Intelligent Systems and Applications in Computer Vision

The book comprehensively covers a wide range of evolutionary computer vision methods and applications, feature selection and extraction for training and classification, and metaheuristic algorithms in image processing. It further discusses optimized image segmentation, its analysis, pattern recognition, and object detection.

Features:

- Discusses machine learning-based analytics such as GAN networks, autoencoders, computational imaging, and quantum computing.
- Covers deep learning algorithms in computer vision.
- Showcases novel solutions such as multi-resolution analysis in imaging processing, and metaheuristic algorithms for tackling challenges associated with image processing.
- Highlight optimization problems such as image segmentation and minimized feature design vector.
- Presents platform and simulation tools for image processing and segmentation.

The book aims to get the readers familiar with the fundamentals of computational intelligence as well as the recent advancements in related technologies like smart applications of digital images, and other enabling technologies from the context of image processing and computer vision. It further covers important topics such as image watermarking, steganography, morphological processing, and optimized image segmentation. It will serve as an ideal reference text for senior undergraduate, graduate students, and academic researchers in fields including electrical engineering, electronics, communications engineering, and computer engineering.

Intelligent Systems and Applications in Computer Vision

Edited by
Nitin Mittal
Amit Kant Pandit
Mohamed Abouhawwash
Shubham Mahajan

CRC Press
Taylor & Francis Group
Boca Raton London New York

CRC Press is an imprint of the
Taylor & Francis Group, an **informa** business

Front cover image: Blackboard/Shutterstock

First edition published 2024
by CRC Press
2385 NW Executive Center Dr, Suite 320, Boca Raton, FL 33431

and by CRC Press
4 Park Square, Milton Park, Abingdon, Oxon, OX14 4RN

CRC Press is an imprint of Taylor & Francis Group, LLC

© 2024 selection and editorial matter, Nitin Mittal, Amit Kant Pandit, Mohamed Abouhawwash and Shubham Mahajan individual chapters

ISBN: 978-1-032-39295-0 (hbk)
ISBN: 978-1-032-59187-2 (pbk)
ISBN: 978-1-003-45340-6 (ebk)

DOI: 10.1201/9781003453406

Typeset in Sabon
by Newgen Publishing UK

Contents

About the editors

Nitin Mittal, received his B.Tech and M.Tech degrees in Electronics and Communication Engineering (ECE) from Kurukshetra University, Kurukshetra, India in 2006 and 2009, respectively. He completed his PhD in ECE from Chandigarh University, Mohali, India in 2017. He worked as a professor and assistant dean of research, ECE Department at Chandigarh University. Presently, he is working as skill assistant professor, department of Industry 4.0 at Shri Vishwakarma Skill University. His research interests include Wireless Sensor Networks, Image Segmentation, and Soft Computing.

Amit Kant Pandit is working as an associate professor and is Ex-Hod, DECE, in Shri Mata Vaishno Devi University (SMVDU), Katra (India). He is a senior member of IEEE and MIR labs member and has 19 years of academic experience.

Mohamed Abouhawwash received the BSc and MSc degrees in statistics and computer science from Mansoura University, Mansoura, Egypt, in 2005 and 2011, respectively. He finished his PhD in Statistics and Computer Science, 2015, in a channel program between Michigan State University, and Mansoura University. He is at Computational Mathematics, Science, and Engineering (CMSE), Biomedical Engineering (BME) and Radiology, Institute for Quantitative Health Science and Engineering (IQ), Michigan State University. He is an assistant professor with the Department of Mathematics, Faculty of Science, Mansoura University. In 2018, Dr. Abouhawwash is a Visiting Scholar with the Department of Mathematics and Statistics, Faculty of Science, Thompson Rivers University, Kamloops, BC, Canada. His current research interests include evolutionary algorithms, machine learning, image reconstruction, and mathematical optimization. Dr. Abouhawwash was a recipient of the best master's and PhD thesis awards from Mansoura University in 2012 and 2018, respectively.

Shubham Mahajan (Member, IEEE, ACM, IAENG) received the B.Tech. degree from Baba Ghulam Shah Badshah University, the M.Tech. degree from Chandigarh University, and the Ph.D. degree from Shri Mata Vaishno Devi University (SMVDU), Katra, India. He is currently employed as an Assistant Professor at Ajeenkya D Y Patil University, Pune. Dr. Mahajan holds ten Indian, one Australian, and one German patent in the field of artificial intelligence and image processing. He has authored/coauthored over 66 publications, including peer-reviewed journals and conferences. His research interests primarily lie in image processing, video compression, image segmentation, fuzzy entropy, nature-inspired computing methods, optimization, data mining, machine learning, robotics, optical communication, and he has also received the "Best Research Paper Award" from ICRIC 2019 (Springer, LNEE). Furthermore, Dr. Mahajan has been honored with several accolades, including the Best Student Award-2019, IEEE Region-10 Travel Grant Award-2019, 2nd runner-up prize in IEEE RAS HACKATHON-2019 (Bangladesh), IEEE Student Early Researcher Conference Fund (SERCF-2020), Emerging Scientist Award-2021, and IEEE Signal Processing Society Professional Development Grant-2021. In addition to his academic pursuits, he has served as a Campus Ambassador for IEEE, representing prestigious institutions such as IIT Bombay, Kanpur, Varanasi, Delhi, and various multinational corporations. Dr. Mahajan actively seeks collaboration opportunities with foreign professors and students, demonstrating his enthusiasm for international research partnerships.

Contributors

Sanjo J. Adwin
PSG College of Technology,
  Coimbatore, Tamil Nadu, India

Mamta Arora
Chandigarh University,
  Gharuan, Punjab, India

Mitali Arya
National Institute of Technology
  Hamirpur,
  Himachal Pradesh, India

Abhishek Bharane
Department of Information
  Technology
Pimpri-Chinchwad College of
  Engineering
Pimpri-Chinchwad, Maharashtra,
  India

Vinay Bhosale
Department of Information
  Technology
Pimpri-Chinchwad College of
  Engineering
Pimpri-Chinchwad, Maharashtra,
  India

Pranav Bongulwar
Department of Information
  Technology

Pimpri-Chinchwad College of
  Engineering,
Pimpri-Chinchwad, Maharashtra,
  India

Jasmine Hazel Crasta
St Joseph Engineering College,
Vamanjoor, Mangalauru,
  Karnataka, India

Soma Debnath
Amity Institute of Information
  Technology, Amity University,
  New Town, Kolkata, India

Priyanka Dhanasekaran
Department of Information
  Science and Technology,
Anna University,
  Chennai, Tamil Nadu, India

Chinmay Dixit
Department of Computer
  Engineering and IT,
College of Engineering,
  Pune, Maharashtra, India

Pulkit Dubey
Department of Computer Science,
Manav Rachna International
  Institute of Research and Studies,
  Faridabad, Haryana, India

S. P. Gautham
Department of Computer Science
  and Engineering,
St Joseph Engineering College,
  Vamanjoor, Mangalauru,
  Karnataka, India

A. V. Geetha
Department of Information
  Science and Technology,
Anna University,
  Chennai, Tamil Nadu, India

G. Gopisankar
Department of Computer
  Science and Engineering,
PSG College of Technology,
  Coimbatore, Tamil Nadu, India

Ashulekha Gupta
Department of Management
  Studies,
Graphic Era (Deemed to be
  University)
Dehradun, Uttarakhand, India

Manoj Kumar Gupta
School of Computer Science and
  Engineering,
Shri Mata Vaishno Devi
  University, Katra, Jammu and
  Kashmir, India

Medha Gupta
Amity Institute of Information
  Technology,
Amity University,
  Kolkata, Tamil Nadu, India

Nidhi Gupta
National Institute of Technology
  Hamirpur,
Himachal Pradesh, India

Pranesh Gupta
Indian Institute of Information
  Technology,
Una, Himachal Pradesh, India

H. N. Gurudeep
St Joseph Engineering College,
Vamanjoor, Mangaluru,
  Karnataka, India

Pai H. Harikrishna
St Joseph Engineering College,
Vamanjoor, Mangaluru,
  Karnataka, India

Anandaram Harishchander
Amrita School of Engineering,
Coimbatore Amrita Vishwa
  Vidyapeetham, Tamil Nadu, India

P. Muhamed Ilyas
Sullamussalam Science College,
Mallapuram, Kerala, India

A. Jayabharathi
PSG College of Technology,
Coimbatore, Tamil Nadu, India

Amit D. Joshi
Department of Computer
  Engineering and IT,
College of Engineering,
  Pune, Maharastra, India

Kapil Joshi
Department of CSE,
Uttaranchal Institute of Technology,
Uttaranchal University,
Dehradun, Uttarakhand, India

Ashima Kalra
Department of ECE, Chandigarh
  Engineering College,
Landran, Mohali, Punjab, India

Divneet Singh Kapoor
Kalpana Chawla Center for
  Research in Space Science and
  Technology,
Chandigarh University,
  Mohali, Punjab, India

K. Sheril Kareem
Sullamussalam Science College,
  Malappuram, Kerala, India

K. Karthik
St Joseph Engineering College,
Vamanjoor, Mangaluru,
  Karnataka, India

Nitish Katal
Indian Institute of Information
  Technology,
Una, Himachal Pradesh, India

Bobbinpreet Kaur
ECE Department, Chandigarh
  University,
Gharuan, Punjab, India

Talha Fasih Khan
Department of Computer Science,
Manav Rachna International
  Institute of Research and Studies,
  Faridabad, Haryana, India

Amit Kumar Kohli
Thapar Institute of Engineering and
  Technology,
Patiala, Punjab, India

Konda Hari Krishna
Koneru Lakshmaiah Education
  Foundation,
Vaddeswaram, Guntur District,
Andhra Pradesh, India

Rajesh Kumar
Meerut Institute of
  Technology,
Meerut, Uttar Pradesh, India

Vivek Kumar
Department of CSE,
  Uttaranchal Institute of
  Technology,
Uttaranchal University,
  Dehradun, Uttarakhand, India

Jyoti Lele
Dr. Vishwanath Karad MIT World
  Peace University,
Pune, Maharashtra, India

T. Mala
Anna University,
Chennai, Tamil Nadu, India

Onakar Mulay
Department of Information
  Technology
Pimpri-Chinchwad College of
  Engineering
Pimpri-Chinchwad, Maharashtra,
  India

Veronica Naosekpam
Indian Institute of
  Information Technology
  Guwahati, Assam, India

Naman Palliwal
Dr. Vishwanath Karad MIT World
  Peace University,
Pune, Maharashtra, India

Tanuja S. Patankar
Department of Information
  Technology

Pimpri-Chinchwad College of
  Engineering
Pimpri-Chinchwad,
  Maharashtra, India

Anuradha C. Phadke
Dr. Vishwanath Karad MIT World
  Peace University,
Pune, Maharashtra, India

M. Preethi
Department of Information
  Technology,
Sri Ramakrishna Engineering
  College,
Coimbatore, Tamil Nadu, India

Arjun Puri
Model Institute of Engineering and
  Technology,
Jammu, Jammu and Kashmir, India

R. Rajalakshmi
Sri Ramakrishna Engineering
  College,
Coimbatore, Tamil Nadu, India

Sahil Rajurkar
Dr. Vishwanath Karad MIT World
  Peace University,
Pune, Maharashtra, India

R. Ramya
Sri Ramakrishna Engineering
  College,
Coimbatore, Tamil Nadu, India

K. C. Ridhi
PSG College of Technology,
Coimbatore, Tamil Nadu, India

Nilkanta Sahu
Indian Institute of Information
  Technology
Guwahati, Assam, India

T. S. Saleena
Sullamussalam Science College,
Malappuram, Kerala, India

S. Sasank
PSG College of Technology,
Coimbatore, Tamil Nadu, India

Kiran Jot Singh
Kalpana Chawla Center for
  Research in Space Science and
  Technology,
Chandigarh University,
Mohali, Punjab, India

Anshul Sharma
Chandigarh University,
Mohali, Punjab, India

Pooja Sharma
I. K. Gujral Punjab Technical
  University,
Jalandhar, Punjab, India

Sheenam
CSE Deptt.,
Chandigarh University,
Gharuan, Punjab, India

S. Suriya
PSG College of Technology,
Coimbatore, Tamil Nadu, India

Gaurav Tewari
Gautam Buddha University,
Gr. Noida, India

Khushal Thakur
Kalpana Chawla Center for
  Research in Space Science and
  Technology,
Chandigarh University,
Mohali, Punjab, India

Vibor Tomar
Dr. Vishwanath Karad MIT World
  Peace University,
Pune, Maharashtra, India

E. Uma
Department of Information Science
  and Technology, Anna University,
  Chennai, Tamil Nadu, India

Yukti Upadhyay
Manav Rachna International
  Institute of Research and Studies,
Faridabad, Haryana, India

S. Vidhya
Sri Ramakrishna Engineering
  College,
Coimbatore, Tamil Nadu, India

Pavinder Yadav
National Institute of Technology
  Hamirpur,
Himachal Pradesh, India

Chapter 1

A review approach on deep learning algorithms in computer vision

*Kapil Joshi, Vivek Kumar, Harishchander Anandaram,
Rajesh Kumar, Ashulekha Gupta, and
Konda Hari Krishna*

1.1 INTRODUCTION

The topic of "computer vision" has grown to encompass a wide range of activities, from gathering raw data to extracting patterns from images and interpreting data. The majority of computer vision jobs have to do with feature extraction from input scenes (digital images) in order to get information about events or descriptions. Computer vision combines pattern detection and image processing. Image understanding comes from the computer vision process. The field of computer vision, in contrast to computer graphics, focuses on extracting information from images. Computer technology is essential to the development of computer vision, whether it is for image quality improvement or image recognition. Since the design of the application system determines how well a computer vision system performs, numerous scholars have proposed extensive efforts to broaden and classify computer vision into a variety of fields and applications, including assembly line automation, robotics, remote sensing, computer and human communications, assistive technology for the blind, and other technologies [1]. Deep learning (DL) is a member of the AI method family. Artificial Neural Networks (ANNs) get their name from the fact that they receive an input, analyze it, and produce a result. Deep learning is based on ANN. Because of the massive amount of data generated every minute by digital transformation, AI is becoming more and more popular. The majority of organizations and professionals use technology to lessen their reliance on people [2].

In machine learning, the majority of features taken into account during analysis must be picked manually by a specialist in order to more quickly identify patterns. DL algorithms gradually pick up knowledge from high level features. A part of machine learning called "further deep learning" is depicted in Figure 1.1. ANNs, which have similar capabilities to human neurons, are the inspiration for deep learning. The majority of machine

DOI: 10.1201/9781003453406-1

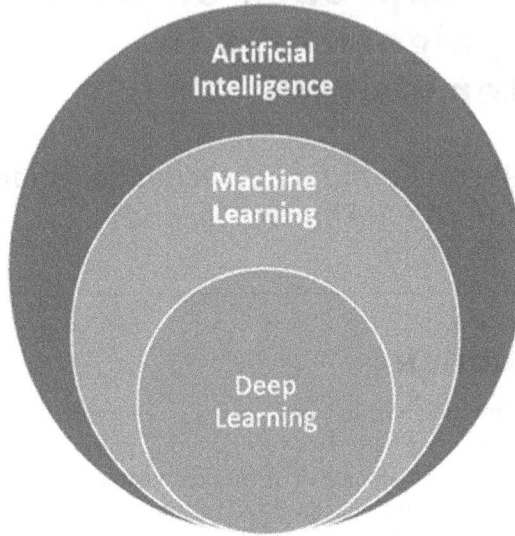

Figure 1.1 Deep learning, machine learning, and artificial intelligence.

learning techniques pale in comparison to ANNs because they can use learning that is supervised, semi-supervised, and unsupervised on a variety of different types of data.

The large family of algorithms known as "deep learning" includes supervised and unsupervised feature learning approaches that include neural networks and hierarchical probabilistic models. Due to their greater performance shown over prior state-of-the-art methods in a number of tasks as well as the volume of complex data from multiple sources, deep learning approaches have recently witnessed an increase in interest. Regarding their applicability in visual understanding, we will concentrate on three one of the key aspects of DL model types in this context: Convolutional Neural Networks (CNN), the "Boltzmann family," which includes Deep Bolzmann Machines, stacked (denoising) autoencoders, and deep belief networks [3]. Robots used in medical applications have been taught to distinguish between scanned and traditional images. Figure 1.2 [4] shows how DL algorithm input is broadly categorized.

For analysis, the DL algorithm needs input. Similar to how human vision works, images are considered in a variety of analysis applications.

Figure 1.3 displays categorization of the input data for the DL phases. But owing to the algorithm's variable inputs, considerable preprocessing is required to reduce the noise. to increase the accuracy of the algorithm.

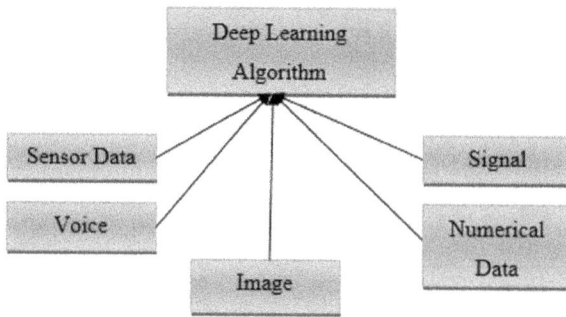

Figure 1.2 Deep learning algorithm's inputs.

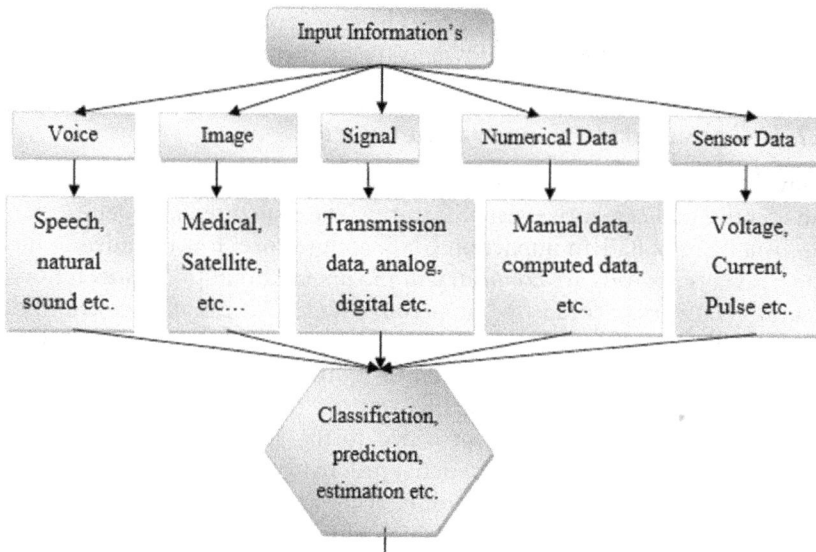

Figure 1.3 Outcome of deep learning algorithms.

1.2 DEEP LEARNING ALGORITHMS

Deep neural networks are challenging to train using back propagation because of the problem of vanishing gradient, which affects training time and precision. As determined by the net difference between the ANN expected output and actual output in the training data, ANNs calculate cost function [5]. After each step, biases and weights are modified based on the

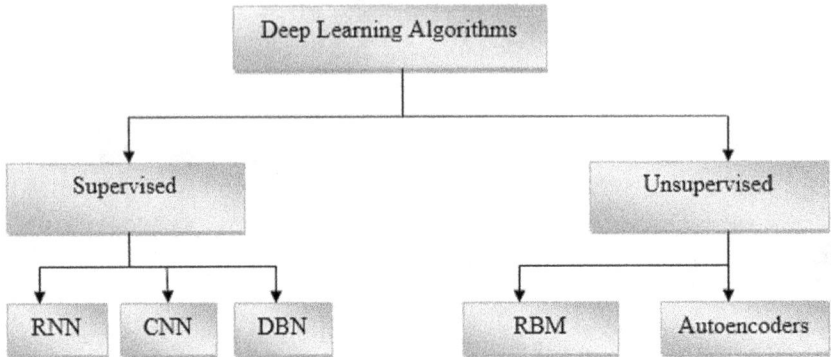

Figure 1.4 Algorithm classifications for deep learning.

cost. The price is as low as it can be. The rate at which cost will alter as a result of weight and biases is known as the gradient.

1.2.1 Convolutional neural networks

Convolutional Neural Network is a DL network for computer vision that can identify and categorize visual features. The structure and operations of the visual cortex had an impact on CNN architecture. It is intended to imitate the ways neurons are connected in the human brain [6]. Convolutional Neural Networks comprise numerous stages, including convolution layer, pooling layer, non-linear processing layer, and sub sampling layers, and it is capable of achieving spatial or temporal correlation in data [7]. Convolutional operations are carried out for feature extraction, and the resulting convolutional is then provided to the activation function. Since non-linearity produces a range of activation patterns for various responses, it is possible to learn the semantic differences across images. CNNs' ability to extract features automatically, which eliminates the requirement for a separate feature extractor, one of its main strengths. Figure 1.5 depicts the architecture of a CNN.

- **Convolution Layers**
  The convolution layer will calculate the scalar product between the weights of the input volume-connected region and the neurons whose output is related to particular regions of the input.

- **Pooling Layers**
  After that, it will simply down sample the input along the spatial dimension, further lowering the number of parameters in that activation [8].

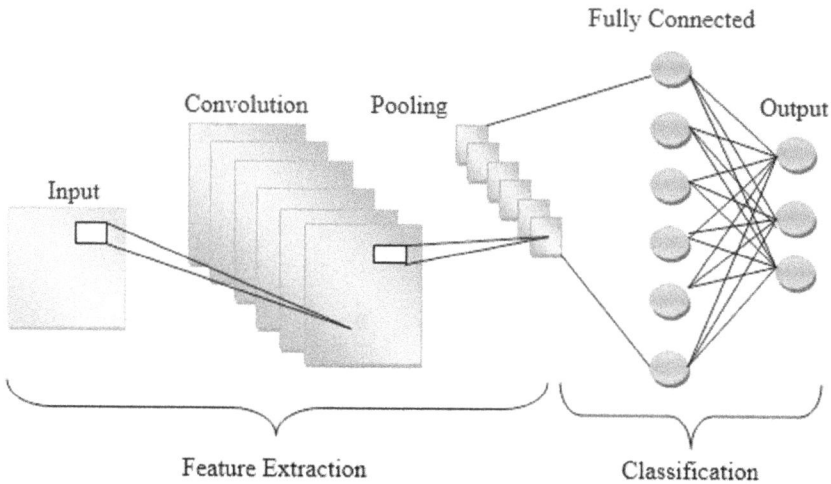

Figure 1.5 CNN architecture.

- **Batch Normalization**
  Batch Normalization is the method through which the activation nodes are scaled and adjusted to normalize the input layer neurons. The output from the preceding is normalized using batch normalization by dividing by the batch standard deviation after subtracting the batch mean [9].

- **Dropout**
  In order to avoid over-fitting, input units are set to 0 at random with a rate of frequency by the "dropout layer" at each training phase. The sum of all inputs is maintained by scaling up non-zero inputs by $\dfrac{1}{1 - \text{rate}}$.

- **Fully Connected Layers**
  After that, it will carry out the same tasks as regular ANNs and try to create categorization scores from the activations. ReLu has also been proposed as a possible application between these layers in order to enhance performance.

1.2.2 Restricted Boltzmann Machines

Such an undirected diagrammatic and modeled depiction of the symmetrical layer, a visible layer, and a hidden layer link among the layers is called a "restricted Boltzmann Machine" (RBM). No relationship exists between

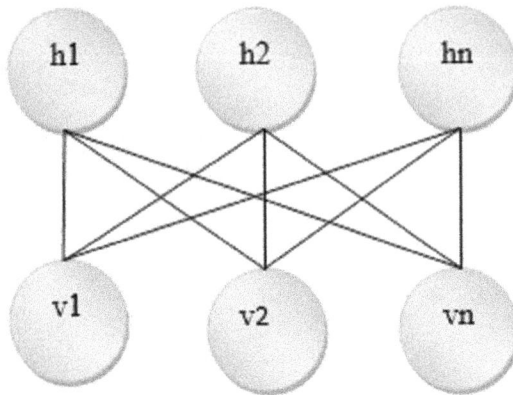

Figure 1.6 Restricted Boltzmann Machine.

an input and the hidden layer in RBM [10]. Restricted Boltzmann Machine exhibits strong feature extraction and representation capabilities. The Restricted Boltzmann machine is a probabilistic network that picks up on the hidden representation, h as well as the probability distribution of its inputs v. The two-layer, typical Restricted Boltzmann Machine method is shown in Figure 1.6. The fundamental benefit of the RBM algorithm is that there are no links between units in the same layer because all components, both visible and concealed, are separate.

The Restricted Boltzmann Machine algorithm seeks to reconstruct the inputs as precisely as possible [11].The input is modified based on the weights and biases throughout the forward stage before beginning to trigger the hidden layer. The hidden layer's activations are then modified based on the weight and biases and transmitted the activation layer's input layer afterward in the following steps: The input layer now searches for the updated activation as a reconstruction of the input, Compare it against the original input.

1.2.3 Deep Boltzmann Machines

In that they use the RBM as a learning module, Deep Boltzmann Machine, are DL models that are members of the "Boltzmann family." The Deep Boltzmann Machine (DBM) has undirected connections between its layers. With the help of the needed data, it accomplishes a layer-by-layer instruction approach that the unlabeled data is trained in, as well as allowing for precise customization. Teaching a stack of RBMs, which are then combined to build a DBM, constitutes pre-training for a DBM with three hidden layers,

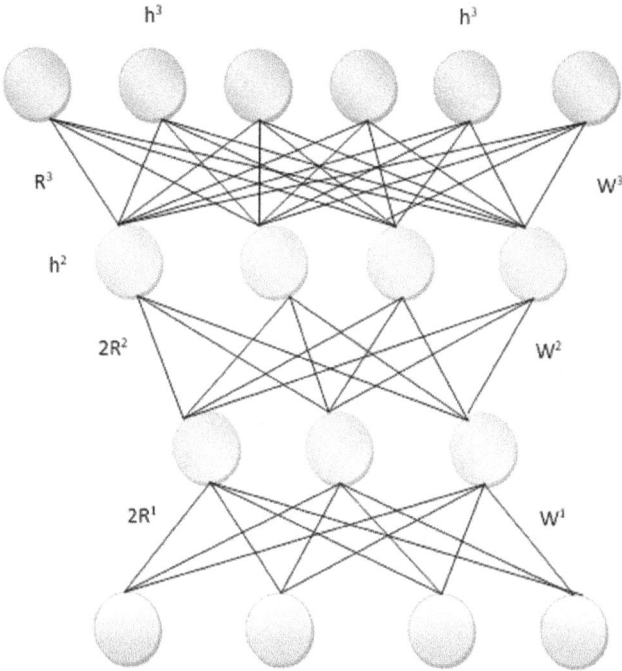

Figure 1.7 Deep Boltzmann Machine.

by defining its energy function to more clearly define the DBM's structure. In relation to the two-layer model as defined by Equation 1.1.

$$E_{DBM}(v, h^{(1)}, h^{(2)}; \theta) = -v^T W h^{(1)} - h^{(1)T} V h^{(2)} - d^{(1)T} h^{(1)}$$
$$- d^{(2)T} h^{(2)} - b^T v \qquad (1.1)$$

Where W, V, d(1), and d(2) are equal to. DBM can be thought of as a bipartite graph with two vertices.

Figure 1.7 DBM's R^1, R^2, and R^3 list the recognition that is intended. The Deep Boltzmann Machine (DBM), a deep generative undirected model, is composed of several hidden layers. In order to affect how lower-level characteristics are learned, it makes use of the top-down connection pattern. R^1, R^2, and R^3 are the recognition model weights, which are increased by two every layer to make up since there is not any top-down feedback [12].

1.2.4 Deep belief networks

A foundation for building models directly from what we see to what we wish to know is provided by DBM. In a way, the layer-by-layer structure

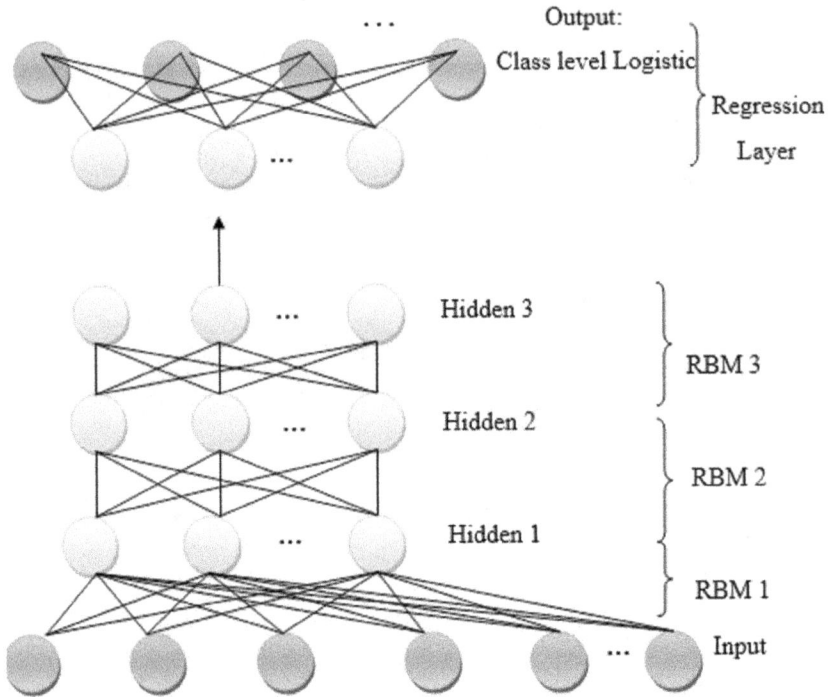

Figure 1.8 Deep Belief Networks.

represents features in a hierarchical manner. The feature-extracting component can be replaced with the self-adaptive network training method. After creating the network, a considerably smaller number of labeled the "back-propagation algorithm," is used to adjust the complete network using a sample set. Large unlabeled samples are used in the layer-by-layer training process [13]. As seen in Figure 1.8, a generative model called "deep belief networks" was produced by stacking several limited Boltzmann machines. An RBM is created by every two neighboring layers. Each restricted Boltzmann machine's visible layer is linked to its predecessor's hidden layer, and the first two levels are directional-less. The top-down between the upper layer and the lower layer with a directed connection are adjusted. In a deep belief network, the various layers of the Restricted Boltzmann Machine are trained in order: the lower Restricted Boltzmann Machines are first trained, followed by the superiors. Back propagation of features to the lowest layers after being extracted by the top Restricted Boltzmann Machine [14].

Pre-training and fine-tuning are the two stages of the deep belief network exercise procedure. Pre-training is a stage of unsupervised training that initializes the model to increase the effectiveness of supervised training. The supervised training stage, which modifies the classifier's prediction to fit the data's ground truth, can be thought of as the fine-tuning process [15].

DBNs can take one of two different forms:

- The deep belief network auto-encoder
- The deep belief network classifier

The auto-encoder deep belief networks are straightforward three-layer neural networks in which the input and output units are joined by a direct connection. Typically, there are a lot fewer hidden units than there are visible units.

There are two steps in the auto-encoding process:

- An input vector is encoded (compressed) to fit in a smaller form.
- It was reconstructed.

The recognition process makes use of the latter's architectural design to produce accurate classification results, with the input data vector represented by the first layer of a Deep Belief Networks' visible layer, the hidden layers the visible layer data, early detectors or reconstructors, and the classification labels represented by the softmax layer, which is the last layer of the Deep Belief Network [16]. Consequently, the classifier to guarantee that the results data is accurately tagged, deep belief network design demands that the last Restricted Boltzmann Machine be discriminative.

1.2.5 Stacked (de-noising) auto-encoders

Similar to how Restricted Boltzmann Machines are a component in Deep Belief Networks, the auto-encoder serves as the foundation of stacked auto-encoders. Therefore, before discussing the DL components of Stacked (Denoising) Autoencoders, it is vital to briefly go over the fundamentals of the autoencoder and its denoising variant.

1.2.5.1 Auto-encoders

A feed-forward neural network that learns a compressed, distributed representation of a dataset is a classic example of an auto-encoder. An auto-encoder is a three-layer neural network trained to reconstruct the inputs by utilizing the output as the input. For the data to be reproducible, it must learn characteristics that capture the variance in the data. If only linear activation functions are utilized and can be used for dimensionality reduction, it can be demonstrated that it is comparable to Principle Component Analysis (PCA). After training, the learned features are employed as the hidden layer activations, and the top layer can be ignored. Contraction, de-noising, and sparseness techniques are used to train auto-encoders.

In auto-encoders, some random noise is injected into the input during de-noising. The original input must be reproduced by the encoder. Regular neural networks will perform better in terms of generalization if inputs are randomly deactivated during training [17]. Setting the hidden layer's number of nodes in contractive auto-encoders to substantially fewer than the number of input nodes drives the network to do dimensionality reduction. As a result, it is unable to learn the identity function since the hidden layer does not have enough nodes to adequately store the input. By giving the weight update function a sparsity penalty, sparse auto-encoders are trained. The connection weights' overall size are penalized, and the majority of the weights have low values as a result. At each stage, old k-1 network hidden layers are used, and a new network with k+ 1 hidden layers is constructed, with the k+ 1th hidden layer using the k+ 1 hidden layer as input. The weights in the final deep network are initialized using the weights from the individual layer training, and the architecture as a

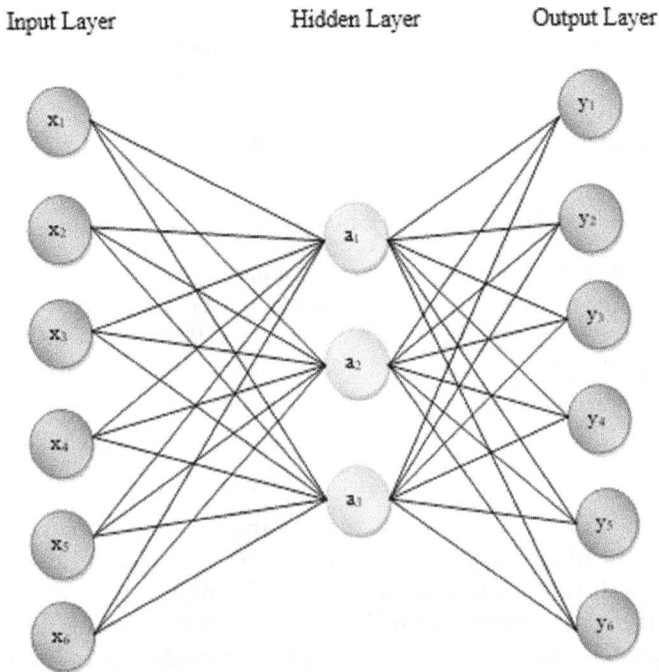

Figure 1.9 Autoencoders.

whole is then tweaked. On the other hand, the network can be tweaked using back propagation by adding an additional output layer on top. Deep networks only benefit from back propagation if the weights are initialized very close to a good solution. This is ensured by the layer-by-layer pre-training. There are also alternative methods for fine-tuning deep networks, such as dropout and maxout.

1.2.5.2 Denoising auto encoders

When given a contaminated input, the denoising auto encoder (DAE) is trained to reassemble a clear, "repaired," version of the input. This is accomplished by first using a stochastic mapping, w qD(w|x), to corrupt the initial input x into w. Then, as with the basic auto-encoder, corrupted input w is mapped to a hidden depicted y = f(w) = s(Ww + b), which we derived reconstruct a z = g(y). A diagram of the process can be found in Figure 1.10. The main modification is that instead of being a deterministic function of x, z is now one of w. The same as before, the reconstruction error is either the squared error loss $L_2(x, z) = \|x - z\|^2$ or the cross-entropy loss L(x, z) = IH(B(x)||B(z)) with an affine decoder. Stochastic gradient descent is used to optimize parameters when they are randomly initialized [18]. It should be noted that qD(w|x) produces a separate corrupted version of each training example x that is shown.

Note that the reconstruction falls among a clean X and its reconstruction against Y is still being minimized by denoising auto-encoders [19, 20]. Therefore, this still entails maximizing a lower constraint on the mutual information between clean input x and representation y [21].The distinction is that y is now produced by using a faulty input with deterministic mapping f. As a result, it forces the acquisition of a mapping that extracts

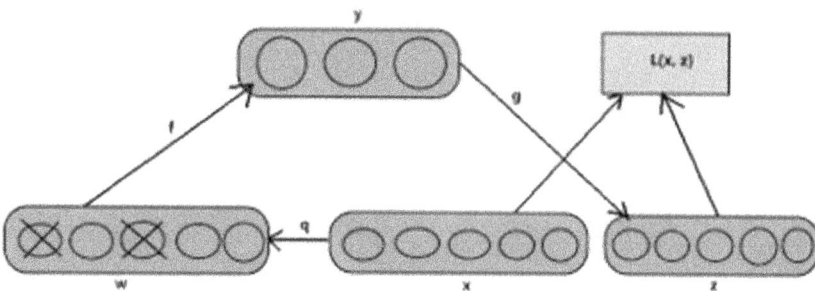

Figure 1.10 The architecture of the denoising autoencoder.

traits helpful for denoising and is significantly more intelligent than the identify [22].

1.3 COMPARISON OF THE DEEP LEARNING ALGORITHMS

Classification of learning, inputting data along with output details and multiple applications are compared in Table 1.1.

1.4 CHALLENGES IN DEEP LEARNING ALGORITHMS

Artificial intelligence, robotics, motion stabilization, virtual reality, automatic panorama stitching, and 3D modeling, scene comprehension, video processing, video stabilization, and motion capture are just a few of the many evolving problems in computer vision that cannot easily be applied in a differentiable system with deep learning [23]. Due to its critical importance in numerous applications, including robotics, surveillance, robotic identification of pedestrians, and real-time vehicle activity tracking, video-scene analysis is a contemporary study area. Despite its widespread use, video-scene analysis is still a difficult problem that calls for more precise algorithms. However, in recent years, improvements in DL algorithms [24] for video-scene study have caused an addressing of the issue of real-time processing.

1.5 CONCLUSION AND FUTURE SCOPE

Deep learning has rendered obsolete many of the computer vision techniques developed over the past 20 years. While machine learning may easily address problems with smaller datasets, deep learning methods are better suited for issues with enormous datasets. We contrast the various models applied to a number of issues, including object detection, object recognition, captioning, and other issues. A few of the most popular DL algorithms, including RBM and auto-encoder, which utilize unsupervised learning, and CNN, and deep belief network, which use supervised learning, are briefly examined. On the basis of their inputs, outputs, and fundamental operation, the algorithms are contrasted. Based on variables including input data, output data, and applications, we compare various algorithms. This study's findings suggest that CNN can successfully solve deep learning challenges involving image inputs. CNN, on the other hand, have substantial computational expenses.

Table 1.1 Comparisons between CNN, RBM, DBM and Auto Encoders

Parameter	Convolutional Neural Networks	Restricted Boltzmann Machines	Deep Belief Networks	Auto encoders
Type of Learning	Supervised	Unsupervised	Supervised	Unsupervised
Input Data	3-D Structured data	Any type of data	Text, Image	Any type of data
Output	Classified, predicted	Reconstructed output	Classified, predicted	Reconstructed output
Application	Image and voice analysis, classification, detection, recognition	Dimensionality Reduction/ Classification	NLP, dimensionality reduction	Dimensionality Reduction

REFERENCES

1. Victor Wiley & Thomas Lucas, 'Computer Vision and Image Processing: A Paper Review'. *International Journal of Artificial Intelligence Research*, Vol. 2, No 1, pp. 28–36, 2018.
2. Savita K. Shetty & Ayesha Siddiqa, 'Deep Learning Algorithms and Applications in Computer Vision'. *International Journal of Computer Sciences and Engineering*, Vol. 7, pp. 195–201, 2019.
3. Ksheera R. Shetty, Vaibhav S. Soorinje, Prinson Dsouza & Swasthik, 'Deep Learning for Computer Vision: A Brief Review'. *International Journal of Advanced Research in Science, Communication and Technology*, Vol. 2, pp. 450–463, 2022.
4. Dr. G. Ranganathan, 'A Study to Find Facts Behind Preprocessing on Deep Learning Algorithms'. *Journal of Innovative Image Processing*, Vol. 3, pp. 66–74, 2021.
5. Swapna G, Vinayakumar R & Soman K. P, 'Diabetes detection using deep learning algorithms'. *ICT Express*, Vol. 4, pp. 243–246, 2018.
6. Dulari Bhatt, Chirag Patel, Hardik Talsania, Jigar Patel, Rasmika Vaghela, Sharnil Pandya, Kirit Modi & Hemant Ghayvat, 'CNN Variants for Computer Vision: History, Architecture, Application, Challenges and Future Scope'. MDPI Publisher of *Open Access Journals*, Vol. 10, p. 2470, 2021.
7. Rachana Patel & Sanskruti Patel, 'A Comprehensive Study of Applying Convolutional Neural Network for Computer Vision'. *International Journal of Advanced Science and Technology*, Vol. 29, pp. 2161–2174, 2020.
8. Keiron O'Shea, Ryan Nash, 'An Introduction to Convolutional Neural Networks'. arXiv, 2015.
9. Shaveta Dargan, Munish Kumar, Maruthi Rohit Ayyagari & Gulshan Kumar, 'A Survey of Deep Learning and Its Applications: A New Paradigm to Machine Learning'. *Archives of Computational Methods in Engineering*, Vol. 27, pp. 1071–1092, 2020.
10. Ali A. Alani, 'Arabic Handwritten Digit Recognition Based on Restricted Boltzmann Machine and Convolutional Neural Networks'. MDPI Publisher of *Open Access Journals*, 2017.
11. Voxid, Fayziyev, Xolbek, Xolyorov, & Kamoliddin, Xusanov, 'Sorting the Object Based on Neural Networks Computer Vision Algorithm of the System and Software'. *ijtimoiy fanlarda innovasiya onlayn ilmiy jurnali*, Vol. 3, No 1, 67–69, 2023.
12. Roy, Arunabha M., Bhaduri, Jayabrata, Kumar, Teerath, & Raj, Kislay, 'WilDect-YOLO: An Efficient and Robust Computer Vision-based Accurate Object Localization Model for Automated Endangered Wildlife Detection'. *Ecological Informatics*, Vol. 75, 101919, 2023.
13. Yang Fu, Yun Zhang, Haiyu Qiao, Dequn Li, Huamin Zhou & Jurgen Leopold, 'Analysis of Feature Extracting Ability for Cutting State Monitoring Using Deep Belief Networks'. CIRP Conference on Modelling of Machine Operations, pp. 29–34, 2015.
14. Weibo Liu, Zidong Wang, Xiaohui Liu, Nianyin Zeng, Yurong Liu & Fuad E. Alsaadi, 'A Survey of deep neural network architectures and their applications'. Neurocomputing, Vol. 234, pp. 11–26, 2017.

15. Jiaojiao Li, Bobo Xi, Yunsong Li, Qian Du & Keyan Wang, 'Hyperspectral Classification Based on Texture Feature Enhancement and Deep Belief Networks'. Remote Sensing MDPI Publisher of *Open Access Journals*, Vol. 10, p. 396, 2018.
16. MehrezAbdellaoui & Ali Douik, 'Human Action Recognition in Video Sequences Using Deep Belief Networks'. *International Information and Engineering Technology Association*, Vol. 37, pp. 37–44, 2020.
17. V. Pream Sudha & R. Kowsalya, 'A Survey on Deep Learning Techniques, Applications and Challenges'. *International Journal of Advance ResearchIn Science and Engineering*, Vol. 4, pp. 311–317, 2015.
18. Parvaiz, Arshi, Khalid, Muhammad Anwaar, Zafar, Rukhsana, Ameer, Huma, Ali, Muhammad, & Fraz, Muhammad Mouzam, 'Vision Transformers in Medical Computer Vision—A Contemplative Retrospection'. *Engineering Applications of Artificial Intelligence*, Vol. 122, 106126, 2023.
19. Malik, Karim, Robertson, Colin, Roberts, Steven A, Remmel, Tarmo K, & Long, Jed A., 'Computer Vision Models for Comparing Spatial Patterns: Understanding Spatial Scale'. *International Journal of Geographical Information Science*, Vol. 37, No 1, 1–35, 2023.
20. Sharma, T., Diwakar, M., Singh, P., Lamba, S., Kumar, P., & Joshi, 'Emotion Analysis for predicting the emotion labels using Machine Learning approaches'. IEEE 8th Uttar Pradesh Section International Conference on Electrical, Electronics and Computer Engineering (UPCON), pp. 1–6, 2021.
21. Joshi, K., Kirola, M., Chaudhary, S., Diwakar, M., & Joshi, 'Multi-focus image fusion using discrete wavelet transform method'. *International Conference on Advances in Engineering Science Management & Technology* (ICAESMT), 2019.
22. Ambore, B., Gupta, A. D., Rafi, S. M., Yadav, S., Joshi, K., & Sivakumar, 'A Conceptual Investigation on the Image Processing Using Artificial Intelligence and Tensor Flow Models through Correlation Analysis'. International Conference on Advance Computing and Innovative Technologies in Engineering (ICACITE) IEEE, pp. 278–282, 2022.
23. Niall O' Mahony, Sean Campbell, Anderson Carvalho, Suman Harapanahalli, Gustavo Velasco Hernandez, Lenka Krpalkova, Daniel Riordan & Joseph Walsh, 'Deep Learning vs. Traditional Computer Vision'. *Computer Vision Conference*, pp. 128–144, 2019.
24. R. Kumar, M. Memoria, A. Gupta, & M. Awasthi, 'Critical Analysis of Genetic Algorithm under Crossover and Mutation Rate'. 2021 3rd International Conference on Advances in Computing, Communication Control and Networking (ICAC3N), 2021, pp. 976–980.

Chapter 2

Object extraction from real time color images using edge based approach

Sheenam, Bobbinpreet Kaur, and Mamta Arora

2.1 INTRODUCTION

Object extraction deals with finding out distinct objects in the image that can further govern the control of some mechanism. The object extraction can be part of some counter-based system wherein, on the basis of count, the system follows the progress [1]. Images are the great source of information and can record the observations. But it is very difficult to process them manually for information extraction. The extraction of objects from the images is one of the most challenging tasks faced in order to make the systems fully automatic [2, 3].

The main principle of object extraction is based on increasing the inter-class relationship and decreasing the intra-class similarity. This can ensure the objects in the image are separated and can be extracted without overlapping. The output of this system can serve as input to the object identification system [4,5].

The object extraction used in object recognition methodology as shown in Figure 2.1. The image captured is the representation of a scenic 3-D scene as a 2-D record. The nontrivial dimensions, usually in numerous recreational objects, are enough to represent the reasons related in most cases and, therefore, play a significant role in different image frameworks, for example for content-based image retrieval systems [6]. Therefore a continuous research is carried out in the direction of designing the automatic systems for extraction of objects from the images. The work is intended to make the systems more efficient in terms of extracting overlapped objects and converting into meaningful information. Further the work is done in order to perform edge linking so that the object boundaries can be connected and form a closed structure [7, 8]. This will help the number and types of objects present in the image. The basic edge-based approaches include application of a mask, which will be done in both x and y directions and then performing element by element multiplication of pixels with the mask coefficients. The image pixels which are mapped to the center of mask will be modified in the

DOI: 10.1201/9781003453406-2

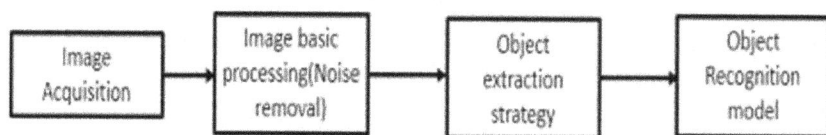

Figure 2.1 Basic Object recognition system.

process and allotted with the updated value as presented in the operation of multiplication.

2.2 APPLICATIONS OF OBJECT EXTRACTION

Numerous applications of object extraction in different contexts have been reported in the literature. We have summarized the few applications below:

1. Vehicular tracking
2. Optical character recognition
3. Tracking people in video frames
4. Ball tracking in different sports
5. Object extraction from satellite images
6. License number plate detection
7. Logo detection from the images
8. Disease detection
9. Medical imaging to detect the tumors in the broad sense
10. Robotics
11. Counter-based applications
12. In agricultural fields to detect any anomalies

The applications listed above are the fields wherein the traditional methods are having different limitations to manage. Each of these application aims at reducing human efforts and providing automated mechanisms by detecting the different objects present in the image. In our proposed model we are going to perform extraction with the help of the edge-filtering approach. One important application is the introduction of cameras at the toll booths to identify automobile number plates through the extraction process and then controlling the gate-opening mechanism. In traditional methods human power was utilized to read and record the number. This process was many times erroneous as huge numbers of cars passed through the toll booths in a day, and the persons on the windows cannot handle so many incoming cars. So the need of automation can be visualized in these kind of scenarios. Thus, a systematic model needs to be developed for managing these kinds of problems [9–12].

2.3 EDGE DETECTION TECHNIQUES

Edge detection can be done by segmentation. Utilizing the changes in the image's gray tones, edge-detection techniques turn original images into edge images. Edge detection in image processing, particularly in computer vision, deals with the localization of significant fluctuations in a gray-level image and the detection of the geometrical and physical characteristics of scene objects. It is a key mechanism to recognize the boundaries between things in a picture, as well as the contours of those objects. The most common method for identifying substantial intensity value discontinuities is edge detection.

Edges are tiny variations in the image's brightness. Edges generally appear where two sections converge that are used by computer vision. Edge detection is employed in object detection, which supports numerous applications such as biometrics and medical image processing. As it enables more advanced image analysis, edge detection is a research topic that is currently being pursued. An image's edges can be used to extract the major characteristics. An important element for picture analysis is edge detection. The gray level contains three various kinds of discontinuities, including points, lines, and edges. All three forms of discontinuities in an image can be discovered using spatial masks. Those techniques are Roberts Edge Detection, Sobel Edge Detection, Prewitt Edge Detection, and Laplacian Edge Detection

2.3.1 Roberts edge detection

This method measures the 2-D spatial gradient of a picture in a straightforward and quick manner. This technique highlights areas of high spatial frequency, which frequently coincide with edges. The most common application of this approach is when both the input and the output are grayscale images [13]. The estimated full amplitude of the spatial gradient of the input image at each place in the output is represented by the pixel values at each location is shown in Figure 2.2.

The gradient is given by:

$$|G| = \sqrt{Gi^2 + Gj^2}$$

Estimated Gradient Calculated By

$$|G| = |Gi^2| + |Gj^2| \text{ compute faster result}$$

2.3.2 Sobel edge detection

Using the Sobel approximation to the derivative, the Sobel method of edge identification for picture segmentation locates edges. At the locations where

-1	0
0	+1

G_i

0	-1
+1	0

$G_j|$

Figure 2.2 Masks used by Roberts Operator.

the gradient is greatest, it comes before the edges. The Sobel method [13] applies a 2-D spatial gradient quantity to a picture, emphasizing edge-corresponding high spatial frequency regions. It is typically used to determine the predicted absolute gradient magnitude at each point in a grayscale input image. As revealed in the table, the operator in hypothesis comprises at least two 3x3 complication kernels. One kernel is just the other kernel turned 90 degrees. The Roberts Cross operator is very similar to this. One kernel is just the other kernel turned 90 degrees. The Roberts Cross operator is very similar to this is shown in Figure 2.3.

The gradient is given by:

$$|G| = \sqrt{G_i^2 + G_j^2}$$

Estimated Gradient Calculated by:

$$|G| = |G_i^2| + |G_j^2| \text{ compute faster result}$$

The angle of orientation of the edge (relative to the pixel grid) giving rise to the spatial gradient is given by:

$$\theta = \arctan(G_i/G_j)$$

2.3.3 Prewitt's Operator

The Prewitt Operator is similar to the Sobel operator and is used for detecting vertical and horizontal edges in images as shown in Figure 2.4.

2.3.4 Laplacian edge detection

The edge detectors previously discussed are different from the Laplacian edge detectors. This technique employs just one filter (also called a kernel). Laplacian [3] detection performs second-order derivatives in a single pass, making it susceptible to noise. Before using this procedure, the image is subjected to Gaussian smoothing to reduce its susceptibility to noise is shown in Figure 2.5.

2.4 RELATED WORK

-1	0	+1
2	0	+2
-1	0	+1

Gi

-1	0	+1
2	0	+2
-1	0	+1

Gj

Figure 2.3 Masks used by Sobel Operator.

-1	0	+1
2	0	+2
-1	0	+1

Gi

-1	0	+1
2	0	+2
-1	0	+1

Gj

Figure 2.4 Masks used by Prewitt's Operator.

-1	-1	-1
-1	8	-1
-1	-1	-1

Gi

0	-1	0
-1	4	-1
0	-1	0

Gj

Figure 2.5 Masks used by Laplacian Operator.

Many researchers have proposed different variants of the noise elimination methods. The main goal of elimination of noise is to achieve a high degree of distinction between edges and the noise affected pixels, thereby making the edge content more visible and strong. The author presented a median filtering approach [15] for noise minimization. In total, six different models of spatial filters are applied and tested for elimination of noise from the images. A spatial domain median calculation based filter is proposed. The experimental results directly illustrate the accuracy levels of minimization in noise through spatial median filtering. Table 2.1 given below, analysis the result of various operator on the basis of complexity and noise sensitivity using edge detection.

An approach toward applying median filtering is utilized in [16] particularly for compound images. The reduction of noise ensures the image quality, thereby making it more meaningful for real time tasks. The compound images contain different levels of information in the form of text, objects, shapes, graphs, and many more attributes. Generally, the sources of

Table 2.1 Analysis result of edge detector

S. No.	Operators	Complexity		Advantages	Disadvantages	Noise Sensitivity
		Time	Space			
1	Roberts	High	High	Diagonal direction points are preserved	Not accurate	Extremely Sensitive to Noise
2	Sobel	Low	High	Simple and time efficient computation	Diagonal direction not fixed	High Sensitive to noise
3	Prewitt's	Low	Lower	Best operator orientation of an image	The magnitude of coefficient is fixed	Least Sensitive to noise
4	Laplacian	Low	Least	There is fixed characteristics in all directions	Error occurred at curved edges	Very Sensitive to Noise

Source: [14].

compound images include the scanned copy of documents and the clicked pictures of the documents that may possess different sorts of information. The median filtered for different scanned images is applied, and the quantitative analysis of the filtration is performed by obtaining the value of the parameters. A similar set of approaches is proposed in [17–19], where direct or a variant of median filter is deployed to minimize the effect of noise. The variants of a median filter will add certain functionalities while making computations of median value as it can add different set of weights to the mask values and while computing the median value some of the pixels will be strengthened as per the mask weight value.

The approach for object extraction from images is developed in [20] where a method and apparatus are proposed for the extraction. The object is having similar intensity values, thereby the process will begin with selecting a seed pixel and then combining the neighbor pixels depending upon the threshold criteria. This joining of pixels to the center pixel will yield a group of pixels, thereby defining an object. Wherever there is a sharp variation in intensity, it will be treated as an edge pixel. Ideally the edge pixels are utilized to extract the threshold and determines the number of threshold values to be used for extraction process to be performed with high level of accuracy.

The edge detection mechanisms and different approaches were discussed and proposed in the literature by different researchers [21–25] with an aim to reduce the number of false edges. The edge detection process is determined, and dependent on, many factors such as presence of noise and similarity between different objects present in the image. The edge will be located through a sharp variation in the intensity where the edge pixels have a different pixel value from its neighbors. As far as object extraction is concerned, it is governed through combining different edge lines obtained from filtering of the image. The edges will form connected structures and then connecting these edges will yield the object boundary. The object extraction makes the image processing capable of finding its applications in real time situations where we need to process the images for extraction of meaningful information.

The detected edges need to be smoothened in order to extract the shape of the required object in the image. Smoothening is directly related to removal of noise pixels. This can be accomplished through suitable spatial filter such as an average filter or a weighted median filter. They directly work on the image pixel by pixel and manipulate the intensity value of the pixel, which is mapped to the center of the image. The other kind of smoothing masks include Gaussian, min-max, adaptive median filter [26–32]. Some Fuzzy Logic based models are proposed through researches based on fuzzification rules [33]. The common goal of achieving edge detection with high accuracy is considered while designing.

This paper proposes a new hybrid technique based on the Aquila optimizer (AO) [38–41] and the arithmetic optimization algorithm (AOA). Both AO and AOA are recent meta-heuristic optimization techniques. They can be used to solve a variety of problems such as image processing, machine learning, wireless networks, power systems, engineering design, and so on. The impact of various dimensions is a standard test that has been used in prior studies to optimize test functions that indicate the impact of varying dimensions on AO-AOA efficiency.

The findings from the literature are as follows:

1. The edge content is highly modified with the presence of noise.
2. The value of PSNR and other metrics like structural similarity will determine the noise minimization effect. The quality assessment requires monitoring, both subjectively as well as objectively.
3. The similar objects are hard to distinguish.
4. The edge pixels are characterized by a sharp change in the intensity pixels.
5. The edge separates the different objects.

2.5 PROPOSED MODEL

The proposed model is shown in Figure 2.6. The proposed model is developed using the basic mechanisms grasped from the literature. The input images captured in the real time environment are prone to noise depending upon

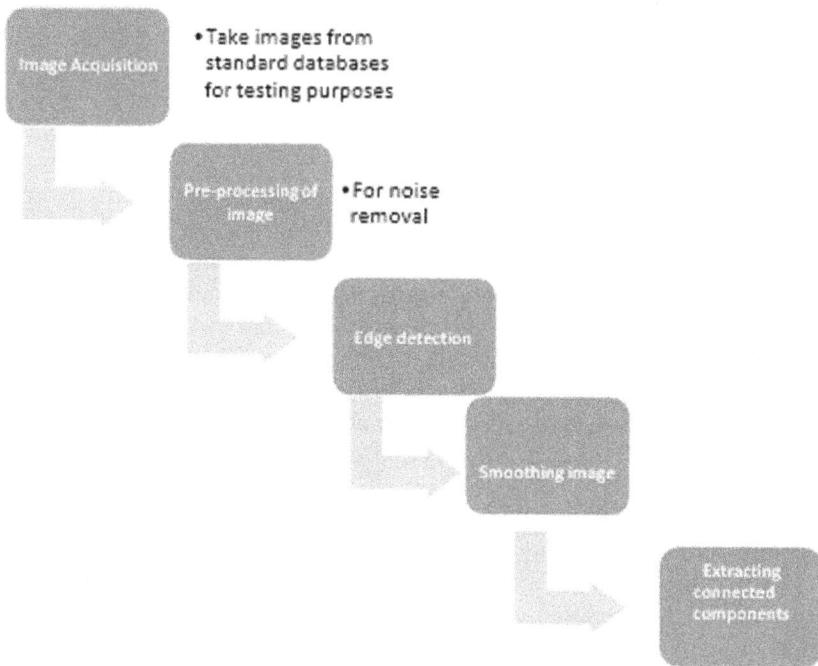

Figure 2.6 The proposed model.

many factors such as sensor abnormalities, lightening conditions and so forth. The acquired image need to be pre-processed in order to extract the objects accurately. In our work we have deployed a median filter at the pre-processing stage in order to remove the noise. The median filter is one in-class filter, which removes noise as well as preserves the edge contents [34]. Since the main portion of this work will be finding the edges, the edges boosted by median filter will give an additional advantage [35]. The appropriate selection of pre-processing filter will ensure no smoothening effect and thereby preserving the information stored in edges. This will be followed by applying an edge detection mechanism. In our work we have used Sobel operator which is giving highly appreciable results. Similar approaches like Perwitts and canny detection can also be used in the extended applications [36]. We have used a mask with size 3*3 and will be applied on image through convolution [37]. After edge detection the smoothing mask is applied in order to reduce the nearly approximate connected components. Thus with the appropriate connected components we can detect all the components completely. The Sobel operator is a standard mask that defines the edges as the points at which the gradient value of the image is maximum. This is a first derivative operator.

Figure 2.6 describes the proposed model for edge based object extraction. Since we are proposing a model for object extraction catering real time situations, the effect of noise needs to be considered and neutralized in order to perform edge detection more accurately. Edges are obtained by looking for abrupt discontinuities in the values of pixels, and thereby affected by noise to a large extent as the noise can blur the edges, which makes it difficult to detect the object boundaries. Therefore introduction of this pre-processing stage will improve the accuracy of the proposed model.

2.6 RESULTS AND DISCUSSION

The proposed model is tested using a MATLAB® environment with an image processing toolbox. To test the variability and generality of the proposed model we have taken four different types of images – standard test image, medical domain image, plant leaf image and a car in a natural environment. All these images are acquired from free Internet sources.

Figure 2.7 shows the output for the standard test image, Lena. The number of connected components is found to be 168, and the edges are connected. Few of the edges are not connected to form the boundary of the image. The Lena image is the standard test image that contains all kinds of image properties, thus proving the variability of the proposed model

Figure 2.8 shows the output for the Leaf image with disease. The output clearly marked the edges of the disease lesions, and thus can serve as an approach for detection of disease. The number of connected components is found to be 385.

Figures 2.9 and 2.10 similarly show the results for the car image and the medical image. The value of connected components is found to be 37 and

Figure 2.7 Sample 1 (L to R) – The input image; Grayscale converted image; output image.

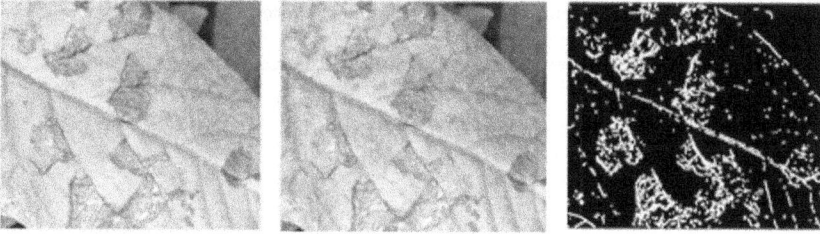

Figure 2.8 Sample 2 (L to R) The input image; Grayscale converted image; output image.

Figure 2.9 Sample 3 (L to R) The input image; Grayscale converted image; output image.

38, respectively. The algorithm is not able to detect the edges, which are merged or are similar to the background.

The analysis of the results obtained can be done through visual inspection of the output image. From the outputs achieved a few observations can be drawn out and comparison shown in Table 2.2:

1. The algorithm is applicable for both colored and grayscale images.
2. The maximum edges are connected and identify the object's boundaries.
3. A small number of edges are not detected.
4. The edges similar to background intensity or merged with background are not detected.

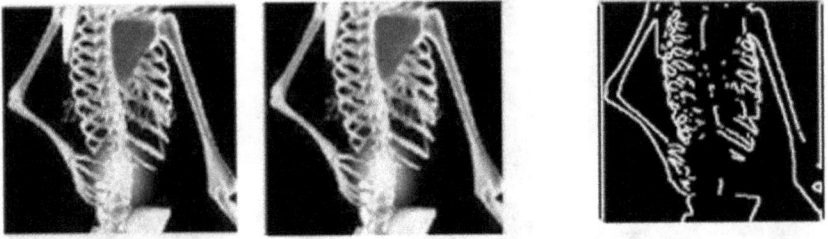

Figure 2.10 Sample 4 (L to R) The input image; Grayscale converted image; output image.

Table 2.2 Connected components for different images

Image	Connected Components
Lena	168
Leaf	385
Car	36
Spine	35

2.7 CONCLUSION

In this chapter, we have presented an approach for image-object extraction using edge detection approach. The broad process and purpose of the image-object extraction and recognition have been described. The objects formed through connected lines and points can be segregated from the background by detecting the edge values in the image and then joining those edges to extract different shapes present in the image. The future applications of object detection requires highly efficient systems for object extraction. In our work we have designed the system to cancel the effects of noise that have been added to the image during the acquisition stage. The spatial filter for noise removal is selected in order to remove the noise as well as preserve the edge strength. Thus, a blurring affect is reduced. Though few of the edges have failed to detect through the system, the overall edges are preserved in a good number. We will be looking forward in future to extending these outputs to serve the object recognition models, which are the backbone of many real time applications.

REFERENCES

1. Buttler, David, Ling Liu, and Calton Pu. "A fully automated object extraction system for the World Wide Web." In *Proceedings 21st International Conference on Distributed Computing Systems*, pp. 361–370. IEEE (2001).

2. Wu, Shaofei. "A traffic motion object extraction algorithm." *International Journal of Bifurcation and Chaos* 25, no. 14 (2015): 1540039.
3. Kim, Sungyoung, Soyoun Park, and Kim Minhwan. "Central object extraction for object-based image retrieval." In *International Conference on Image and Video Retrieval*, pp. 39–49. Springer: Berlin, Heidelberg (2003).
4. Liang, Cheng-Chung, and Thomas Moeller. "System for interactive 3D object extraction from slice-based medical images." U.S. Patent 6,606,091, issued August 12, 2003.
5. Gevers, Theo, and Arnold W. M. Smeulders. "Color-based object recognition." *Pattern Recognition* 32, no. 3 (1999): 453–464.
6. Belongie, Serge, Jitendra Malik, and Jan Puzicha. "Shape matching and object recognition using shape contexts." *IEEE transactions on pattern analysis and machine intelligence* 24, no. 4 (2002): 509–522.
7. Fan, Jianping, Xingquan Zhu, and Lide Wu. "Automatic model-based semantic object extraction algorithm." *IEEE Transactions on circuits and systems for video technology* 11, no. 10 (2001): 1073–1084.
8. Idris, Fayez, and Sethuraman Panchanathan. "Review of image and video indexing techniques." *Journal of Visual Communication and Image Representation* 8, no. 2 (1997): 146–166.
9. Xu, Haifeng, Akmal A. Younis, and Mansur R. Kabuka. "Automatic moving object extraction for content-based applications." *IEEE Transactions on Circuits and Systems for Video Technology* 14, no. 6 (2004): 796–812.
10. Tsuchikawa, Megumu, Atsushi Sato, Akira Tomono, and Kenichiro Ishii. "Method and apparatus for moving object extraction based on background subtraction." U.S. Patent 5,748,775, issued May 5, 1998.
11. Shang, Yanfeng, Xin Yang, Lei Zhu, Rudi Deklerck, and Edgard Nyssen. "Region competition based active contour for medical object extraction." *Computerized Medical Imaging and Graphics* 32, no. 2 (2008): 109–117.
12. Baltsavias, Emmanuel P. *Object extraction and revision by image analysis using existing geospatial data and knowledge: State-of-the-art and steps towards operational systems*. ETH Zurich, 2002.
13. Muthukrishnan, Ranjan, and Miyilsamy Radha. "Edge detection techniques for image segmentation." *International Journal of Computer Science & Information Technology* 3, no. 6 (2011): 259.
14. Katiyar, Sunil Kumar, and P. V. Arun. "Comparative analysis of common edge detection techniques in context of object extraction." arXiv preprint arXiv:1405.6132 (2014).
15. Church, James C., Yixin Chen, and Stephen V. Rice. "A spatial median filter for noise removal in digital images." In *IEEE SoutheastCon 2008*, pp. 618–623. IEEE (2008).
16. Church, James C., Yixin Chen, and Stephen V. Rice. "A spatial median filter for noise removal in digital images." In *IEEE SoutheastCon 2008*, pp. 618–623. IEEE, 2008.
17. Dong, Yiqiu, and Shufang Xu. "A new directional weighted median filter for removal of random-valued impulse noise." *IEEE signal processing letters* 14, no. 3 (2007): 193–196.

18. Kumar, N. Rajesh, and J. Uday Kumar. "A spatial mean and median filter for noise removal in digital images." *International Journal of Advanced Research in Electrical, Electronics and Instrumentation Engineering* 4, no. 1 (2015): 246–253.

19. Wang, Gaihua, Dehua Li, Weimin Pan, and Zhaoxiang Zang. "Modified switching median filter for impulse noise removal." *Signal Processing* 90, no. 12 (2010): 3213–3218.

20. Kamgar-Parsi, Behrooz. "Object extraction in images." U.S. Patent 5,923,776, issued July 13, 1999.

21. Shrivakshan, G. T., and Chandramouli Chandrasekar. "A comparison of various edge detection techniques used in image processing." *International Journal of Computer Science Issues (IJCSI)* 9, no. 5 (2012): 269.

22. Sharifi, Mohsen, Mahmood Fathy, and Maryam Tayefeh Mahmoudi. "A classified and comparative study of edge detection algorithms." In *Proceedings. International conference on information technology: Coding and computing*, pp. 117–120. IEEE, 2002.

23. Nadernejad, Ehsan, Sara Sharifzadeh, and Hamid Hassanpour. "Edge detection techniques: evaluations and comparisons." *Applied Mathematical Sciences* 2, no. 31 (2008): 1507–1520.

24. Middleton, Lee, and Jayanthi Sivaswamy. "Edge detection in a hexagonal-image processing framework." *Image and Vision Computing* 19, no. 14 (2001): 1071–1081.

25. Ziou, Djemel, and Salvatore Tabbone. "Edge detection techniques – an overview." *Pattern Recognition and Image Analysis C/C of Raspoznavaniye Obrazov I Analiz Izobrazhenii* 8 (1998): 537–559.

26. Lee, Jong-Sen. "Digital image smoothing and the sigma filter." *Computer vision, graphics, and image processing* 24, no. 2 (1983): 255–269.

27. Ramponi, Giovanni. "The rational filter for image smoothing." *IEEE Signal Processing Letters* 3, no. 3 (1996): 63–65.

28. Meer, Peter, Rae-Hong Park, and K. J. Cho. "Multiresolution adaptive image smoothing." *CVGIP: Graphical Models and Image Processing* 56, no. 2 (1994): 140–148.

29. Kačur, Jozef, and Karol Mikula. "Solution of nonlinear diffusion appearing in image smoothing and edge detection." *Applied Numerical Mathematics* 17, no. 1 (1995): 47–59.

30. Hong, Tsai-Hong, K. A. Narayanan, Shmuel Peleg, Azriel Rosenfeld, and Teresa Silberberg. "Image smoothing and segmentation by multiresolution pixel linking: further experiments and extensions." *IEEE Transactions on Systems, Man, and Cybernetics* 12, no. 5 (1982): 611–622.

31. Tottrup, C. "Improving tropical forest mapping using multi-date Landsat TM data and pre- classification image smoothing." *International Journal of Remote Sensing* 25, no. 4 (2004): 717–730.

32. Fang, Dai, Zheng Nanning, and Xue Jianru. "Image smoothing and sharpening based on nonlinear diffusion equation." *Signal Processing* 88, no. 11 (2008): 2850–2855.

33. Taguchi, Akira, Hironori Takashima, and Yutaka Murata. "Fuzzy filters for image smoothing." In *Nonlinear Image Processing V*, vol. 2180, pp. 332–339. International Society for Optics and Photonics, 1994.

34. Chen, Tao, Kai-Kuang Ma, and Li-Hui Chen. "Tri-state median filter for image denoising." *IEEE Transactions on Image Processing* 8, no. 12 (1999): 1834–1838.

35. Arce, McLoughlin. "Theoretical analysis of the max/median filter." *IEEE transactions on acoustics, speech, and signal processing* 35, no. 1 (1987): 60–69.

36. Canny, John. "A computational approach to edge detection." *IEEE Transactions on pattern analysis and machine intelligence* 6 (1986): 679–698.

37. Marr, David, and Ellen Hildreth. "Theory of edge detection." *Proceedings of the Royal Society of London. Series B. Biological Sciences* 207, no. 1167 (1980): 187–217.

38. Mahajan, S., Abualigah, L., Pandit, A. K. et al. "Fusion of modern meta-heuristic optimization methods using arithmetic optimization algorithm for global optimization tasks." *Soft Comput* 26 (2022): 6749–6763.

39. Mahajan, S., Abualigah, L., Pandit, A. K. et al. "Hybrid Aquila optimizer with arithmetic optimization algorithm for global optimization tasks." *Soft Comput* 26 (2022): 4863–4881.

40. Mahajan, S. and Pandit, A.K. "Hybrid method to supervise feature selection using signal processing and complex algebra techniques." *Multimed Tools Appl* (2021).

41. Mahajan, S., Abualigah, L. & Pandit, A.K. "Hybrid arithmetic optimization algorithm with hunger games search for global optimization." *Multimed Tools Appl* 81 (2022): 28755–28778.

42. Mahajan, S., Abualigah, L., Pandit, A. K. et al. "Fusion of modern meta-heuristic optimization methods using arithmetic optimization algorithm for global optimization tasks." Soft Comput 26 (2022): 6749–6763.

Chapter 3

Deep learning techniques for image captioning

R. Ramya, S. Vidhya, M. Preethi, and R. Rajalakshmi

3.1 INTRODUCTION TO IMAGE CAPTIONING

Creating a textual caption for a set of images is known as *image captioning*. This translates images, which are seen as a sequence of pixels to a sequence of words, making it an end-to-end sequence to sequence challenge. Both the language or statements and the visuals must be processed for this reason. NVIDIA has developed a tool to assist those with poor or no vision using image captioning technology. It makes it easier for persons who are visually challenged to understand what is going on in a picture. Image captioning comes with an encoder-decoder structure. The image feature extractions, object detection comes to encoder part, consecutively language modelling comes under decoder.

The picture captioning paradigm streamlines and expedites the close captioning procedure for the creation, editing, delivery, and archiving of digital information. For creating great captions for both photos and movies, well-trained models replace manual efforts. Millions of images are distributed internationally by the media in the form of magazines, emails, social media, and so on. The picture captioning model expedites the production of subtitles and frees executives to concentrate on more crucial activities.

Artificial intelligence is transitioning from discussion forums to underpinning systems for classifying and detecting gigabytes of media content in social media. This makes it possible for analysts to create commercial plans and for community managers to keep an eye on interactions.

3.1.1 How does image recognition work?

A digital image is one that is composed of picture elements, often known as pixels, and each of which has a definite, finite amount of numeric representation for its degree of intensity. The computer interprets a picture as the numerical values of these pixels, thus it must recognize patterns and regularities in this data to identify a particular image.

DOI: 10.1201/9781003453406-3

The technique of image recognition often entails building a neural network that analyses each image pixel. To "train" these networks how to recognize related images, we feed them as many pre-labeled photos as we can. We feed these networks as many pre-labeled photos as possible in order to "train" them how to detect comparable images.

A digital image is a numerical matrix, as was already said. This number represents the data associated with the image pixels. The location and intensity of every pixel in the image are supplied as data to the recognition system. The position and intensity of every pixel in the image is transmitted as data to the recognition system. Using this information, we might train the algorithm to identify patterns and connections between various photographs.

Later, when the training process is completed, we evaluate the system's performance using test data. To increase system accuracy and deliver precise results for picture identification, periodic weights in neural networks are adjusted. As a consequence, neural networks process this numerical data using the deep learning method. Three widely used picture identification methods are Speed Up Robust Features (SURF), PCA (Principal Component Analysis) and Scale-invariant Feature Transform (SIFT).

The encoder is linked to a dense or completely connected layer that generates confidence scores for each potential label. The output of confidence scores from image recognition models for each label and input image must be noted in this context. It is important to take notice of the output of confidence ratings from image recognition models for each label and input picture. In order to get a single prediction for single-class image recognition, we choose the label with the greatest confidence score. Final labels are only assigned when multi-class recognition is used, and each label's confidence score exceeds a predetermined threshold. One more thing is regarding accuracy. The major portion of image recognition systems are bench-marked on common datasets using consistent accuracy standards. Top-1 accuracy is the proportion of images for which the model output class with the greatest confidence score corresponds to the actual label of the image. Top-5 accuracy is the proportion of images for which the actual label is one of the 5 most reliable model outputs.

3.2 INTRODUCTION TO DEEP LEARNING

Deep Learning is a subcategory of Artificial Intelligence. The DL depends generally on Artificial Neural Networks (ANNs) to figure worldviews enlivened by the working of human knowledge. According to the knowledge of the human, there are of various processing cells or "neurons," which all play out a basic activity as well as cooperate with one another to take the choice/determination [1,2]. Deep learning is deals with precisely getting the knowledge "assigning the rating" between the other layers of a brain network, proficiently, and without any supervision. The present interest is to empower progressions in handling hardware equipment [3]. Self-association and collaborations between little units have demonstrated better performance compared to control of the central unit, especially the multipart non-direct cycle model that is to tolerate the fault and versatility for achieving the new information.

To acquire the basic comprehension of deep learning, there is a considerable distinction between illustrative examination and prescient examination. Illustrative examination includes characterizing a conceivable numerical model that depicts the distinctiveness. This involves gathering the information about steps, shaping speculations on designing the information, and approving that these speculations contrast with the result of illustrative examination models [4].

Prescient examination includes the revelation of step-by-step procedure along with rules that underlie a peculiarity and that one is structured as a prescient examination model, which limit the faults/mistakes between the real and the anticipated results based on all possible/conceivable variables. AI dismisses the conventional method on worldview because the training network framework can replace the issue and the framework is taken care of countless preparation designs (bundle of data sources are used to get accurate results) that is acquires and utilized to produce the exact result.

3.2.1 Pros of the deep learning algorithm

In this present decade, there are fast improvements in DL algorithm and upgrades in gadget capacities which, including registering power, storage limit, utilization of power, and picture sensor, has enhanced the presentation as well as effective cost and then stimulated the spread of "vision-based application." When comparing customary CV with DL, the DL algorithm empowers accomplishing more noteworthy precision in undertakings like picture characterization, semantic division, object discovery and Simultaneous Localization and Mapping. Meanwhile the brain networks utilized in DL are *prepared* as opposed to *customized*, applications utilizing this DL-based methods like, picture characterization, semantic division, object discovery methods frequently unneeded master investigation and calibrating and take advantage of the colossal measure of video information accessible in the present frameworks [5,6]. DL moreover gives prevalent adaptability in light of the fact that CNN representations and structures can be re-prepared by utilizing traditional information for any utilization case, in spite of CV algorithm, which will generally be more area explicit.

3.2.2 Customary / traditional CV methodology

The customary methodology is to utilize deep rooted CV algorithm like element descriptors ("SIFT," "SURF," "BRIEF," and so on) for object discovery. Earlier development of DL, a stage called highlight extraction was done for errands like picture grouping. Highlights are little "fascinating," illustrative or useful patches in pictures. A few CV calculations, like edge discovery, corner location or limit division might be associated with this step. However, as many elements as practicable are extricated from pictures, and these highlights structure a definition (known as a sack of-words) of each item class. At the sending stage, these definitions are looked for in different pictures. On the off chance that a critical number of elements from one pack of-words are in another picture, the picture is delegated containing that particular item (for example seat, horse, and so forth).

The trouble with this customary methodology is that picking, which is fundamental highlights, are significant in each given picture [7]. As the quantity of classes to characterize increments, highlight extraction turns out to be increasingly unwieldy. It really depends on the CV specialist's judgment and a long experimentation interaction to conclude which elements best depict various classes of items. Besides, each element definition requires managing with plenty of boundaries, which must all be tweaked by the CV specialist. DL acquainted the idea of end-with end, realizing where the machine is simply given a dataset of pictures which have been commented on with what classes of item are available in each picture. Consequently a DL model is "prepared" on the given information, where brain networks find the basic examples in classes of pictures and consequently works out the most distinct and striking highlights regarding every particular class of item for each article. It has been deep rooted that DNNs offer far superior performances than conventional calculations, but with compromises regarding registering prerequisites and preparing time. With all the cutting edge approaches in CV utilizing this system, the work process of the CV specialist has changed emphatically where the information and aptitude in removing hand-made highlights has been supplanted by information and skill in emphasizing through profound acquiring structures.

The improvement of CNNs has had an enormous impact in the field of CV in recent years and is liable for a major leap in the capacity to perceive objects. This burst in progress has been empowered by an expansion in registering power, as well as an expansion in how much information is accessible for preparing brain organizations. The new blast in and broad reception of different profound brain network structures for CV is clear in the way that the fundamental paper ImageNet Classification with Deep CNN [8,9] has been referred to more than three thousand times.

CNNs utilize portions (otherwise called channels), to distinguish highlights (for example edges) all through a picture. A part is only a lattice of values, called loads, which are prepared to identify explicit elements. As their name demonstrates, the primary thought behind the CNNs is to spatially convolve the part on a given info picture check if the element it is intended to recognize is available. To offer a benefit addressing how sure it is, just an explicit component is available, a convolution activity is done by processing the dab result of the piece and the information region where a portion is covered (the region of the unique picture the part is taking a gander at is known as the responsive field).

To work with the learning of piece loads, the convolution layer's result is added with a predisposition term and afterward taken care of to a non-straight initiation capability. Enactment capabilities are normally non-straight capabilities like Sigmoid, TanH and ReLU (Rectified Straight Unit). Contingent upon the idea of information and arrangement undertakings, these enactment capabilities are chosen likewise [10]. For instance, ReLUs are known to have a more natural portrayal (neurons in the mind either fire or they do not). Accordingly, it yields great outcomes for picture acknowledgment undertakings, as it is less powerless to the disappearing slope issue and it produces sparser, more proficient portrayals.

To accelerate the preparation interaction and lessen how much memory is consumed by the organization, the convolutional layer is in many cases followed by a pooling layer to eliminate overt repetitiveness present in the information highlight. For instance, max pooling moves a window over the information and just results in the most extreme worth in that window really decreasing to the significant pixels in a picture [7]. The profound CNNs may have a few sets of convolutional and pooling layers. At long last, a fully connected layer smooths the past layer volume into an element vector and afterward a result layer, which figures the scores (certainty or probabilities) for the result classes/highlights through a thick organization. This result is then passed to a relapse capability.

3.2.3 Limitations/challenges of traditional CV methodology

DL is in some cases over the top excess as frequently conventional CV strategies can tackle an issue considerably more productively and in less lines of code than DL. Calculations like SIFT and, indeed, even basic tone threshold and pixel counting calculations are not class-explicit – that is, they are exceptionally broad and play out no differently for any picture. Conversely, highlights gained from a profound brain net are intended for your preparation data set which, while possibly not greatly developed, likely will not perform well for pictures quite the same as the preparation set. Thus, SIFT and different calculations are frequently utilized for applications like picture sewing/3D cross section reproduction, which do not need explicit class information. These assignments have been demonstrated to be feasible via preparing huge datasets, but this requires a colossal examination exertion and it is not viable to go through this work for a shut application. One necessitates the rehearsed presence of mind with regard to picking which course to take for a given CV application. For instance, to characterize two classes of items on a mechanical production system transport line, one with red paint and one with blue paint, a profound brain net will work given that enough information can be gathered to prepare from. Nonetheless, the equivalent can be accomplished by utilizing a straightforward variety threshold. A few issues can be handled with less difficult and quicker methods [11].

These days, the customary methods are utilized when the issue can be rearranged so that they can be sent on minimal expense micro-controllers or to restrict the issue for profound learning strategies by featuring specific elements in information, enlarging information or helping with data set commentary. The number of picture change procedures can be utilized to work on your brain net preparation. At long last, there are a lot additional difficult issues in CV, for example, Robotics, expanded reality, programmed display sewing, computer generated reality, demonstrating, movement assessment, video adjustment, movement catch, video handling, and scene understanding ,which ca not just be handily executed in a differential way with profound advancement yet benefit from arrangements utilizing "conventional" strategies [12,13].

There are clear compromises between customary CV and profound learning-based approaches. Exemplary CV calculations are deep rooted, straightforward, and improved for execution and power proficiency, while DL offers more prominent exactness and flexibility at the expense of a lot of figuring assets. Cross breed approaches combine conventional CV and profound learning and present the advantages of the two procedures. They are particularly reasonable in high execution frameworks, which should be carried out rapidly. For instance, in a security camera, a CV calculation can productively identify faces or different elements or moving objects in the scene. These identifications can then be passed to a DNN for character check or article arrangement. The DNN need just be applied on a little fix of the picture, saving huge registering assets and preparing exertion contrasted with what would be expected to deal with the whole casing.

3.2.4 Overcome the limitations of deep learning

There are likewise difficulties presented by DL. The most recent DL approaches might accomplish considerably better exactness; but this bounce comes at the expense of billions of extra number related activities and an expanded prerequisite for handling power. DL requires these processing assets for preparing [14] and less significantly for deduction. It is fundamental to have devoted equipment or preparing and AI speeded-up stages: for example, VPUs for deduction for designers of AI.

Vision handling results utilizing DL are additionally subject to the picture goal. Accomplishing sufficient execution in object arrangement, for instance, requires high resolution pictures or video – with the ensuing expansion in how much information should be handled, put away, and moved. Picture goal is particularly significant for applications in which it is important to distinguish and characterize objects somewhere far off, for example in surveillance camera film. The edge decrease methods talked about already, for example, utilizing SIFT highlights or optical stream for moving articles to first recognize a district of interest are helpful concerning picture goals and furthermore with regard to decreasing the time and information expected for preparing.

Preparing a DNN consumes most of a day. Contingent upon registering equipment accessibility, preparing can require merely hours or days. Also, preparing for some random application frequently requires numerous emphases as it involves experimentation with various preparation boundaries. The most widely recognized procedure to diminish preparing time is move learning. Concerning conventional CV, the discrete Fourier change is another CV strategy that once experienced significant prevalence however much it presently appears to be dark [15]. The calculation can be utilized to accelerate convolutions as shown and consequently may again happen to substantial significance. It should be said that simpler, more space-explicit undertakings than general picture characterization would not need as much information (in that frame of mind, of hundreds or thousands as opposed to millions). This is as yet a lot of information, and CV procedures are in many cases used to support preparing information through information expansion or to diminish the information down to a specific kind of component through other pre-handling steps.

Pre-handling involves changing the information (ordinarily with conventional CV procedures) to permit connections/examples to be all the more effectively deciphered prior to preparing your model. Information expansion is a typical pre-handling task which is utilized when there is restricted preparing information. It can include performing arbitrary turns, shifts, shears, and so forth on the pictures in your preparation set to really build the quantity of preparing pictures. Another methodology is to feature highlights of interest prior to passing the information to a CNN with CV-based techniques like foundation deduction and division.

3.3 DEEP LEARNING ALGORITHMS FOR OBJECT DETECTION

Deep CNN may be a special variety of Neural Networks, which has shown exemplary performance on many competitions associated with computer vision and image processing. A number of the exciting application areas of CNN embrace image classification and segmentation, object detection, video process, linguistic communication process, and speech recognition. The powerful mentality of deep CNN is primarily because of the employment of multiple feature extraction stages that may mechanically learn representations from the info. The availability of an outsized quantity of information and improvement within the hardware technology has accelerated the analysis in CNNs, and recent attention-grabbing deep CNN architectures are reported. Many ennobling concepts to bring advancements in CNNs are explored, such as the use of various activation and loss functions, parameter optimization, regularization and architectural innovations. However, the numerous improvements within the naturalistic capability of the deep CNN are achieved through subject area innovations. Notably, the concepts of exploiting spatial and channel info, depth and dimension of design and multi-path information processing have gained substantial attention.

Similarly, the thought of employing a block of layers as a structural unit is additionally gaining quality. As a result, this survey focuses on the intrinsic taxonomy present in the recently proposed deep CNN architectures and consequently classifies the recent innovations in CNN architectures into seven completely different classes. These seven classes are based on abstraction exploitation, depth, multi-path, width, feature-map exploitation, channel boosting, and a focus. To boot, the elementary understanding of CNN parts, current challenges, and applications of CNN are provided [16].

One of the tasks that deep neural networks (DNNs) excel at is image recognition. To recognize pattern, neural networks were designed. This architecture consists of three parts. One is input layer; the second is hidden layer and the third is output layer. The signals are received from input layer, the role of hidden layer is processing and the last decision about input data is made by hidden layer. For doing computation process, each layer should be interconnected with nodes.

Detecting objects is a one kind of computer vision technique. In order to obtain the information of image and videos, we can use computer vision. Another computer vision technique is object classification. The main difference between object detection and classification is that one deals with *where* the object is in the frame and other deals with finding out *what* the object is. Combining the object detection and classification techniques play an integral part of many applications. Consider an example for detection and classification is used in autonomous cars. Object detection helps to detect any obstacles that are presented in the path and role of object classification helps in understanding the nature of the things.

3.3.1 Types of deep models for object detection

CNN plays a major role in the field of object detection. We can now identify different kinds of objects and even their movement in films thanks to the LeNet-5. The MNIST dataset's handwritten digits were detected using the LeNet, which was the first network to use CNN for object detection. Only dense and convolutional layers were present in the LeNet. The architectural complexity of LeNet is five layers with approximately 0.06M parameters. The ReLU activation function and the maxpooling layers were introduced by the AlexNet.

The inception block was first implemented in Google LeNet (inception v1). The output of various size kernels is stacked together as a block by the inception block. This enables the network to collect many properties without being restricted to a specific kernel size.

The network captures different kinds of options without fixing on a single kernel size. The other name of ResNet building block is residual block or identity block. An identity or residual block is a one which is simply the activation of a layer that is connected to a deeper layer.

While training the deep CNN, that time the residual blocks solves a drawback called the vanishing gradient problem. According to this theory, when more layers are added to the network, we can decrease the loss. Selectively, at a certain point the loss can decrease and then focus on increasing as more layers are added.

The analysis carried out inside the DL algorithm for picture capture is emphasized networks. The subsequent networks area unit chosen as a result of the AlexNet is little and its input size good for the chosen data set. The inception net was chosen because the block that creates it in various forms of some other networks. Due to hardware restrictions the recent networks were not selected. Now current networks perform well but they have expensive computational structures [17, 18].

3.4 HOW IMAGE CAPTIONING WORKS

The computer vision community has taken a keen interest in image captioning, which aims to provide machine-generated natural language descriptions for an image. The encoder-decoder framework with the visual recognition mechanism is used by nearly all image captioning models as a result of the success of deep learning techniques in machine translation. The encoder converts the input images into fix-length vector features, while the decoder converts the image features back into word-by-word descriptions.

3.4.1 Transformer based image captioning

The majority of approaches to image captioning in the literature use a visual encoder and a linguistic decoder. This raises the challenge of what part of the image to translate word by word. Transformer architecture employs machine translation, which is the process of automatically translating text from one natural language into another while maintaining the intended meaning and creating fluid writing in the target language. Depending on the distance, a word in a text may be either to the left or to the right of another word. Images provide for more leeway in the relative spatial relationship between semantic components than phrases. This achieves a cutting-edge performance by indirectly relating informative parts of the image via dot-product attention.

Given the input $X \in R^{A \times B}$ for each layer of transformer where A is the entries and B is the number of features, entry represents word feature in a sentence and also region in an image. Through multi-head dot product attention, the transformer's primary job is to refine each entry in relation to other entries. Now the input is converted into queries S, keys T, and values U as per Equations 3.1, 3.2 and 3.3 respectively and B^K and B^U is the dimension of the key vector and value vector.

$$S = XW_S, W_s \in R^{B \times B_T} \tag{3.1}$$

$$T = XW_T, W_T \in R^{B \times B_T} \tag{3.2}$$

$$U = XW_U, W_X \in R^{B \times B_U} \tag{3.3}$$

The dot product attention for S, T, and U is given in Equation 3.4,

$$Attention(S, T, U) = Soft\max\left(\frac{ST^T}{\sqrt{B_T}}\right)U \tag{3.4}$$

The inner architecture of transformer is modified as sub-transformers to decode the variety of information in image regions and encode spatial relationships between image regions [19]. The attention in Equation 3.4 gives a weighted average for each S generating irrelevant information sometimes. An attention on attention (AOA) module alleviates this problem by doing element-wise multiplication between an attention gate and information vector [20].

3.4.2 Visual scene graph based image captioning

Recent methods for picture labeling include using attention to learn the relationships between regions of the image, successfully encoding their context in the image. The scene graph (SG) holds the organized semantic data of a picture, such as awareness of any visible objects, their characteristics, and bilateral associations. Graph Convolution Networks (GCN) gives spatial features of object and Long Short-Term Memory (LSTM) representing semantic features, combined to known as GCN-LSTM architecture to work as an image encoder [21]. GCN uses Faster region with CNN (R-CNN) is employed to predict the possible objects in the given input region. Semantic and spatial relation graph is constructed for the detected objects in the input object. Now GCN-LSTM is trained with full image with the constructed graph. Figure 3.1 shows the typical CNN based encoder and recurrent neural network (RNN)- LSTM based decoder for image captioning.

Auto- encoder based SG is to learn language dictionary for sentence reconstructing by its own. Initially a syntactic tree is constructed and then it is transformed to SG based on rules [22].

For an input image A, the encoder E fetches the n visual representations Z, then the sentence h generated by the decoder D is [23]

$$Z = E(A) \tag{3.5}$$

$$h = D(Z) \tag{3.6}$$

In the image A, Z is obtained as a structure with combinations of object unit (O) which are the discrete objects, its properties named as attribute units (A) and communication between object twins called as relationship unit R. Z in relationship with O, A and R is given by,

$$V = V_o \cup V_a \cup V_r \tag{3.7}$$

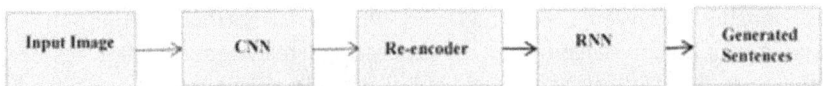

Figure 3.1 Conventional encoder- decoder based image captioning.

Multi-layer perceptron (MLP) along with Soft-max is employed to classify A [23]. Visual semantic units (VSU) use geometry, visual appearance, and semantic embedding cues to obtain the features in each node. The features from the VSU are embedded by GCN. The relationship between VSU and linguistic words can be (1) A word in the caption and a VSU can both be put into one of the following three groups: O, A and R. A phrase could frequently be aligned with one of the image's VSUs, which express the same information in many modalities.

Using a soft-attention mechanism and a context gated attention module (CGA), each caption word is hierarchically aligned with the VSUs. The higher category level of VSUs then performs the gated fusion operation. A two layer LSTM is used for sentence decoder. The concatenation of the current word's embedding, the mean-pooled image feature, and the prior hidden state of the second LSTM serves as the input vector to the

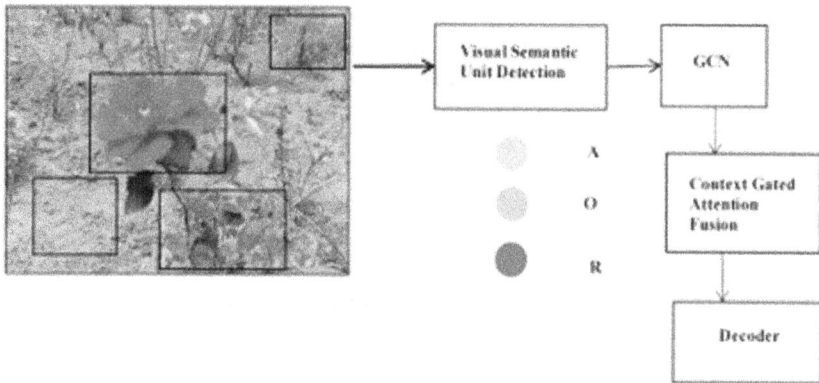

Figure 3.2 Visual Semantic Units and Caption Generation.

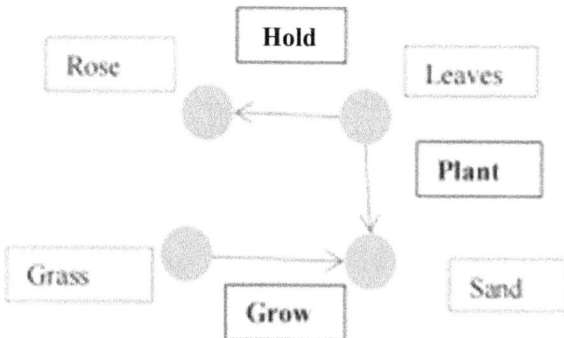

Figure 3.3 Objects and relationships.

first LSTM at each time step. Figure 3.2 shows the VSU and decoder for caption generator. Figure 3.3 shows the O and R of VSU acted on the given input image.

The decoder can frame sentences as "leaves hold rose," "grass grow sand," "rose planted in sand." Hierarchy parsing architecture can be used for image captioning and functions as an image encoder to read the hierarchical structure in images. In order to further improve sentence generation, tree-structured topology have been added to all instance-level, region-level, and image-level features [24].

3.4.3 CHALLENGES IN IMAGE CAPTIONING

As the frequently generated captions in a sequential fashion – that is, the next generated word – depends on both the previous word and the image attribute, traditional captioning systems suffer from a lack of naturalness. The problem with these systems is that they have difficulty generalizing situations when the same objects appear in scenes with unknown contexts because the trained models are over-fitted to the common objects that co-occur in a common context (such as bed and bedroom) (e.g., bed and forest). The examination of the caliber of created captions presents the third difficulty. Utilizing automated measures, while somewhat beneficial, is still inadequate because they ignore the image [25].

3.5 CONCLUSION

This chapter provides the knowledge on DL techniques involved in image captioning along with what we have discussed: the deep models for object detection, differences, limitations of DL, and traditional image feature methods. The conventional translation captioning approaches use word-by-word decoding, and it may change the meaning of the caption. An attention mechanism works well for this problem. Also out of A, O, R units the O model gives improvement in performance rather than using combined A and R units. Using A and R increases the computational load because of the residual connections. The increase in the relationship unit will uplift the Consensus-Based Image Description Evaluation (CIDE) score. Finally we provided the common challenges faced by captioning systems. Utilizing automated measures, while somewhat beneficial, is still inadequate because they ignore the image. When scoring various and descriptive captions, their scores frequently remain insufficient and perhaps even misleading.

REFERENCES

[1] Bonaccorso, G. (2018) Machine Learning Algorithms. *Popular Algorithms for Data Science and Machine Learning*, Packt Publishing Ltd, 2nd Edn.

[2] O'Mahony, N., Murphy, T., Panduru, K., et al. (2017) Real-time monitoring of powder blend composition using near infrared spectroscopy. In: 2017 Eleventh International Conference on Sensing Technology (ICST). IEEE.

[3] Lan, Q., Wang, Z., Wen, M., et al. (2017) High Performance Implementation of 3D Convolutional Neural Networks on a GPU. *Comput Intell Neurosci* (2017).

[4] Diligenti, M., Roychowdhury, S., Gori, M. (2017) Integrating Prior Knowledge into Deep Learning. In: 2017 16th IEEE International Conference on Machine Learning and Applications (ICMLA). IEEE.

[5] Zeng, G., Zhou, J., Jia, X., et al. (2018) Hand-Crafted Feature Guided Deep Learning for Facial Expression Recognition. In: 2018 13th IEEE International Conference on Automatic Face and Gesture Recognition (FG 2018). IEEE, pp. 423–430.

[6] Li, F., Wang, C., Liu, X., et al. (2018) A Composite Model of Wound Segmentation Based on Traditional Methods and Deep Neural Networks. *Comput Intell Neurosci* 2018.

[7] AlDahoul, N., Md. Sabri, A. Q., Mansoor, A. M. (2018) Real-Time Human Detection for Aerial Captured Video Sequences via Deep Models. *Comput Intell Neurosci* 2018.

[8] Alhaija, H. A., Mustikovela, S. K., Mescheder, L., et al. (2017) Augmented Reality Meets Computer Vision: Efficient Data Generation for Urban Driving Scenes, International Journal of Computer Vision.

[9] Tsai F. C. D. (2004) Geometric hashing with line features. *Pattern Recognit* 27:377–389.

[10] Rosten, E., and Drummond, T. (2006) *Machine Learning for High-Speed Corner Detection*. Springer: Berlin, Heidelberg, pp. 430–443.

[11] Horiguchi, S., Ikami, D., Aizawa, K. (2017) Significance of Soft-max-based Features in Comparison to Distance Metric Learning-based Features, IEEE Xplore.

[12] Karami, E., Shehata, M., and Smith, A. (2017) Image Identification Using SIFT Algorithm: Performance Analysis against Different Image Deformations.

[13] Dumoulin, V., Visin, F., Box, G. E. P. (2018) *A Guide to Convolution Arithmetic for Deep Learning.*

[14] Wang, J., Ma, Y., Zhang, L., Gao, R. X. (2018) Deep learning for smart manufacturing: Methods and applications. *J Manuf Syst.*

[15] Tsai F. C. D. (1994) Geometric hashing with line features. *Pattern Recognit* 27:377–389. https://doi.org/10.1016/0031-3203(94)90115-5

[16] Khan, A., Sohail, A., Zahoora, U. et al. (2020) A survey of the recent architectures of deep convolutional neural networks. *Artif Intell Rev* 53, 5455–5516.

[17] Alom, Md. Zahangir, Taha, Tarek, Yakopcic, Christopher, Westberg, Stefan, Hasan, Mahmudul, Esesn, Brian, Awwal, Abdul & Asari, Vijayan. (2018). The History Began from AlexNet: A Comprehensive Survey on Deep Learning Approaches.

[18] Wu, Xiaoxia, Ward, Rachel, and Bottou, Léon. (2018). *WNGrad: Learn the Learning Rate in Gradient Descent.*

[19] He, S., Liao, W., Tavakoli, H. R., Yang, M., Rosenhahn, B., and Pugeault, N. (2021). Image Captioning Through Image Transformer. In: Ishikawa,

H., Liu, CL., Pajdla, T., and Shi, J. (eds) Computer Vision – ACCV 2020. Lecture Notes in Computer Science, Vol. 12625.

[20] Huang, L., Wang, W., Chen, J., and Wei, X. Y. (2019). Attention on attention for image captioning. In: Proceedings of the IEEE International Conference on Computer Vision. 4634–4643.

[21] Yao, T., Pan, Y., Li, Y., and Mei, T. (2018) Exploring visual relationship for image captioning. In: Proceedings of the European Conference on Computer Vision (ECCV), 684–699.

[22] Yang, X., Tang, K., Zhang, H., and Cai, J. (2019) Auto-encoding scene graphs for image captioning. In: Proceedings of the IEEE Conference on Computer Vision and Pattern Recognition. 10685–10694.

[23] Guo, L., Liu, J., Tang, J., Li, J., Luo, W., Lu, H. (2019) Aligning linguistic words and visual semantic units for image captioning. In: Proceedings of the 27th ACM International Conference on Multimedia. 765–773.

[24] Yao, T., Pan, Y., Li, Y., Mei, T. (2019) Hierarchy parsing for image captioning. In: Proceedings of the IEEE International Conference on Computer Vision. 2621–2629.

[25] www.ibm.com/blogs/research/2019/06/image-captioning/ (accessed on 15 July 2022).

Chapter 4

Deep learning-based object detection for computer vision tasks

A survey of methods and applications

Priyanka Dhanasekaran, E. Uma, A. V. Geetha, and T. Mala

4.1 INTRODUCTION

Computer vision is a field in which a 3D scene can be recreated or interpreted using basic 2D images. The subject of computer vision has been fast evolving due to the continual advancement of sophisticated technologies such as Machine Learning (ML), Deep Learning (DL), and transformer neural networks. Figure 4.1 represents the overall learning process of ML and DL.

In ML, handcrafted features are used with proper feature selection techniques [16], whereas DL models can directly extract salient information from images or videos [22]. Thus, advances in DL have made computer vision technologies more precise and trustworthy. The Convolutional Neural Networks (CNN) in DL have made it appropriate for many industrial applications and a trustworthy technology to invest in for businesses wishing to automate their work and duties.

DL enables computational models with several processing layers to learn and represent data at different levels of abstraction, simulating how the brain processes and comprehends multimodal information and implicitly capturing complex structures of big data. Further, the DL model uses different optimization algorithms [15] to have an impact on accuracy and training speed. A wide range of unsupervised and supervised feature learning techniques are included in the DL family, which also includes neural networks and hierarchical probabilistic models. DL techniques perform better than prior state-of-the-art techniques because of a huge volume of input from different sources such as visual, audio, medical, social, and sensor. With the help of DL, significant progress has been made in several computer vision issues, including object detection, motion tracking, action recognition, human posture estimation, and semantic segmentation [17]. CNN act as a mainstream network in the field of computer vision is shown in Figure 4.2. The development of deep network emerged for computer vision tasks is shown in Table 4.1.

DOI: 10.1201/9781003453406-4

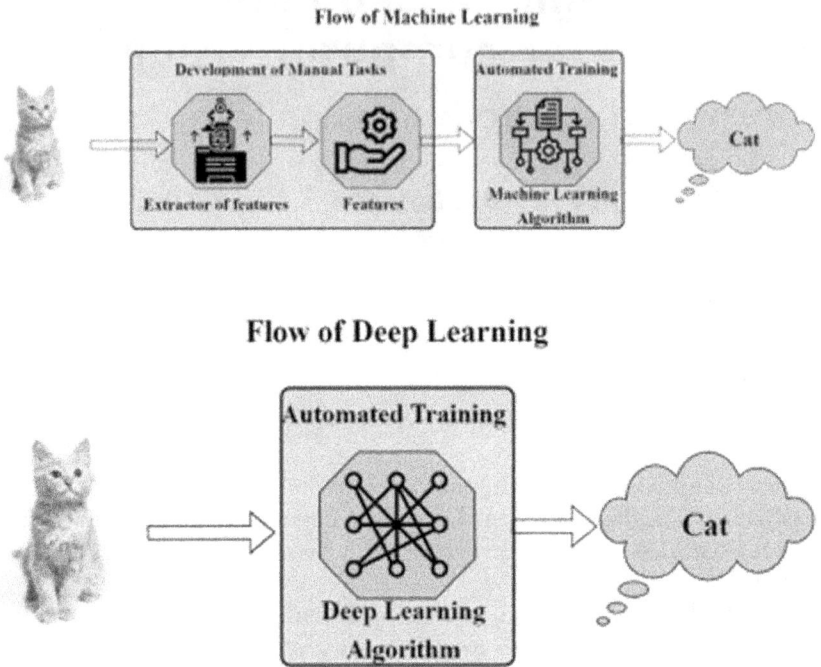

Figure 4.1 Flow of learning process.

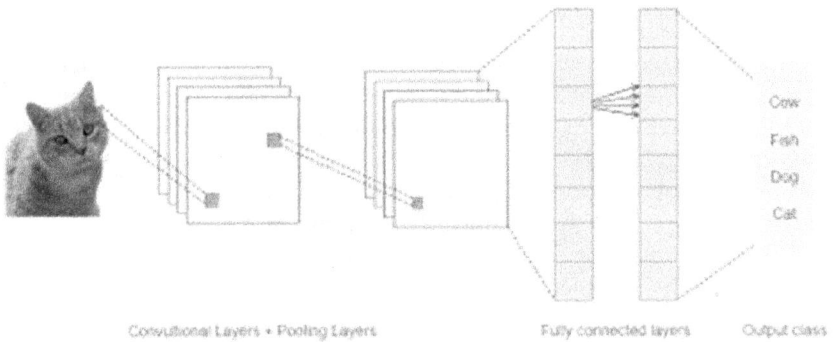

Figure 4.2 Overall view of CNN architecture.

Table 4.1 Evolution of deep learning models

Architecture	Key Characteristics
AlexNet [11]	8 layers, 5 convolution layers, 3 Fully Connected (FC) layers, ReLU nonlinearity used instead of tanh function
VGGNet [21]	Increased depth with 16 or 19 layers, convolutions with 3x3 filters, trains longer due to large parameter space
GoogLeNet [23]	Inception layers to increase depth and width while keeping computational budget constant, 22 layers with 1 x 1 convolution kernels. Inception V2 improved efficiency through 3 x 3 filters. Inception v3 further improved V2 through Batch Normalization
ResNet [6]	152 layers, multiple parameter layers to learn residuals, Residual connections mitigate vanishing gradient problem & reduces parameter space
DenseNet [8]	Interconnects layers with each other, L(L+1)/2 connections, mitigates vanishing gradient problem and reduces parameter space
Mobile Nets [7]	Uses depthwise separable convolution layer for faster and simpler deep neural network, small size and low parameter space for use in mobile and embedded devices

4.2 OBJECT DETECTION

Object detection is an essential task in computer vision that involves identifying and localizing objects within an image or video. The primary goal of object detection is to provide machines with the ability to perceive and understand their surroundings by detecting and recognizing the objects present in them. This capability serves as a foundation for various other computer vision tasks, such as instance segmentation, object tracking, and image captioning.

The traditional methods for object detection, such as the Viola-Jones face detector, utilized techniques such as Adaboost with cascade classifiers, integral images, and the Haar wavelet. The Histogram of Oriented Gradients (HOG) and Deformable Part Models (DPM) were also introduced as powerful feature descriptors. However, the performance of these methods reached a saturation point before the development of deep learning techniques. Recent advancements in deep learning, particularly in CNN, have revolutionized the field of object detection. DL-based object detection methods employ supervised learning, where a model is trained on annotated images to detect objects. These models can handle complex scenes with varying illumination, occlusions, and object orientations. Although collecting a significant amount of annotated data for training deep learning models is challenging, the availability of benchmark datasets like MS-COCO, PASCAL VOC, KITTI, openImage, and ILSVRC with annotated images for object detection has been instrumental in the advancement of DL-based object detection methods.

Object detection models are classified into two types: one-stage detectors and two- stage detectors. One-stage detectors combine both the tasks of finding instances and classifying them with a single network. On the other hand, two-stage detectors have separate modules for each task. The structure and characteristics of each model belonging to a particular object detection category are discussed in detail in the following sections.

4.3 TWO-STAGE OBJECT DETECTORS

4.3.1 R-CNN

Region-based Convolutional Neural Network (R-CNN) extracts the object proposals (region boxes) by merging similar pixels into regions. R-CNN provides nearly two thousand object proposals and identifies the regions having the probability of being an object using a selective search algorithm [25]. Each selected region is reshaped to a fixed size (warped) and inputted to the backbone CNN architecture to extract the features. Thus, each region proposal is rescaled and processed by the CNN, due to the fixed size input representation of the Fully Connected (FC) layer. Further, the classifier and regressor process the feature vector to obtain the class label and bounding box respectively. Figure 4.3 depicts the structure of R-CNN model.

However, R-CNN faces certain issues, such as a slow processing rate in extracting candidate proposals using selective search and redundant CNN feature computation due to overlapped region proposals. Moreover, training time is increased due to the fixed process in extraction of candidate proposals and shows a high prediction time of 47 seconds per image.

4.3.2 SPPNet

Spatial Pyramid Pooling Network (SPPNet) [5] is a modification of R-CNN that can handle images of arbitrary size and aspect ratio. SPPNet processes

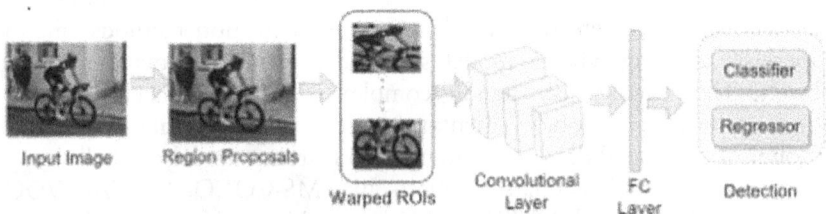

Figure 4.3 Architecture of R-CNN.

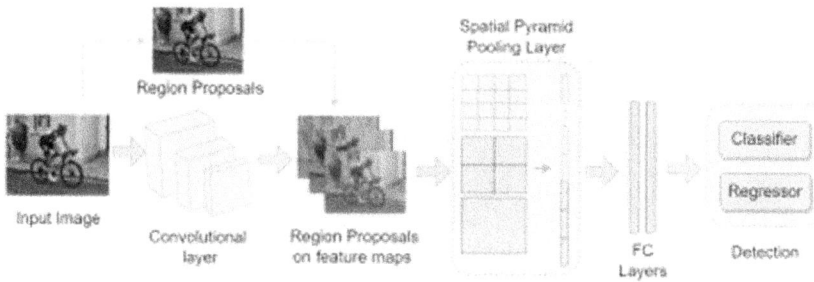

Figure 4.4 Architecture of SPPNet.

the entire image with the CNN layer and adds a pooling layer before the FC layer. The region proposal is extracted using selective search, and candidate regions are mapped onto the feature maps of the last convolutional layer. Next, the candidate feature maps are inputted to the spatial pooling layer and then the FC layer. Finally, classification and regression are performed. SPPNet addresses the warping-based overlapped CNN computation issue by fine-tuning the FC layer. However, the previous layers based on region proposal selection are still not addressed, leading to an increase in training and prediction time. SPPNet architecture is shown in Figure 4.4.

4.3.3 Fast RCNN

Fast RCNN addresses the issue of training multiple region proposals separately as in R-CNN and SPPNet, by utilizing the single trainable system [3]. In Fast RCNN entire image is inputted to the convolutional layer to obtain the feature. The candidate region proposals are obtained using a selective search algorithm; such regions are called Region of Interest (ROI). Such region proposals are mapped onto the final feature maps of the CNN layer. Further, ROI pooling concatenates the feature maps of corresponding region proposals. Thus, a feature map is obtained for every region proposal and then feed to the FC layer. The final layer of classification and regression is performed for object detection.

4.3.4 Faster RCNN

Faster RCNN introduced the Region Proposal Network (RPN) to generate candidate region proposals instead of selective search. RPN makes use of an anchor, a fixed bounding box with different aspect ratios to localize the object [20]. The RPN module consists of a fully convolutional network with a classifier and a bounding box regressor to provide an objectness score. The image is inputted to the CNN part to obtain the feature maps, which are

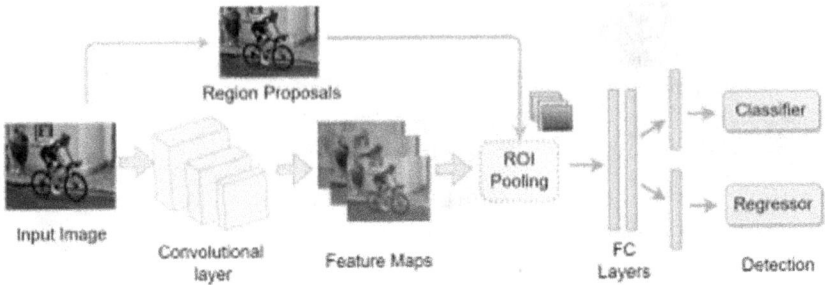

Figure 4.5 Architecture of Faster RCNN.

provided as input to the RPN module. Anchor boxes are selected and predict the object score, removing those with low objectness scores. RPN utilizes multi-task loss optimization for classification and regression. The convolutional feature maps and predicted region proposal are concatenated using ROI pooling. Faster RCNN addresses the issue of slow selective search with a convolutional RPN model, which makes the network learn region proposal along with object detection. The prediction time of Faster RCNN is improved to five frames per second. Figure 4.5 shows the network structure of Faster RCNN model.

4.3.5 R-FCN

The Region-based Convolutional Neural Network (RCNN) model had utilized the fully connected (FC) layer before the object detection layer, which made localization difficult due to the translation-invariant property of CNN. To overcome this limitation, Jifeng et al. [2] modified the FC layer with a fully convolutional layer. However, the performance of the model did not improve significantly. Thus, the Region-based Fully Convolutional Network (R-FCN) was introduced, which includes the position-sensitive score to capture the spatial information of the object, and localization is performed by pooling. The R-FCN model as shown in Figure 4.6, uses ResNet-101 CNN to extract feature maps, and the position-sensitive score map is combined with RPN output for classification and regression. While it has a faster detection speed than other models, its improvement in accuracy is not substantial compared to Faster RCNN.

4.3.6 FPN

Feature Pyramid Network (FPN) addresses the issue of capturing the small objects in the image, which is faced by the Faster RCNN model [12]. FPN

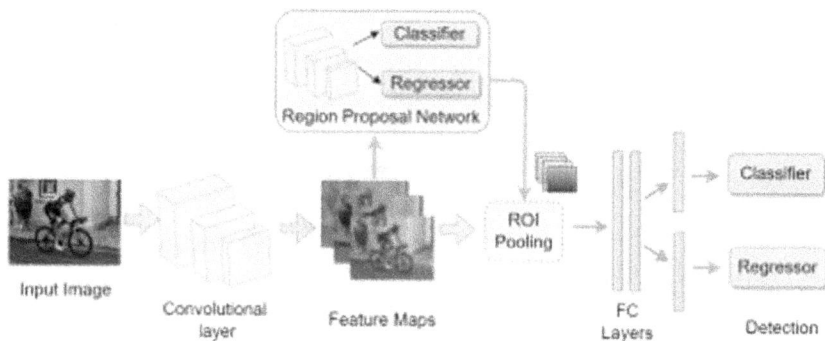

Figure 4.6 Architecture of R-FCN.

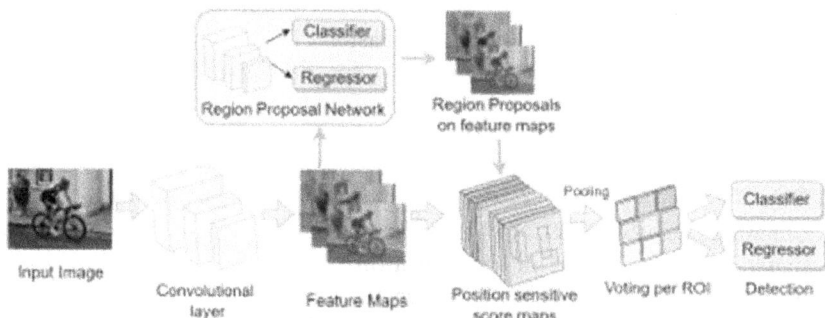

Figure 4.7 Architecture of FPN.

follows either a bottom-up pathway based on kernel hierarchy at different scales or a top-down pathway using upsampling of feature maps from high-level to high-resolution features. An image pyramid is generated by scaling the images at different aspect ratios. Each scale is sent to the detector to obtain the predictions and they are combined using different methods. The top-down architecture with lateral connections aids in extracting the high-level features at different scales as shown in Figure 4.7.

4.3.7 Mask RCNN

Mask RCNN is an extension of Faster RCNN, and the structure is depicted in Figure 4.8. Mask RCNN includes a branch for the prediction of pixel-wise object segmentation in parallel with existing object detection [4]. The fully convolutional layer is applied to the final region proposal output to obtain the object mask. The ROI pooling layer is modified with ROI alignment to

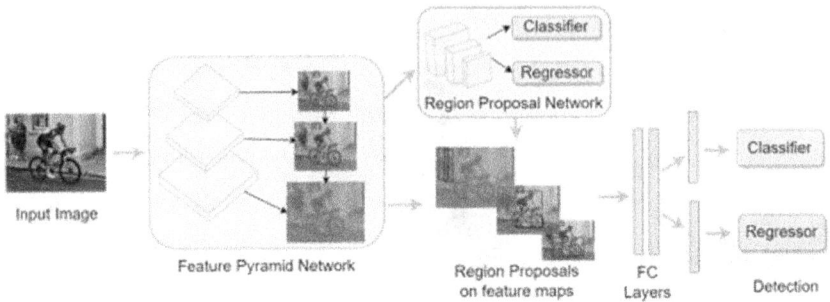

Figure 4.8 Architecture of Mask RCNN.

resolve the pixel-wise misalignment while performing spatial quantization. The additional branch has a little overhead in computation and the prediction time of Mask RCNN is similar to Faster RCNN. Thus, Mask-RCNN lacks real-time requirement of prediction speed. The authors of [10] used Mask-RCNN in tennis for action recognition.

4.3.8 G-RCNN

Granulated RCNN (G-RCNN) is an improved version of Faster RCNN designed for video-based object detection [18]. G-RCNN utilizes a network similar to AlexNet model, which includes 5 convolutional layers, 3 pooling layers, and 3 fully connected layers. Additionally, it incorporates a granulation layer, ROI generation, and anchor process. To extract region proposals in an unsupervised manner, granules (clusters) are formed after the first pooling layer. G-RCNN effectively combines spatial and temporal granules, obtained from static images and video sequences, to capture spatio-temporal information. The granules are processed through the AlexNet layer, anchored for region proposals, and fed to the classifier and regressor for detecting class labels and bounding boxes. Figure 4.9 depicts the detailed view of the G-RCNN model.

4.4 ONE-STAGE OBJECT DETECTORS

4.4.1 YOLO

You Only Look Once (YOLO) is a single CNN model that predicts object classes and their bounding boxes simultaneously on the full image [19]. It divides the image into $K*K$ grid cells and assigns each cell the responsibility of detecting objects that it contains. YOLO uses anchor boxes to provide multiple bounding boxes for each grid cell based on the aspect ratio of

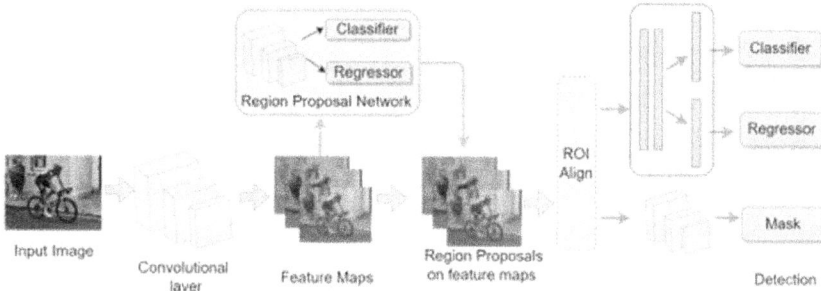

Figure 4.9 Architecture of G-RCNN.

different classes. The predicted bounding box is represented by six values: confidence score for being an object, bounding center coordinates (b_x & b_y), bounding box height and width, and class score.

YOLO achieves high speed but low localization accuracy compared to two-stage detectors. Its limitations include detecting small or clustered objects and difficulty in determining the number of objects in a grid cell. The latest YOLO models focus on addressing these challenges through optimization techniques. The generic structure of the YOLO model is shown in Figure 4.10(a). YOLOv2, YOLOv3 and YOLOv4 are popular object detection models that have been widely used in computer vision tasks. Furthermore, YOLOv4 model incorporates various techniques such as Bag of Freebies and Bag of Specials methods to reduce training and prediction times [1].

4.4.2 CenterNet

A new perspective of object detection is performed by modeling objects as points instead of bounding boxes [27]. CenterNet uses the stacked hourglass-101 model as a backbone for feature extraction, which is pre-trained on the ImageNet dataset. The network provides three outputs as shown in Figure 4.10(b), namely: (1) keypoint heatmap to detect the center of the object; (2) offset to correct the location of an object; and (3) dimension to determine the object aspect ratio. The model training is fine-tuned using the multitask loss of three outputs. Computationally expensive Non-Maximum Suppression (NMS) technique is not required due to detection of the object points instead of boxes. The prediction of a bounding box is generated using the offset output. The network achieves high accuracy with less prediction time compared with previous models. However, it lacks in generalization ability to have different backbone architectures.

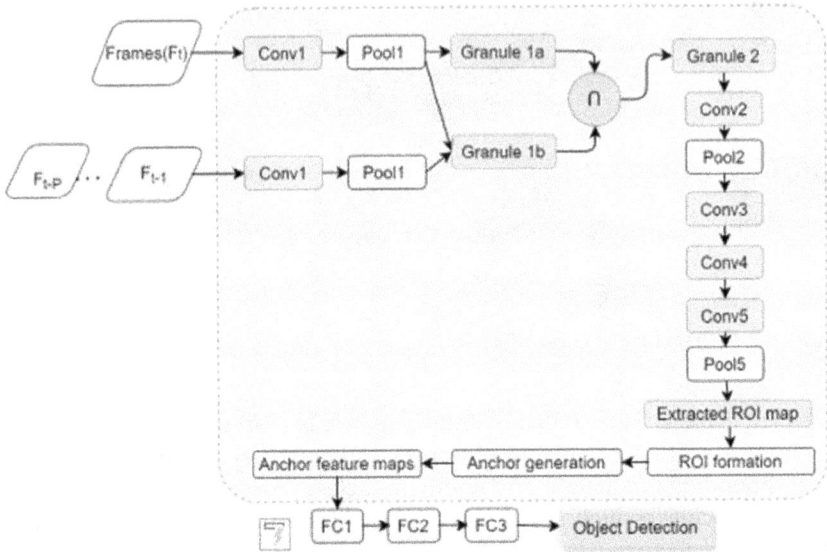

Figure 4.10 Architecture of YOLO and CenterNet.

4.4.3 SSD

Single Shot Multi-box Detector (SSD) is an object detection model proposed by Szegedy et al. [14], which outperforms Faster RCNN and YOLO in terms of average precision and object localization. The model uses VGG16 as a backbone and adds multi-scale feature layers to detect objects at different scales, including small objects. The multi-scale layer provides the offset of default boxes and specific height and weight. The model optimizes using weighted average localization and confidence loss, and applies NMS to remove duplicate predictions.

Although SSD enables real-time object detection, it has difficulty detecting small objects, which can be improved by using VGG19 and Resnet models as backbones. Figure 4.11(a) illustrates the SSD architecture. The authors used SSD for multiple real-time object identification [9].

4.4.4 RetinaNet

The performance reduction of the one-stage detector compared with the two-stage detector is predominantly due to the high-class imbalance between foreground and background objects. Lin et al [13] proposed a RetinaNet model using a new loss function named Focal loss, it provides lower loss for easy misclassified samples and the detector focuses on the

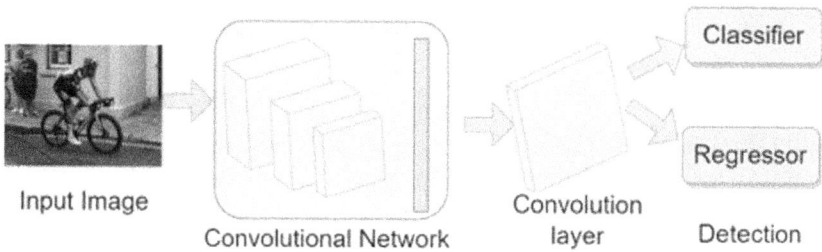

Figure 4.11 Architecture of SSD and RetinaNet.

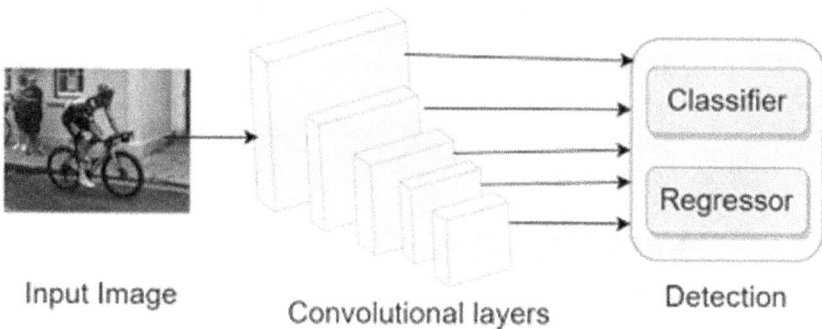

Figure 4.12 Architecture of EfficientDet.

hard misclassified samples. RetinaNet uses Resnet and FPN model as the backbone to extract the features and two sub-networks of fully convolutional layers for classification and regression. Each pyramidal scale layer of FPN is processed by the subnetworks to detect the object class and bounding box in different scales. The diagrammatic representation of RetinaNet is shown in Figure 4.11(b). Thus, RetinaNet, which is simple, fast, and easy to implement and train, has outperformed previous models and paved the way for enhancing model optimization through a new loss function.

4.4.5 EfficientDet

EfficientDet is a model that improves detection accuracy and speed by scaling in different dimensions [24]. It uses multi-scale features, Bi-directional FPN layers, and model scaling. The backbone network is EfficientNet, and multiple BiFPN layers are used to extract features. As in Figure 4.12, the final output is processed by a classifier and regressor network. EfficientDet uses a compounding coefficient to jointly scale the network in all dimensions, resulting in high accuracy and low computational cost.

4.4.6 YOLOR

YOLOR, a novel object detection model proposed by Wang et al. [26], combines explicit and implicit knowledge to create a unified representation. It uses architecture called the scaled YOLOv4 CSP model for object detection and performs multi-task detection using implicit deep learning that generalizes to different tasks. YOLOR achieves significant performance and speed compared to current state-of-the-art object detection models by using modifications like kernel alignment, manifold space reduction, feature alignment, prediction refinement, and multitask learning in a single model. These modifications ensure appropriate kernel space is selected for different tasks, kernels are translated, rotated, and scaled to match the appropriate output kernel space. The model achieves significant performance and speed compared with current state-of-the-art models, making it a promising new approach to object detection.

4.5 DISCUSSION ON MODEL PERFORMANCE

In this section, the object detectors are evaluated for their real-time performance in terms of accuracy and speed. The benchmark datasets like Pascal Visual Object Class (VOC) 2012 and Microsoft Common Objects in Context (MS-COCO) are utilized for model performance comparison. Pascal VOC 2012 and COCO contain 11k training images with more than 27k labeled objects and 2 million images with 7.7 objects per image, respectively. The huge volume of training data supports the deep learning models for feature extraction instead of handcrafted features.

The commonly used metric for comparison of object detectors is precision, recall, and frames per second (fps). Moreover, mean Average Precision (mAP) is the most common metric of evaluation. Intersection over Union (IoU) is used to measure the ratio of bounding box region of overlap and union between the predicted and ground truth. True positives are defined as a correct prediction if IoU is more than a threshold. False positives are predictions that have IoU below the threshold. Thus, precision is defined as the ratio of true positive and all the observations, whereas recall is calculated using the ratio of true positive and ground truth. Further, the average precision is computed for every class separately. Thus, mAP is calculated based on the mean of average precision of all the classes.

Table 4.2 shows the performance of the object detection model on the MS-COCO dataset. The $AP_{0.5}$ represents the average precision computed from IoU with a threshold of 50 percent. The object detectors and their respective backbone detectors for feature map extraction are also highlighted with

Table 4.2 Performance of different object detector on MS COCO dataset

Models	Backbone architecture	$AP_{0.5}$	fps	Year
R-FCN	Resnet-101	53.20%	≈ 3	2016
SSD	VGG-16	41.20%	46	2016
YOLOv2	DarkNet-19	44.00%	81	2016
FPN	ResNet-101	59.10%	5	2017
Mask R-CNN	ResNeXt-101-FPN	62.30%	5	2018
RetinaNet	ResNet-101-FPN	49.50%	12	2018
YOLOv3	DarkNet-53	51.50%	45	2018
CenterNet	Hourglass-104	61.10%	7.8	2019
EfficientDet-D2	Efficient-B2 768	62.30%	41.7	2020
YOLOv4	CSPDarkNet-53	64.90%	31	2020
YOLOR	Scaled YOLOv4 CSP	73.3%	30	2021

Table 4.3 Performance of different object detector on Pascal VOC 2012 dataset

Models	Backbone architecture	$AP_{0.5}$	fps	Year
SPP-Net	ZF-5	59.20%	≈ 0.23	2015
R-CNN	AlexNet	58.50%	≈ 0.02	2016
Fast R-CNN	VGG-16	65.70%	≈ 0.43	2015
Faster R-CNN	VGG-16	67.00%	5	2016

their mAP and fps. Similarly, Table 4.3 shows the object detector performance on the Pascal VOC 2012 dataset. The authors of G-RCNN have not discussed the $AP_{0.5}$, instead providing the mAP value as 80.9 percent on Pascal VOC 2012 dataset with AlexNet as backbone architecture.

The performance of object detectors is mainly based on the input size, training method, optimization, loss function, feature extractor, and so on. Therefore, a common benchmark dataset is required to analyze the model improvement in terms of accuracy and inference time. Thus, the study utilized the standard benchmark dataset like PASAL VOC and MS COCO datasets. From the analysis, it is inferred that for real-time object detection YOLOv4 and YOLOR perform better concerning average precision and inference time.

4.5.1 Future trends

Despite the development of various object detectors, the field of object detection has plenty of room for improvement.

- Lack of lightweight object detectors with reasonable accuracy and speed is the major requirement of object detectors for embedding in edge devices.

- Semi-supervised learning can be a promising approach for improving object detection performance in scenarios with limited availability of annotated images, enabling the use of domain transfer learning for effective object detection with less training data.
- Increasing the ability of object detectors for multi-task and 3D object detection is utilized to increase the capability of detectors.
- Object detection in video sequences offers numerous applications and poses unique challenges compared to static image detection.

4.6 CONCLUSION

This chapter offers a comprehensive review of deep learning-based object detection methods. It categorizes the object detection methods into single-stage and two-stage deep learning algorithms. Recent algorithmic advancements and their architecture are covered in depth. The chapter primarily discusses developments in CNN-based methods because they are the most widely used and ideal for image and video processing. Most notably, some recent articles have shown that some CNN-based algorithms have already become more accurate than human raters.

However, despite the encouraging outcomes, more development is still required – for instance, the current market demand to develop a high-precision system using lightweight models for edge devices. This work highlights the ongoing research in improving deep neural network-based object detection, which presents various challenges and opportunities for improvement across different dimensions, such as accuracy, speed, robustness, interpretability, and resource efficiency.

REFERENCES

[1] Bochkovskiy, A., Wang, C.Y., Liao, H.Y.M.: *Yolov4: Optimal speed and accuracy of object detection.* arXiv preprint arXiv:2004.10934 (2020).

[2] Dai, J., Li, Y., He, K., Sun, J.: R-fcn: Object detection via region-based fully convolutional networks. *Advances in neural information processing systems.* P. 29 (2016).

[3] Girshick, R.: Fast r-cnn. In: Proceedings of the IEEE international conference on computer vision. pp. 1440–1448 (2015).

[4] He, K., Gkioxari, G., Dollár, P., Girshick, R.: Mask r-cnn. In: Proceedings of the IEEE international conference on computer vision. pp. 2961–2969 (2017).

[5] He, K., Zhang, X., Ren, S., Sun, J.: Spatial pyramid pooling in deep convolutional networks for visual recognition. IEEE transactions on pattern analysis and machine intelligence 37(9), 1904–1916 (2015).

[6] He, K., Zhang, X., Ren, S., Sun, J.: Deep residual learning for image recognition. In: 2016 IEEE Conference on Computer Vision and Pattern Recognition (CVPR). pp. 770–778 (2016).

[7] Howard, A., Zhu, M., Chen, B., Kalenichenko, D., Wang, W., Weyand, T., Andreetto, M., Adam, H.: Mobilenets: Efficient convolutional neural networks for mobile vision applications (2017).

[8] Huang, G., Liu, Z., Van Der Maaten, L., Weinberger, K.Q.: Densely connected convolutional networks. In: 2017 IEEE Conference on Computer Vision and Pattern Recognition (CVPR). pp. 2261–2269 (2017).

[9] Kanimozhi, S., Gayathri, G., Mala, T.: Multiple real-time object identification using single shot multi-box detection. In: 2019 International Conference on Computational Intelligence in Data Science (ICCIDS). pp. 1–5. IEEE (2019).

[10] Kanimozhi, S., Mala, T., Kaviya, A., Pavithra, M., Vishali, P.: Key object classification for action recognition in tennis using cognitive mask rcnn. In: Proceedings of International Conference on Data Science and Applications. pp. 121–128. Springer (2022).

[11] Krizhevsky, A., Sutskever, I., Hinton, G.E.: Imagenet classification with deep convolutional neural networks. *Advances in neural information processing systems* 25 (2012).

[12] Lin, T.Y., Dollár, P., Girshick, R., He, K., Hariharan, B., Belongie, S.: Feature pyramid networks for object detection. In: Proceedings of the IEEE conference on computer vision and pattern recognition, pp. 2117–2125 (2017).

[13] Lin, T.Y., Goyal, P., Girshick, R., He, K., Doll'ar, P.: Focal loss for dense object detection. In: Proceedings of the IEEE international conference on computer vision. pp. 2980–2988 (2017).

[14] Liu, W., Anguelov, D., Erhan, D., Szegedy, C., Reed, S., Fu, C.Y., Berg, A.C.: Ssd: Single shot multibox detector. In: European conference on computer vision. pp. 21–37. Springer (2016).

[15] Mahajan, S., Abualigah, L., Pandit, A.K., Nasar, A., Rustom, M., Alkhazaleh, H.A., Altalhi, M.: Fusion of modern meta-heuristic optimization methods using arithmetic optimization algorithm for global optimization tasks. *Soft Computing* pp., 1–15 (2022).

[16] Mahajan, S., Pandit, A.K.: Hybrid method to supervise feature selection using signal processing and complex algebra techniques. *Multimedia Tools and Applications*, pp. 1–22 (2021).

[17] Mahajan, S., Pandit, A.K.: Image segmentation and optimization techniques: a short overview. *Medicon Eng Themes* 2(2), 47–49 (2022).

[18] Pramanik, A., Pal, S.K., Maiti, J., Mitra, P.: Granulated rcnn and multi-class deep sort for multi-object detection and tracking. IEEE Transactions on Emerging Topics in Computational Intelligence 6(1), 171–181 (2022).

[19] Redmon, J., Divvala, S., Girshick, R., Farhadi, A.: You only look once: Unified, real-time object detection. In: Proceedings of the IEEE conference on computer vision and pattern recognition. pp. 779–788 (2016).

[20] Ren, S., He, K., Girshick, R., Sun, J.: Faster r-cnn: Towards real-time object detection with region proposal networks. *Advances in neural information processing systems* 28 (2015)

[21] Simonyan, K., Zisserman, A.: *Very deep convolutional networks for large-scale image recognition.* arXiv preprint arXiv:1409.1556 (2014).

[22] Sulthana, T., Soundararajan, K., Mala, T., Narmatha, K., Meena, G.: Captioning of image conceptually using bi-lstm technique. In: *International Conference on Computational Intelligence in Data Science*. pp. 71–77. Springer (2021).

[23] Szegedy, C., Liu, W., Jia, Y., Sermanet, P., Reed, S., Anguelov, D., Erhan, D., Van- houcke, V., Rabinovich, A.: Going deeper with convolutions. In: 2015 IEEE Conference on Computer Vision and Pattern Recognition (CVPR). pp. 1–9 (2015).

[24] Tan, M., Pang, R., Le, Q.V.: Efficientdet: Scalable and efficient object detection. In: Proceedings of the IEEE/CVF conference on computer vision and pattern recognition. pp. 10781–10790 (2020).

[25] Uijlings, J.R., Van De Sande, K.E., Gevers, T., Smeulders, A.W.: Selective search for object recognition. *International Journal of Computer Vision* 104(2), 154–171 (2013).

[26] Wang, C.Y., Yeh, I.H., Liao, H.Y.M.: You only learn one representation: Unified network for multiple tasks. arXiv preprint arXiv:2105.04206 (2021).

[27] Zhou, X., Wang, D., Krähenbühl, P.: *Objects as points*. arXiv preprint arXiv:1904.07850 (2019).

Chapter 5

Deep learning algorithms for computer vision

A deep insight into principles and applications

Medha Gupta and Soma Debnath

5.1 INTRODUCTION

The exponential rise in the availability of information and big data over the past few years drives the motivation to filter and extract high and very specific information from raw sensor data – for example, speech progression, images, videos, and so forth. We know that computers do not perceive images as the human eye does. They are naturally capable of understanding the numeric notation. To perceive images in machine readable format, the first and foremost step of any computer is to convert the information contained in an image in understandable and readable form for machines [1,2,3]. Since images are constructed of a grid of pixels that cover every tiny part of the image, each pixel can be considered to be a "spot" of a singular color. The greater content of pixels in an image represents higher resolution if the image. It is known that the human brain associates some important "features" (size, shape, color, etc.) with each object, which helps one focus solely on those features to recognize those objects correctly [4,5,6,7]. It succeeds in delivering highly accurate results when put to use to extract certain particular "features" from images and identify each feature to an individual category of objects. A convolution matrix identifies the patterns or "features" that need to be extracted from the raw visual data, further helping in image identification. A neural network on the other hand, defines a succession of algorithms that aim to conclude the rudimentary relationships in synchronized data in a method that largely imitates the way a human brain would deliver on the principal relationship in the set of data. In the true sense, a neural network cites an arrangement of artificial "neurons" that utilizes a convolution matrix to break down the visual input and recognize the key "features" needed to be extracted for image categorization and thus concluding information from the same.

Some of the different libraries which are utilized in computer vision are TensorFlow, Keras, MATLAB®, and so forth. These libraries which depend largely on GPU-accelerated libraries deliver soaring multi-GPU-accelerated training. Apart from being computationally efficient and reducing the input

DOI: 10.1201/9781003453406-5

images into a form which is quite easier to process without the loss of any important feature, CNNs have the advantage of detecting and extracting important features from any visual input without any human intervention. This non-involvement of human-interaction gives it the added advantage when compared to its predecessors.

Deep learning in computer vision has risen high in the evolving world. From object detection to identifying whether the X-Ray is indicative of presence of cancer, deep learning methodologies when appropriately implemented in the domain of computer vision can flourish to be helpful to the mankind as well. The chapter delves into the preliminary concepts of deep learning along with a detailed overview of applied deep learning algorithms used in computer vision. The following sections discuss miscellaneous tools, libraries, and frameworks of deep learning in computer vision. Computer vision has proved to be versatile, with its applications in various industrial sectors, such as the transportation sector, the manufacturing unit, healthcare, retail, agriculture, construction, and so forth. The penultimate section includes these flexible industrial applications. The chapter concludes with a reference to a few prospects of the same domain, which could be further progressed accordingly.

5.2 PRELIMINARY CONCEPTS OF DEEP LEARNING

Deep learning is rightly defined as a subset of machine learning, which teaches computerized machines to learn and do the primary job of any human, that is, of learning by example. Deep learning is essentially only a neural network with multiple layers. It is called "DEEP" Learning owing to the presence of these multiple layers, which help us to learn from the data. These multiple layers are made of numerous interconnected neurons making the entire network mimic the architecture and arrangement of "neurons" in the human brain.

Deep learning functions entirely on Neural Networks primarily called neural nets, which make the computerized systems accumulating inspiration from the biological neural network that make up the human brain [5,8,9]. There are mainly three vivid types of neural networks that build the fundamentals for more than half the pre-trained models in deep learning, namely,

- Artificial Neural Networks (ANNs)
- Convolution Neural Networks (CNNs)
- Recurrent Neural Networks (RNNs)

As seen in Figure 5.1, the foundational structure of any neural network consists of a number of intermediate hidden layers that take input from the supervised and unsupervised data into the input nodes; the network works

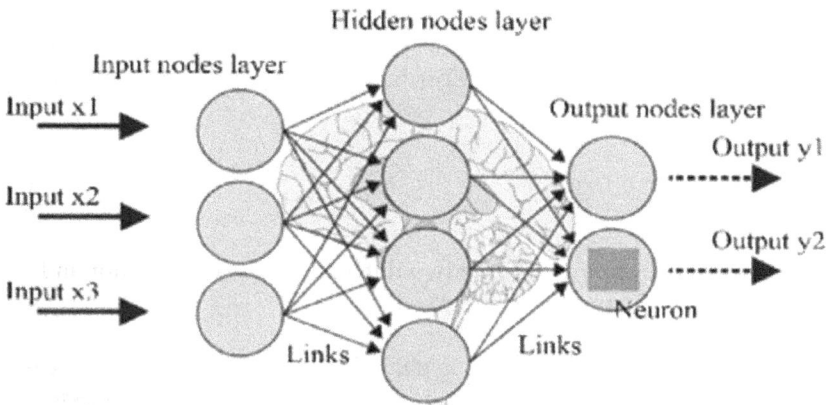

Hidden nodes layer

Input nodes layer

Input x1

Input x2

Input x3

Output nodes layer

Output y1

Output y2

Neuron

Links

Links

Figure 5.1 Fundamental structure of a Neural Network.

in these hidden layers and redirects the results to the output nodes. Each input node is connected to every other hidden layer present in the model, and every hidden layer is connected to each output node forming an interconnected structure.

5.2.1 Artificial Neural Network

Artificial Neural Networks (ANNs) evolve by programming daily computers to function like the human brain. This is an attempt to stimulate the network of artificial "neurons" that would enable the computers to behave and make decisions as a human being would do.

The key component on which CNNs and RNNs are built is the Feed Forward Neural Network, also known as Deep Feed Forward Networks or Multi-layer Perceptrons.

In the Feed Forward Neural Network, the nodes' connections do not form a loop that results in a unidirectional flow of information. Looking at Figure 5.1, we can say Feed Forward Neural Networks data enters the input nodes, flow through the hidden layers, and exit through the output nodes. The absence of any loop formation in the network eliminates the probability of the output information flowing backwards into the network. Feed Forward Neural Networks were built with the aim to approximate functions.

Cost Function in Feed Forward Neural Networks: In order to upgrade the delivered performance of the neural networks, minor adjustments are done to the neurons weights and biases using a cost function which is given by

$$C(w,b) \equiv \frac{1}{2n} \sum_x \|y(x) - a\|^2 \tag{5.1}$$

Loss Function in Feed Forward Neural Networks: The Loss Function in any Neural Network is used to cross-check whether the learning process requires any updating. This function is used to exhibit the difference between the predicted and actual distributions of probabilities. The function is given by:

$$L(\theta) = -\sum_{i=1}^{k} y_i \log(\hat{y}_i) \qquad (5.2)$$

Phases of Operation: There are two different phases of operation in Feed Forward Neural Networks, namely,

- Learning Phase: In the starting phase of the network operation the network weights are adjusted to make sure the output unit possesses the largest value.
- Classification Phase: Though the network weights remain intact, in the classification phase the input pattern becomes modified in every layer until it is mapped to the output layer. Classification is done based on selected categories mapped to the output unit with the largest value.

The classification phase is faster by comparison with the learning phase.

5.2.2 Convolution Neural Network (CNNs)

A Convolution Neural Network is a sub-domain of ANNs and a super sub-division of deep learning, which has proved its dominance in computer vision. CNNs were primarily designed to process and classify images (pixel

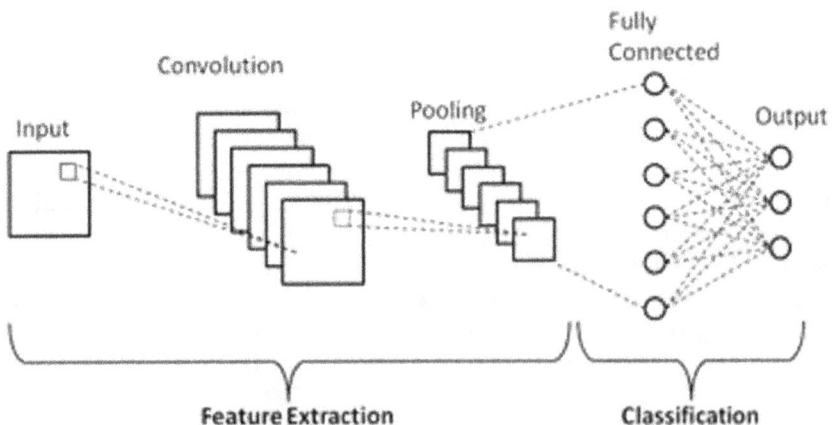

Figure 5.2 Building block of CNN.

data) [10,11,12]. CNNs are used to reduce the input of large sized images into a form that is understandable and easier to process by the computerized machines without losing the distinct features of the image.

A Convolution Neural Network is built of multiple building blocks such as convolution layers (kernel/filter), pooling layers, and fully connected layers. The amalgamation of all these layers makes up the CNN that is used for automatic and progressive learning of spatial hierarchies of features present in the input image through an algorithm that supports back-propagation [13,14].

5.3 RECURRENT NEURAL NETWORK (RNNs)

Recurrent Neural Networks are a subset of neural networks where the output of the current step is fed as an input for the next step. Unlike the other neural networks, where all the inputs and outputs are independent of each-other, RNNs have all of them interconnected.

In tasks where one needs to predict the outcome of the next sequence, such as a text sequence, where the next word would be dependent on the previous words of the sentence, RNNs would be implemented. The technique through which RNNs are able to exhibit this phenomenon is the presence of Hidden Layers, which are intermediate layers remembering information and generating the next output based on the previously fed inputs.

For example: If the input for the neural network is [1,2,3] and it has been asked to predict the next outcome. The outcome could be anything – any number, any alphabet, or any symbol. A typical normal neural network would not be able to predict an accurate result there is no connection between the inputs and outputs. If the same input is fed to a RNN based on the previous inputs, it would predict the output as [4].

RNNs have a memory that stores all the previously calculated information. It increases the network efficiency and reduces the network complications since it uses those similar parameters for each input thereby performing the same calculations on the hidden layers.

Working Principle of RNN:

Unlike in other neural networks, the hidden layers would have their own weights and biases, which would be different from one-another and thus would not be inter-connected. In an RNN, all the intermediate hidden layers possess the same weights and biases for which the entire setup would work like a memory or one recurrent hidden layer remembering the inputs of each layer. The output for one-layer acts like an input for the next. This feature gives them the added advantage for working in time sequence prediction problems or any sequential problem such as sentence generation.

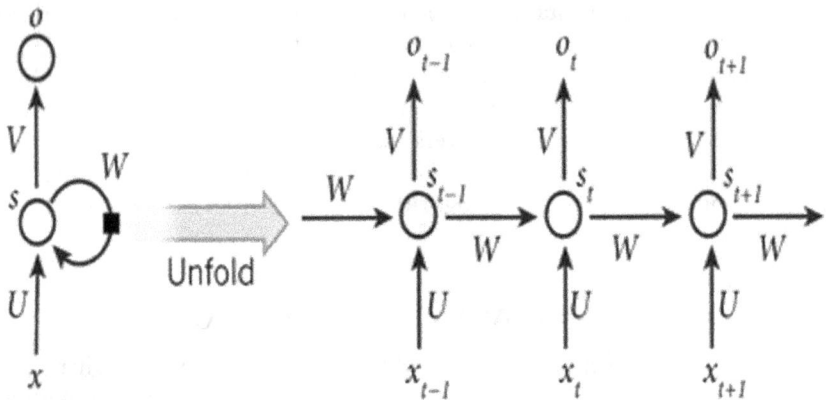

Figure 5.3 Structure of RNN.

The formula for calculating a hidden state in a recurrent neural network is

$$h_t = f\left(h_{t-1}, x_t\right) \tag{5.3}$$

Formula for the application of activation function (tanh):

$$h_t = \tanh\left(W_{hh}h_{t-1} + W_{xh}x_t\right) \tag{5.4}$$

Formula for calculating the output:

$$h_t = \tanh\left(W_{hh}h_{t-1} + W_{xh}x_t\right) \tag{5.5}$$

Despite the given advantages, RNNs possess few disadvantages, such as the vanishing gradient problem or the exploring gradient problem. With the intention to conquer the same, LSTM (Long Short-Term Memory) has been introduced.

5.4 OVERVIEW OF APPLIED DEEP LEARNING IN COMPUTER VISION

Computer vision is the field of study that defines how computerized machines interpret visual content of data such as images and videos. Computer vision algorithms minutely study different segments of the visual input, thereby applying interpretations to predictive or decision deriving tasks. Computer vision and image processing are not the same thing. Image processing involves enhancement or modification of images, such as optimizing brightness or contrast, subtracting noise or blurring sensitive information to derive a new result altogether. Image processing also does not require the identification of visual content. On the other hand, computer vision is solely about identifying and classifying visual input to translate the visual data based on the evaluative information identified during the training phase.

Deep learning has been activated in a number of fields of computer vision and has emerged to be one of the most developing and promising strengths in the area of computer applications. Some of these fields are as follows:

1. Image Classification: Image Classification [15] involves the assignment of a certain type of label to the input pictorial data. Often referred to as "object classification" or "Image Classification," examples include:
   • Analysing an X-Ray image or a CT scan image and labelling whether the same is cancer or not.
   • Organizing a hand-written digit.

Image Classification with localization results in assigning class labels to the pictorial input and thereby showcasing the position of the object in the image.

2. Object Detection: Object Detection is the process of classifying the image with the process of localization (pinpointing the position of the particular object in the image) in spite of the image consisting of multiple other objects that may be required to undergo the process of localization and classification further. A few examples of object detection:
   • Drawing a box, labelling and thus classifying each object in the given image irrespective of it being an indoor or an outdoor photograph.
3. Object Segmentation: Also referred to as *semantic segmentation*, this process involves the segmentation and further classification of every object that has been identified in the image separately [16].
4. Image Colorization: Image Colorization refers to the conversion of a greyscale image to a full color image. Often referred to as "neural colorization," this technique can be thought of as the application of a filter to a black and white image. This process does not have any involvement of object evaluation.
5. Image Reconstruction: Image Reconstruction refers to the process of substituting corrupted or missing parts of an image.

6. Image Synthesis: Image Synthesis includes bringing about certain small-scale modification to the visual input generating thought of modifications of already existing images. This includes:
   - Addition or deletion of an object to the scene;
   - Bringing a change in the style of an existing object in the scene.
7. Image Super-Resolution: This process involves the generation of a brand-new version of the input image, which is gradually more detailed and of a higher resolution than the existing visual input data.

5.6 INDUSTRIAL APPLICATIONS OF COMPUTER VISION

Computer vision has escalated widely to a new height with its innovative initiations and implemented deep learning algorithms. It has brought changes in the industrial sector by

- Inventing innovative solutions to minimize undertaken risks, the total time required for the accomplishment of any job;
- Strengthening operator productivity;
- Executing automatic control, which is qualitative in nature for better product quality, which also helps in an exponential increase in productivity; and
- Reduction in human involvement which saves the lives of many.

A few of the fascinating industrial applications of computer vision are as follows:

- Intuitive monitoring solutions: Drone assisted systems enable companies to bring about remote inspections of their sites without having to be present physically. This application helps in the mining industry, which is hazardous to humans where miners need to descend into the deep mines for visual data collection.
- Prognostic supporting systems: Prognostic Supporting Systems when assisted with computer vision technology and sensors have paved the way for easy tracking of the condition of critical infrastructure and call for maintenance. FANUC's Zero Down Time is such an example, which collects image related metadata to reveal any probable abnormalities in the machines.
- Robot Palletizing structures: These structures which are rightly guided by computer vision, load and unload boxes automatically after correctly identifying them.
- Benchmark quality control: Improved computer-vision powered cameras can record every minute detail happening around us, which has helped companies to keep a check on everything. Instruments like the WebSPECTOR, which is a surface inspector, rightly pinpoints

defects in objects, thereby taking snapshots and collecting image related metadata, which help them to classify errors and their types accordingly.

- Safety Monitoring solutions: Using the technology of computer vision helps to monitor any ill person anywhere in public, monitoring manufacturing and construction sites that can ultimately protect people working in unsafe environments by alerting to machine alarms for dangerous events.
- Warehouse management systems: Helps to reduce inventory time from hours to minutes, thereby saving huge amounts on operational costs. Pegasus robot technology used by Amazon, guarantees to improve storing accuracy by 50 percent at the storing centres.

5.7 FUTURE SCOPE IN COMPUTER VISION

Computer vision can create a system that can replicate the complex functioning of the human visual system, bridging the gap between the digital and physical worlds. Computer vision has already gained popularity owing to its simple and discrete methodologies and techniques. The future of computer vision is bright, with many companies and tech giants working for its implementation to make lives simpler. The future computer vision scope in different industries could be as follows:

- Autopilot system in the automotive industry, which can drive an individual from one place to another with only a simple command. An array of vision sensors could also be employed, which would add advanced driver assistance functions, such as accident prevention systems, road sign detection, and so forth.
- The healthcare industry faces an ever-growing challenge of enormous patient data production every year. Microsoft is working towards introducing a tool that puts artificial intelligence and computer vision to use to collect this data, and thereby analyze and process it, recommending medical treatments to the respective patients based on these reports, their medical history.
- For the agriculture sector, computer vision offers solutions that monitor and maintain farm equipment and livestock efficiently. Utilizing Deep Learning Methodologies along with satellite images can help farmers estimate anticipated seasonal yield from their smartphones directly. Proper implementation of computer vision can facilitate various tasks in the cycle of production of crops, such as planting to harvesting periodically and automatically.
- Efficient Fraud Management, Fake Bill Detection, authenticating users through facial recognition while drawing money from an ATM machine can be easily done by computer vision, saving a lot of time.

- Computer vision, when coupled with the right sensors, can do wonders. For the manufacturing industry, product inspection can be revolutionized by the correct implementation of computer vision. It can automate the entire process of quality control and error detection before packaging the products. The same technology can be used for rectifying errors in 3D printing where implementing computer vision with artificial intelligence enables a machine to predict the behavior of any material in various physical conditions.

5.8 CONCLUSION

Owing to the continuous expansion of vision technology, it can be said that in the immediate future, computer vision would prove to be the primary technology giving solutions to almost every real-world problem. The technology is capable of optimizing businesses, strengthening security, automating services, and thus seamlessly bridging the gap between tech and the real world. Integrating deep learning methodologies in computer vision has taken vision technology to a new level that will lead to the accomplishment of various difficult tasks.

REFERENCES

1. Wilson, J.N., and Ritter, G.X. *Handbook of Computer Vision Algorithms in Image Algebra.* CRC Press; 2000 Sep 21.
2. Hornberg, A., editor. *Handbook of Machine Vision.* John Wiley; 2006 Aug 23.
3. Umbaugh, S.E. *Digital Image Processing and Analysis: Human and Computer Vision Applications with CVIPtools.* CRC Press; 2010 Nov 19.
4. Tyler, C.W., editor. *Computer Vision: From Surfaces to 3D Objects.* CRC Press; 2011 Jan 24.
5. Guo, Y., Liu, Y., Oerlemans, A., Lao, S., Wu, S., and Lew, M.S. *Deep Learning for Visual Understanding: A Review. Neurocomputing.* 2016 Apr 26;187:27–48.
6. Debnath, S. and Changder, S. Automatic detection of regular geometrical shapes in photograph using machine learning approach. In 2018 Tenth International Conference on Advanced Computing (ICoAC) 2018 Dec 13 (pp. 1–6). IEEE.
7. Silaparasetty, V. *Deep Learning Projects Using TensorFlow 2.* Apress; 2020.
8. Hassaballah, M. and Awad, A.I., editors. *Deep Learning in Computer Vision: Principles and Applications.* CRC Press; 2020 Mar 23.
9. Srivastava, R., Mallick, P,K,, Rautaray, S.S., and Pandey, M., editors. *Computational Intelligence for Machine Learning and Healthcare Informatics.* Walter de Gruyter GmbH & Co KG; 2020 Jun 22.
10. de Campos Souza, P.V. Fuzzy neural networks and neuro-fuzzy networks: A review the main techniques and applications used in the literature. *Applied soft computing.* 2020 Jul 1;92:106275.

11. Rodriguez, L.E., Ullah, A., Espinosa, K.J., Dral, P.O., and Kananenka, A.A. A comparative study of different machine learning methods for dissipative quantum dynamics. arXiv preprint arXiv:2207.02417. 2022 Jul 6.
12. Debnath, S., and Changder, S.. Computational approaches to aesthetic quality assessment of digital photographs: state of the art and future research directives. *Pattern Recognition and Image Analysis*. 2020 Oct; 30(4):593–606.
13. Debnath, S., Roy, R., and Changder, S. Photo classification based on the presence of diagonal line using pre-trained DCNN VGG16. *Multimedia Tools and Applications*. 2022 Jan 8:1–22.
14. Ajjey, S.B., Sobhana, S., Sowmeeya, S.R., Nair, A.R., and Raju, M. Scalogram Based Heart Disease Classification Using Hybrid CNN-Naive Bayes Classifier. In 2022 International Conference on Wireless Communications Signal Processing and Networking (WiSPNET). 2022 Mar 24 (pp. 345–348). IEEE.
15. Debnath, S., Hossain, M.S., Changder, S. Deep Photo Classification Based on Geometrical Shape of Principal Object Presents in Photographs via VGG16 DCNN. In: Proceedings of the Seventh International Conference on Mathematics and Computing. 2022 (pp. 335–345). Springer, Singapore.
16. Abdel-Basset, M., Mohamed, R., Elkomy, O.M., & Abouhawwash, M. Recent metaheuristic algorithms with genetic operators for high-dimensional knapsack instances: A comparative study. *Computers & Industrial Engineering*. 2022 166: 107974.

Chapter 6

Handwritten equation solver using Convolutional Neural Network

Mitali Arya, Pavinder Yadav, and Nidhi Gupta

6.1 INTRODUCTION

It is a difficult task in image processing to use a Convolutional Neural Network (CNN) to create a robust handwritten equation solver. Handwritten mathematical expression recognition is one of the most difficult problems in the domain of computer vision and machine learning. In the field of computer vision, several alternative methods of object recognition and character recognition are offered. These techniques are used in many different areas, such as traffic monitoring [3], self-driving cars [9], weapon detection [17], natural language processing [11], and many more.

Deep learning is subset of machine learning in which neural networks are used to extract increasingly complex features from datasets. The deep learning architecture is based on data understanding at multiple feature layers. Further, CNN is another core application of deep learning approach, consisting of convolutions, activation functions, pooling, densely linked, and classification layers. Over the past several years, deep learning has emerged as a dominating force in the field of computer vision. When compared to classical image analysis problems, CNNs have achieved the most impressive outcomes.

Deep learning is becoming increasingly important today. Deep learning techniques are now being used in several fields like handwriting recognition, robotics, artificial intelligence, image processing, and many others. Creating such a system necessitates feeding our machine data in order to extract features to understand the data and make the possible predictions. The correction rate of symbol segmentation and recognition cannot meet its actual requirements due to the two-dimensional nesting assembly and variable sizes. The primary task for mathematical expression recognition is to segment and then classify the characters. The goal of this research is to use a CNN model that can distinguish handwritten digits, characters, and mathematical operators from an image and then set up the mathematical expression and compute the linear equation.

DOI: 10.1201/9781003453406-6

The purpose of this study lies in designing a deep learning model capable of automatic recognizing handwritten numerals, characters, and mathematical operations when presented with an image of the handwriting. In addition, the purpose extends to build a calculator that is capable of both setting up the mathematical statement and computing the linear equation.

The chapter is divided into different sections. Section 2 presents a thorough summary of current handwritten character recognition research studies in recent years. Section 3 goes through each component of the CNN in depth. Section 4 describes the proposed deep learning algorithms for handwritten equation recognition as well as the dataset used. Section 5 discusses the comparative analysis of different technical approaches. In addition, future scope and conclusion are provided in Section 6.

6.2 STATE-OF-THE-ART

There are variety of methods that have been developed to recognize handwritten digits. Handwritten digit recognition has many applications such as bank checks, postal mail, education and so forth. In past years, many methods have been used to recognize handwritten digits such as Support Vector Machines (SVM), Naive Bayes, CNN, K-Nearest Neighbors and so forth. In a few decades, CNN has achieved good performance in handwritten digit recognition.

For offline handwritten character recognition (HCR), Agarwal et al. [4] employed CNN and Tensorflow. They divided the HCR system into six stages: image data collection, image preprocessing for enhancement, image segmentation, feature extraction using the CNN model, classification, and postprocessing for detection. Furthermore, Softmax Regression was used to assign probabilities to handwritten characters because it produces values ranging from 0 to 1 and sums to 1. The use of normalization in conjunction with feature extraction resulted in higher accuracy results as achieved more than 90 percent. However, the study did not provide any information on the specific comparative outcomes and the dataset that was examined.

Bharadwaj et al. [6] used Deep Convolution Neural Networks for effective handwritten digit detection on Modified National Institute of Standards and Technology (MNIST) dataset. The dataset consists of 250 distinct forms of writing and 70,000 digits. The proposed technique includes the steps of preprocessing, model construction and compilation, training the model, evaluation of the trained model, and detection of digits. Both the computer-generated and handwritten digits were recognized by the model. They predicted real-world handwritten digits with 98.51 percent accuracy and 0.1 percent loss. However, the model could only identify the characters from clear and good quality images. The most difficult component is dealing with the images that are blurred or have noise in the real-world images.

Thangamariappan et al. [16] used a variety of machine learning techniques for handwritten digit recognition, including Naive Bayes, Random Forest, SVM and others.

The model was trained using a multi-layer perceptron neural network model on the MNIST dataset. They achieved 98.50 percent accuracy in Digit Recognition with MNIST dataset. Meanwhile, the test accuracy on the same dataset was 88.30 percent, which was relatively low when compared to the training accuracy.

Chen et al. [14] used CNN to recognize four basic arithmetic operations: addition, division, subtraction, and multiplication. On the MNIST dataset, the CNNs performance was trained and evaluated. The improved CNN model is tested in handwritten digit recognition and four arithmetic operations. The convergence speed of CNN model has been reduced and observed 91.20 percent accuracy. The authors only experimented on clear images in their proposed model, which was trained on the MNIST dataset. The trained model was unable to recognize characters in noisy or blurry images.

Gawas et al. [8] proposed a system for recognizing handwritten digits and symbols that consist of addition, subtraction, and multiplication and for solving basic equations containing these operations. They used CNN and created a front-end interface that allows the user to write the equation, which is then identified and solved. They used libraries such as OpenCV, Keras, and Flask to deploy the model. The trained model could only perform fundamental mathematical operations, but failed to solve linear equations.

The authors created their own dataset for the handwritten equation solver and trained it using deep learning models. By constructing the equation solver calculator, researchers not only identify the characters and symbols, but also has solved the linear mathematical problem.

6.3 CONVOLUTIONAL NEURAL NETWORK

CNN (in certain cases referred to as ConvNet) is a type of deep neural network that specializes in image processing and has a grid-like topology [5]. A CNN is a feed-forward neural network with multiple layers. It is formed by assembling many layers on top of one another in the sequence that can be seen in Figure 6.1. CNN trains the model using raw pixel image input, then extracts features for better categorization.

Convolutional layers, fully connected layers, and pooling layers comprise the three different kinds of layers that are included in a deep neural network.

6.3.1 Convolution layer

The first layer utilized to extract various information from input images is the Convolutional Layer. The dot product is computed using an array of

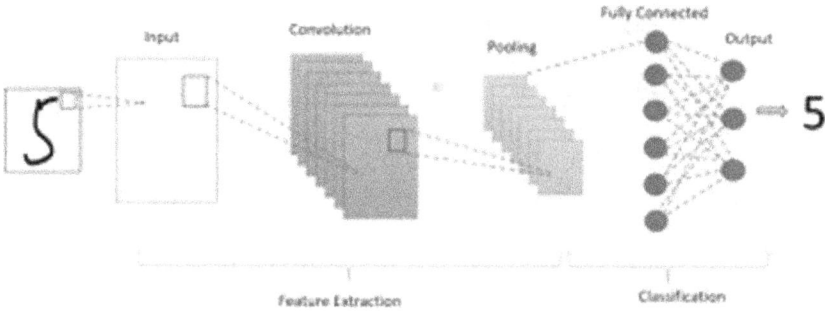

Figure 6.1 CNN architecture for handwritten images.

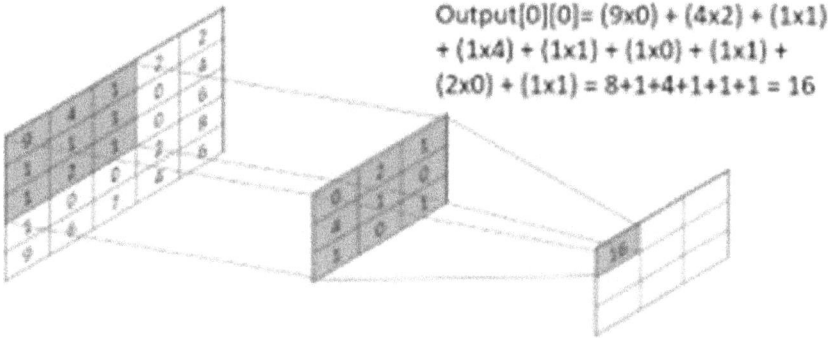

Output[0][0]= (9x0) + (4x2) + (1x1)
+ (1x4) + (1x1) + (1x0) + (1x1) +
(2x0) + (1x1) = 8+1+4+1+1+1 = 16

Figure 6.2 Convolution layer.

data input and a two-dimensional array of weighted parameters known as a kernel or filter, as shown in Figure 6.2.

6.3.2 Pooling layer

This layer is generally used to make the feature maps smaller. It reduces the number of training parameters, which speeds up computation. There are mainly three kinds of pooling layers: Max Pooling. It chooses the maximum input feature from the feature map region as shown in Figure 6.3, Average Pooling. It chooses the average input feature from the feature map region, and Global Pooling. This is identical for employing a filter with the dimensions h x w, that is, the feature map dimensions.

6.3.3 Fully connected layer

The last few layers of the neural network are Fully Connected Layers. As shown in Figure 6.4, if the preceding layer is entirely linked, every neuron

10	54	43	34
45	34	41	12
34	32	89	65
20	12	23	22

Max Pool ⟶ Stride-2

54	43
34	89

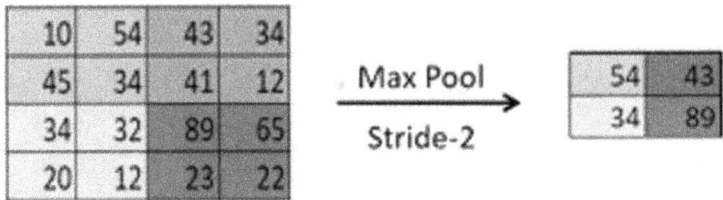

Figure 6.3 Max pooling.

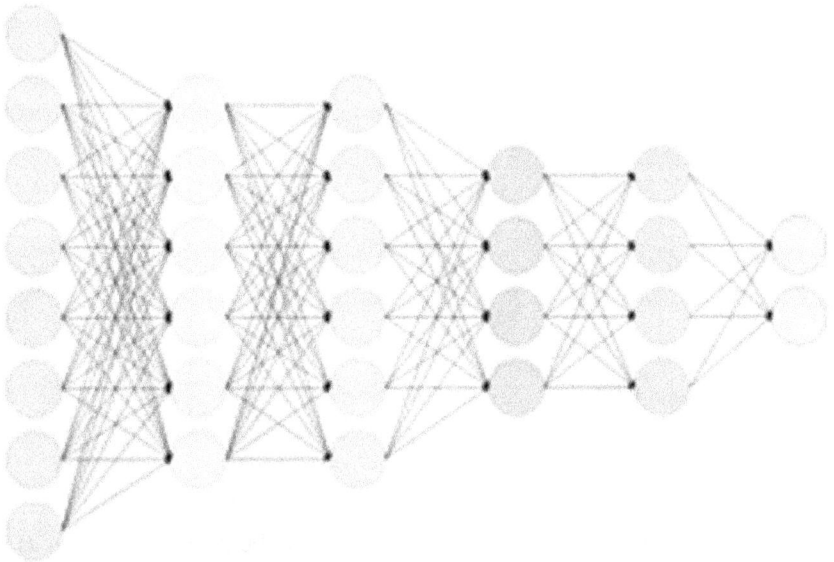

Figure 6.4 Fully connected layer.

in that layer is coupled to every other neuron in the layer below it. In our proposed method, two fully connected layers in CNN are employed followed by the classification layer.

6.3.4 Activation function

In simple words, Activation Function, shown in Figure 6.5, activates the neurons. It helps in deciding whether or not a neuron should fire and determining the output of the convolution layer. These are the most common activation functions: Sigmoid [12], ReLU [13], Leaky ReLU [7], and Softmax [10].

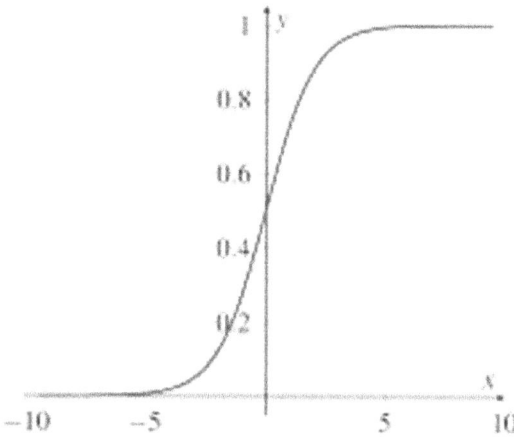

Figure 6.5 Several different kinds of activation functions.

The sigmoid and SoftMax activation functions are represented in Equations 6.1 and 6.2, respectively.

$$\sigma(x) = {}^1 1 + e^{-x\exp (x_i)} \tag{6.1}$$

$$\text{Softmax } (x_i) = \Sigma \exp (x_j) \tag{6.2}$$

6.4 HANDWRITTEN EQUATION RECOGNITION

6.4.1 Dataset preparation

The first and most important step of any research is dataset acquisition. The numerals and operations data and the character/variable dataset were collected from Kaggle [1, 2]. Then these were augmented to prepare a large dataset. The dataset contains approximately 24,000 images, which has 16 classes, such as 0–9 numerals, variable and five basic mathematical operators/symbols, namely, addition, subtraction, multiplication, equals, and division, as shown in Figure 6.6.

6.4.2 Proposed methodology

The proposed CNN model is used to recognize simple equations that consists of arithmetic operators: addition, subtraction, multiplication and division. It is also used to recognize simple linear equations of the type $x + a = b$ where x is a variable and a, b are constants. The block diagram of the implemented model is illustrated by Figure 6.7.

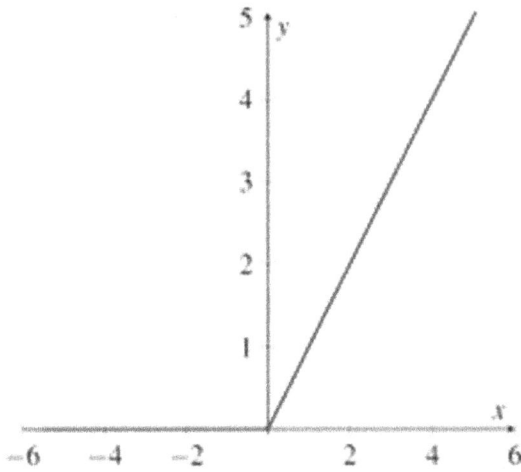

Figure 6.6 Sample images in the dataset.

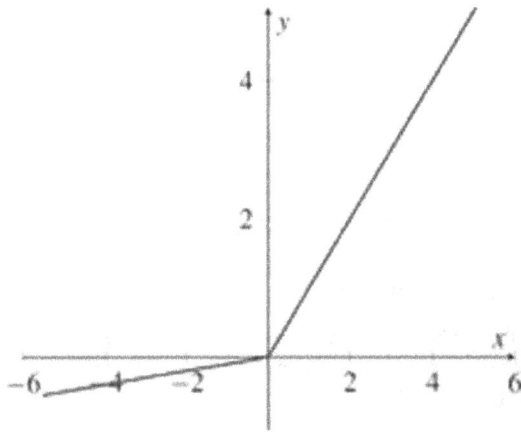

Figure 6.7 Block diagram of proposed scheme for handwritten digits.

6.4.2.1 Dataset acquisition

The dataset contains approximately 24,000 handwritten images divided into two sets – training images and testing images. Number of training images is approximately 1,300 whereas testing images are taken to be approximately 50 for each category. Pre-processing which includes resizing, cropping of images, padding etc. was done to make the dataset uniform. The images have a resolution of 95 x 84 for digits and 94 x 89 for the character M. The images were further resized to 100 x 100 for smooth training and better results.

6.4.2.2 Preprocessing

The goal of preprocessing is to improve the image quality to analyze it more effectively. Im age are preprocessed using several well-known techniques, like image resizing, normalizing and augmentation are a few examples.

(1) **Image Augmentation** is a technique of artificially increasing dataset size. Image augmentation use various techniques of processing or a combination of multiple methods of processing, such as random rotation, shear, shifts, flips, and so on.
(2) **Image Resizing:** In CNN, model accepts all the images of the same sizes only. Therefore, all images need to be resized to the fixed size. The images have been resized to 100 x 100 for sequential model and 224 x 224 for inception model.
(3) **Normalization** is a process that alters the intensity range of pixels. The primary goal of normalization is to make computational task more efficient by reducing values from 0 to 1.
(4) **Label Encoding** is the process of translating labels into a numeric representation that machines can read.

6.4.2.3 Recognition through CNN model

Handwritten datasets were used for training data acquisition after being supplemented with various methods such as shearing, rotating, nearest filling, shifting, and so on. There are approximately 23,000 training sample images and 950 testing sample images in the handwritten numeral dataset. To increase the variability and diversity of training data, we deformed it using several transformations such as rotation, translation, scaling, and vertical and horizontal stretching. By adding a few samples to make the dataset more diverse, the goal is to make it easier for CNN to find handwritten digits when the dataset is used to train CNN.

6.4.2.4 Processing inside CNN model

The model was trained using a sequential approach. The sequential model builds the structure layer by layer. The model contains seven Conv2D layers, four MaxPooling2D layers, and six drop-out layers with the rate 0.2 (Fraction of the input units to drop).

In addition, the activation function of the employed convolution layer was modified from Sigmoid to ReLU Activation function, and subsequently to Leaky ReLU Activation function. Leaky ReLU speeds up training and also has other benefits over ReLU. As it is a multiclass classification issue, the Softmax Activation function was utilized in the final dense layer. The model was then optimized using the Adam optimizer.

Also, the model is trained using inception architecture [15]. Inception-V3 is a convolutional neural network-based deep learning image categorization algorithm. The input layer, 1x1 convolution layers, 3x3 convolution layers, 5x5 convolution layers, max pooling layers, and combining layers are the parts that make up an inception model. The module is simple to unpack and comprehend when broken down into the constituent parts.

6.4.3 Solution approach

The solution approach is shown using the flowchart in Figure 6.8. The handwritten mathematical equation is provided by the user.

Image segmentation is the process of dividing an image information into divisions known as image segments, which helps to minimize the computational complexity and makes the further processing or analysis easier. The segmentation stage of an image analysis system is crucial because it isolates the subjects of interest for subsequent processing such as classification or detection. Image categorization is used in the application to better accurately classify image pixels. Figure 6.9 represents the actual deployment of the proposed methodology. The input image is segmented into well-defined fixed proportions. In the case of simple character recognition provided in

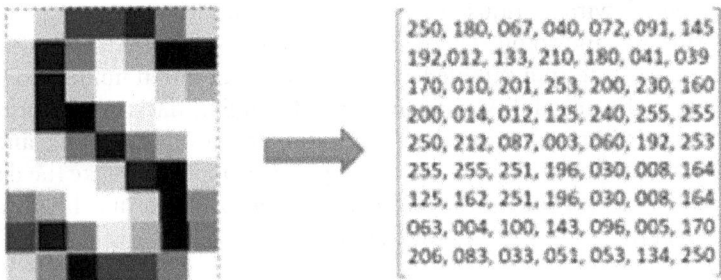

Figure 6.8 Sequence of proposed solution approach.

Figure 6.9 Practical implementation of proposed scheme.

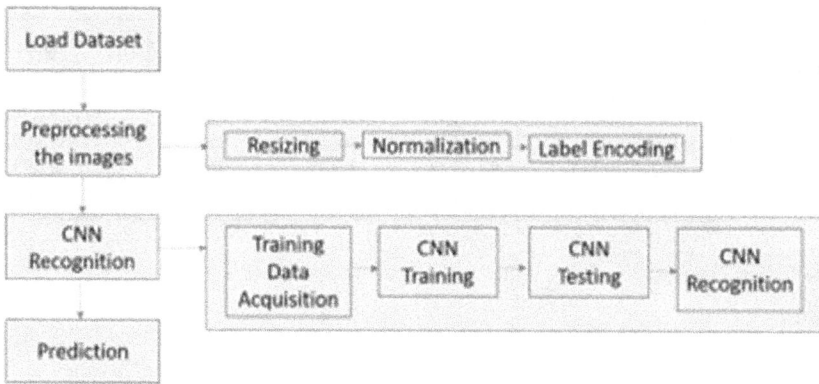

Figure 6.10 Algorithm for handwritten equation solver.

the image, we have segmented it into three parts, that is, into two numerals and one operator. This case is considered in general form of image. The segmentation is a 1:3 ratio. The proposed model works with the constraints being the middle segment should be an operator and the extreme segments should belong to numerals.

Figure 6.10 describes the steps in the algorithm of the proposed model to solve the equations from handwritten images. In both cases, each of the segment is thresholded by Otsu's Algorithm. Later, the segmented binary image is normalized before fed to the model for training. The size of segmented image is further defined by four coordinates, as left, right, top, bottom. Each segment will be now framed into a new image named as *segs*. Each of these segments is resized into 100x100.

Now, the segmented character/variables or operators are extracted and recognized by the trained model. The end goal of the training is to be able to recognize each block after analyzing an image. It must be able to assign a class to the image. Therefore, after recognizing the characters or operators from each image segment, the equation is solved using mathematical formulas on a trained model.

6.5 RESULTS AND DISCUSSION

The results of the experiments show that CNN can correctly segment handwritten typefaces and then combine the results into an equation. It is capable of recognizing fundamental operations. Instead of manually recognizing handwritten digits and symbols, the trained CNN model can effectively recognize basic four arithmetic operations, digits, and characters. In terms of checking mathematical equations, the CNN model has a relatively stable performance. The model was trained through both *Sequential Approach*

Table 6.1 Comparison between sequential model and inception model

Model	Training Accuracy	Training Loss	Validation Accuracy	Validation Loss
Sequential Model(10 epochs)	98.10%	5.95%	74.30%	129.63%
Sequential Model(30 epochs)	99.46%	1.66%	99.20%	3.38%
Inception Model(10 epochs)	97.32 %	9.38%	98.56%	6.39%
Inception Model(30 epochs)	99.422%	1.83%	99.50%	2.70%

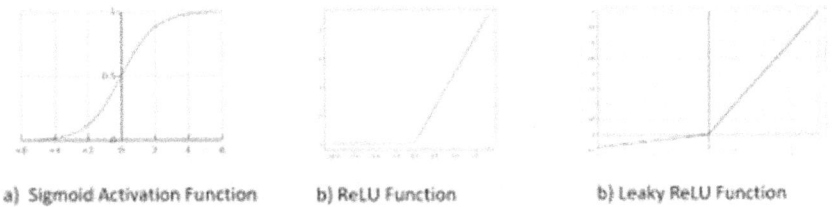

a) Sigmoid Activation Function   b) ReLU Function   b) Leaky ReLU Function

Figure 6.11 Accuracy and loss of sequential model.

Figure 6.12 Accuracy and loss of inception model.

and *Inception Architecture*. The comprehensive outcomes of both models are shown in Table 6.1.

The performance of the proposed model is observed not efficient for 10 epochs for model loss and validation loss. Therefore, both models are further trained for 30 epochs to observe the accurate results.

However, both models gave similar results in case of 30 epochs. The graph for Sequential model for 30 epochs is shown in Figure 6.11. The model accuracy is 99.46 percent with 1.66 percent model loss.

The graph for Inception's model for 30 epochs is shown in Figure 6.12. The accuracy of model is observed 99.42 percent with 1.83 percent model loss.

The proposed model functions correctly on handwritten equations, regardless the handwriting style, that is, whether, good or bad. Even if the equation is written in messy handwriting, the proposed model is able to detect it correctly as shown in Figure 6.13. In this example of poor handwriting, we can see digit 8 is not written in good handwriting, even though the model is detecting it accurately and able to solve the equation. However, in addition to good efficiency, the proposed model possess some limitations as well.

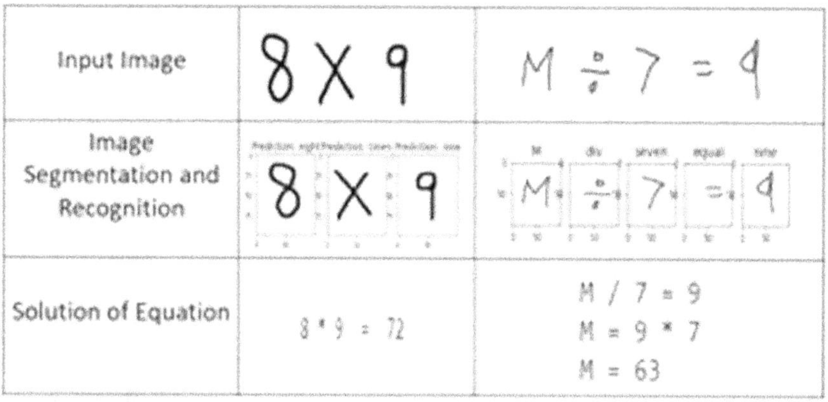

Input Image	8×9	$M \div 7 = 9$
Image Segmentation and Recognition	8×9	$M \div 7 = 9$
Solution of Equation	$8 * 9 = 72$	$M / 7 = 9$ $M = 9 * 7$ $M = 63$

Figure 6.13 Sample of poor handwritten equation.

6.6 CONCLUSION AND FUTURE SCOPE

Handwritten Digit Recognition and Equation Solver have been implemented using Convolutional Neural Networks. By replacing activation functions of the CNN architecture algorithm with a Leaky ReLU, an improved CNN algorithm is proposed through the Sequential Approach. Both the Sequential Model and Inception Model have been used for experimentation to observe the results. Both models produced better results for 30 epochs rather than for 10 epochs. The proposed CNN model with Leaky ReLU is tested for handwritten numeral recognition, where it is used to automatically check four basic arithmetic operations, addition, subtraction, multiplication, and division. Until now, it has been trained to solve simple handwritten linear equations on a single character/variable only.

Also, the handwritten equation solver must be capable of recognizing the characters and operators from the input images as quickly as possible. Hence, there is a need of the Graphics Processing Unit (GPU) for the large dataset to significantly reduce the training as well as testing time. The overall recognition accuracy of the CNN based handwritten recognition model is observed as 99.46 percent. Future work may be extended to solve handwritten quadratic equations from the images. Also, the work may include to solve equations having more than one character/variable.

REFERENCES

[1] Dataset. www.kaggle.com/code/rohankurdekar/ handwritten-basic-math-equation-solver/data.
[2] Dataset. www.kaggle.com/datasets/vaibhao/handwritten-characters.

[3] Mahmoud Abbasi, Amin Shahraki, and Amir Taherkordi. Deep learning for network traffic monitoring and analysis (ntma): A survey. *Computer Communications*, 170:19–41, 2021.

[4] Megha Agarwal, Vinam Tomar Shalika, and Priyanka Gupta. Handwritten character recognition using neural network and tensor flow. *International Journal of Innovative Technology and Exploring Engineering (IJITEE)*, 8(6S4):1445–1448, 2019.

[5] Saad Albawi, Tareq Abed Mohammed, and Saad Al-Zawi. Understanding of a convolutional neural network. In *2017 international conference on engineering and technology (ICET)*, pages 1–6. Ieee, 2017.

[6] Yellapragada, S.S., Bharadwaj, P., Rajaram, V.P., Sriram, S., and Sudhakar and Kolla Bhanu Prakash. Effective handwritten digit recognition using deep convolution neural network. *International Journal of Advanced Trends in Computer Science and Engineering*, 9(2):1335–1339, 2020.

[7] Arun Kumar Dubey and Vanita Jain. Comparative study of convolution neural network's relu and leaky-relu activation functions. In Sukumar Mishra, Yogi Sood, and Anuradha Tomar (eds) *Applications of Computing, Automation and Wireless Systems in Electrical Engineering*, Lecture Notes in Electrical Engineering pages 873–880. Springer, 2019. https://doi.org/10.1007/978-981-13-6772-4_76

[8] Jitesh Gawas, Jesika Jogi, Shrusthi Desai, and Dilip Dalgade. Handwritten equations solver using cnn. *International Journal for Research in Applied Science and Engineering Technology (IJRASET)*, 9:534–538, 2021.

[9] Abhishek Gupta, Alagan Anpalagan, Ling Guan, and Ahmed Shaharyar Khwaja. Deep learning for object detection and scene perception in self-driving cars: Survey, challenges, and open issues. *Array*, 10:100057, 2021.

[10] Ioannis Kouretas and Vassilis Paliouras. Simplified hardware implementation of the softmax activation function. In *2019 8th international conference on modern circuits and systems technologies (MOCAST)*, pages 1–4. IEEE, 2019.

[11] Daniel W. Otter, Julian R. Medina, and Jugal K. Kalita. A survey of the usages of deep learning for natural language processing. *IEEE transactions on neural networks and learning systems*, 32(2):604–624, 2020.

[12] Andrinandrasana David Rasamoelina, Fouzia Adjailia, and Peter Sinčák. A review of activation function for artificial neural network. In *2020 IEEE 18th World Symposium on Applied Machine Intelligence and Informatics (SAMI)*, pp. 281–286. IEEE, 2020.

[13] Johannes Schmidt-Hieber. Nonparametric regression using deep neural networks with relu activation function. *The Annals of Statistics*, 48(4):1875–1897, 2020.

[14] Chen ShanWei, Shir LiWang, Ng Theam Foo, and Dzati Athiar Ramli. A cnn based handwritten numeral recognition model for four arithmetic operations. *Procedia Computer Science*, 192:4416–4424, 2021.

[15] Christian Szegedy, Vincent Vanhoucke, Sergey Ioffe, Jon Shlens, and Zbigniew Wojna. Rethinking the inception architecture for computer

vision. In *Proceedings of the IEEE conference on computer vision and pattern recognition*, pages 2818–2826, 2016.

[16] P. Thangamariappan and J.C. Pamila. Handwritten recognition by using machine learning approach. *International Journal of Engineering Applied Sciences and Technology*, 4(11): pp. 564–567, 2020.

[17] Pavinder Yadav, Nidhi Gupta, and Pawan Kumar Sharma. A comprehensive study towards high-level approaches for weapon detection using classical machine learning and deep learning methods. *Expert Systems with Applications*, p. 118698, 2022.

Chapter 7

Agriware

Crop suggester system by estimating the soil nutrient indicators

S. P. Gautham, H. N. Gurudeep, Pai H. Harikrishna, Jasmine Hazel Crasta, and K. Karthik

7.1 INTRODUCTION

Farming is a cornerstone of the food business and is the foundation of a country's economic growth. As there is an expansion in population in countries like India, China, Syria, Niger, Angola, Benin, and Uganda, food demand also increases, and the agriculture industry needs to be able to meet all those demands for urban and rural areas. Decades ago, without technological progress in the agricultural sector, unanticipated losses were experienced because individuals had lost faith in the agricultural sector. The conventional method of creating assumptions that could not live up to expectations is one of the leading causes of this. Apart from these, natural calamities poison the earth by stealing its substance. The widening spread of contaminated climatic conditions have radically changed the air, soil, and water. The altered soil in nature cannot thus agree with the farmer's hypothesis, resulting in a massive agricultural loss. Hence, to improve the agricultural industry, precision farming must be developed to a greater extent than traditional methods [1]. Most farmers are unfamiliar with precision farming and are unaware of the scientific agricultural techniques. Awareness about modern farming with newer agricultural innovations and farming practices will help farmers increase efficiency, thus reducing the farm risk in terms of production, poor irrigation facilities, human resources, usage of modern techniques in farming. The existing method of soil classification and crop suggestion is manual and time-consuming, leading to human errors when the results are ambiguous.

Using modern technologies such as artificial intelligence in the agricultural sector has seen many positive impacts [2]. Types of crops that can be grown vary in different parts of the world based on the type of the soil, its nutrition, climatic conditions like rainfall, temperature, and so forth. The crop yield and its growth are extremely dependent upon these factors, and every type of crop may not be suitable for that location. Hence, building a system that classifies the soil type and predicts the kinds of crops that can be grown would be of great assistance for farmers. Such systems will assist in the large production of crops in those regions [3].

DOI: 10.1201/9781003453406-7

A major focus of the present work is to design a model for soil classifi-cation and to forecast the types of crops that can be grown using artificial intelligence and computer vision techniques, thereby extending a helping hand to the farmers. The input image is preprocessed first, then feature extraction and optimum feature selection are performed for effective soil classification. Convolutional neural networks have been used for feature extraction, selection, and classification of soil images. Different parameter values of the soil nutrients are then fed into the model to suggest suitable crops. For the final development of the proposed model, VGG16 was used for soil classification and Random Forest for predicting crops.

The rest of the article presented in this chapter is as follows: Section 2 briefs on existing works related to soil classification and prediction of crops. Section 3 details the proposed methodology for classifying the soil and predicting the possible list of crops that can be grown and the precise detail of the model designed for the task. Section 4 describes the experimental analysis and observations, along with the results of the proposed model. A conclusion followed by further improvements in the proposed system and future directions is at the end.

7.2 RELATED WORK

It is noted that many research works are going on, and researchers are applying mod-ern techniques for developing the agricultural sector to have better crop production. Agriculture has helped farmers from information exchange regarding seasonal change, demand, and cultivation. Climate and soil-related data are exceptionally useful for the farmers to defeat the misfortunes that might happen due to inappropriate cultivation.

Zubair et al. [4] proposed a model to predict seasonal crops based on the locality, and the overall process is built based on the regional areas of Bangladesh. For building up the model Seasonal Autoregressive Integrated Moving Average (SARIMA) was used for forecasting the rainfall and tem-perature, random forest regression for predicting the crops production. The model finally suggests crops that can have top production based on the season and location. However, the model focused only on the crops that can be grown in Bangladesh and did not focus on the varieties of the soil. Another approach for crop selection and yield prediction by T. Islam et al. [5] used 46 parameters for the prediction process. Along with the deep neural network model, support vector machine, logistic regression, and random forest algorithms were considered for comparing the accuracy and error rate. However, it is noticed that the study was limited and focused on the region of Bangladesh. Kedlaya et al. [6] proposed a pattern matching technique for predicting crops using historical data that relies on different parameters like weather conditions and soil property. The system suggests the farmers to plant an appropriate crop on the basis of the season and area or region of sowing. Area here speaks about the place, land of sowing. It is noted that such a system was implemented only for two regional districts of Karnataka, and proper classification of soil was not included as a part of their work.

Indian farmers face a common problem as they do not choose the right crop based on their soil requirements. A solution to this has been identified by S. Pudumalar et al. through precision agriculture [7]. Precision agriculture is a cutting edge cultivating strategy that suggests the right crop based on their specific parameters increasing productivity and thus reduces the wrong choice of a crop. S. Khaki and L. Wang [8] proposed a crop yield prediction using deep neural networks and found that their model had a predominant expectation accuracy with a root-mean-square-error of 12 percent. It also performed a feature selection using the trained DNN model, which effectively reduced the input space dimension without any drop in the prediction results. Finally, the outcome showed that the environmental parameters greatly improved crop yield and productivity.

S. Veenadhari et al. [9] presented a software tool called "Crop Advisor" is a user-friendly web application for predicting the climatic factors on the selected crops of the Madhya Pradesh districts. However, other agro parameters responsible for crop yield were not included in this product tool, as those input parameters vary in individual fields based on the area and seasonal conditions. Nevavuori et al. [10] proposed a CNN-based model for soil image classification tasks that showed outstanding performance in crop yield prediction on Normalized Difference Vegetation Index (NDVI) and Red Green Blue (RGB) data. Significantly, the CNN model showed better performance with RGB data compared to the NDVI data.

Y. J. N. Kumar et al. [11] in the agriculture sector developed a supervised machine learning prediction approach for a better crop yield from past historical factors, including temperature, humidity, ph, and rainfall. The model used the Random Forest algorithm to attain the best accurate value for crop prediction. N. Usha Rani and G. Gowthami [12] developed a smart crop suggester recommendation android-based application that assists farmers in choosing a preferable crop to yield higher production. It has a user-friendly interface that helps the farmers to get a suitable crop suggestion that is most suitable based on location, season, soil type, and rainfall analyzed considering last year's agriculture data.

The above studies have made their attempts for crop prediction using advanced technologies that can be applied in the agricultural sector for better yield of crops considering several parameters. In each of these research works, it can be noticed that most of the effort was either based on region-wise conditions or included only crop suggestions or soil classification. It is observed that very limited works have both soil classification and crop suggestion. So for the betterment of agriculture and to help farmers, we have come across a crop suggestion application after soil classification. Once a soil image is classified and then different parameters based on the soil and climatic conditions are considered, the application suggests a suitable crop that can have a better yield based on the inputs parameters. To achieve this, we have used neural network techniques and, in the next section, we discuss the proposed approach that has been developed for soil classification and crop suggestion.

7.3 PROPOSED METHODOLOGY

A soil classification system is a machine learning-based application designed to help farmers to classify the soil type and to predict the crops based on different parameters. Image processing methods and convolutional neural networks are combined to categorize soil images. Later, based on the input parameters using a random forest algorithm, the type of crop is suggested. Once a user uploads a soil image using the soil classification model, the soil is classified into its category. After the classification result is obtained, the user needs to enter the parameters of the soil and weather information; the input data then gets processed and compared with the model and, finally, suggests the crops based on the soil and the given input parameters. The sequence of tasks involved in the overall progress of the proposed classification and crop suggestion model using the network is shown in Figure 7.1.

For the soil classification task, we have considered four types of soil images – Black, Clay, Red and Sandy. With VGG16 [13] CNN architecture, the input soil image classification was carried out. The reason for

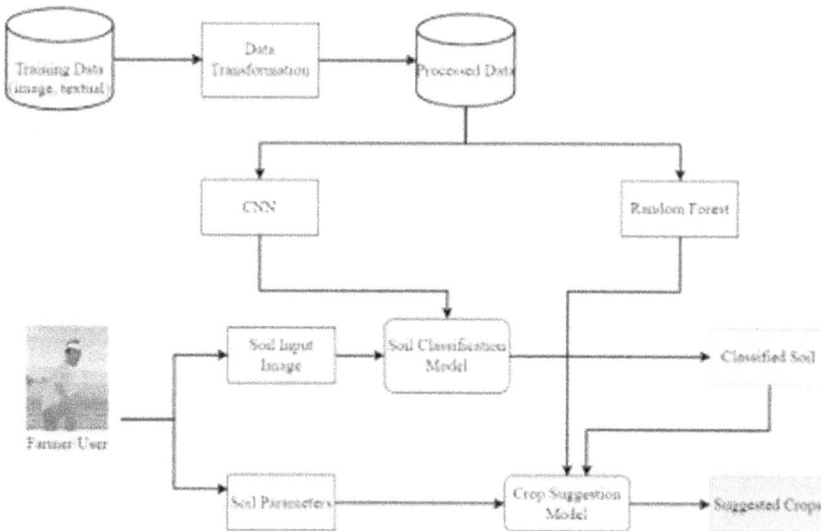

Figure 7.1 Proposed approach for soil image classification and crop prediction.

selecting VGG16 is that it is a fantastic vision model architecture to date. One of the uniqueness of VGG16 is that instead of having a large number of hyper-parameter, it focuses on having convolution layers of a 3x3 filter with a stride=1 and consistently uses a similar padding. The max pooling layer of the 2x2 filter with stride=2 can be seen in the architectural model. This convolution and max pool layer arrangement is followed consistently throughout the whole architecture. At the end of the network, two fully connected layers followed by a softmax layer are present for classifying the soil image.

Soil images of size m × n are fed into the neural network for training. Images in the dataset are of different dimensions; therefore, resizing was performed before feeding it into the network. Augmentation operations like random flip and translation along the vertical and horizontal axis were performed on the training dataset, preventing the network from overfitting. While training deep neural network architectures using a transfer learning approach, the final layers – fully connected, Softmax, and classification output layers – need to be configured according to the dataset used. The replaced and fully connected layer parameters are specified according to the dataset used for a new classification model. Increasing the WeightLearnRateFactor and BiasLearnRateFactor values helps the network learn features faster, with the addition of the new layers. Additionally, while preparing the model, a few hyper-parameter values like mini-batch, epochs, batch normalization, learning rate, regularizations, optimizers and activators were varied and prepared to find the best suitable values in building the final model for the soil image classification task. Here, adam optimiser, loss function as categorical cross entropy with learning rate of 0.001, were set during training the neural network.

Once the soil type is obtained from the neural network model, using a random forest algorithm, crop suggestions are predicted. Soil parameters like nitrogen, potassium, phosphorus, pH, and weather information such as temperature, humidity, and rainfall, are fed into the crop suggester model. The random forest algorithm creates a forest as a collection of decision trees, increasing randomization as the trees grow. The technique seeks the best features from the random subset of features when splitting a node, adding more diversity and improving the model. Finally, the best crop that can be grown is suggested to the farmers based on the type of soil obtained from the soil classification model using a random forest algorithm.

7.4 EXPERIMENTAL RESULTS AND DISCUSSION

For this experimental work, we used a dataset extracted from Kaggle,[1] consisting of 400 soil images in total, split into training and testing sets. For crop prediction model seven parameters of soil such as nitrogen, phosphorus, potassium, temperature, humidity, pH and rainfall data were

Table 7.1 A sample showing four categories of soil images

Category Name	Sample Image 1	Sample Image 2
Black		
Clay		
Red		
Sandy		

1 www.kaggle.com/datasets/prasanshasatpathy/soil-types
2 www.kaggle.com/datasets/shashankshukla9919/crop-prediction

Table 7.2 Test cases for integration testing

Test No.	Description	Expected Result	Status
1	An image uploaded by a user to a web page is processed in the back-end by the CNN model.	The soil is classified	Pass
2	As soil parameters are entered in the web page, the random forest algorithm is used to process them in the back-end.	Type of crop is suggested	Pass

considered from Kaggle.[2] A set of sample images that are used for this crop prediction model is shown in Table 7.1.

Integration testing is performed by connecting the front end to the back end. Integration testing is where individual units are merged and tested as a group. This level of testing aims to expose incompatibilities, faults, and irregularities in the interaction between integrated modules. When the image is uploaded, it is processed by the CNN model that classifies the soil type. After entering the soil parameters and seasonal data values in the crop suggestion system, it is processed by the Random Forest algorithm that suggests the suitable crop that can be grown. Test cases that were validated using a test image for soil classification and prediction of crop are shown in Table 7.2.

The proposed CNN models' performance are measured using the standard metrics like accuracy, loss, validation accuracy, and validation

Table 7.3 Observed soil classification performance

No.of Epochs	Accuracy	Loss	Val_accuracy	Val_loss
5	0.8450	0.3660	0.9750	0.1394
10	0.9250	0.1899	0.9500	0.1066
15	0.9400	0.1798	0.9875	0.0740
20	0.9725	0.0732	0.9500	0.1408
25	0.9725	0.0827	0.9375	0.0977

Table 7.4 Observed crop prediction performance

Algorithm	Accuracy	Precision	Recall
Logistic Regression	0.7906	0.6491	0.7580
Support Vector Machine	0.8829	0.5025	0.6774
K Nearest Neighbors	0.9108	0.4281	0.5832
Random Forest	0.9595	0.9375	0.7528

loss. Accuracy is computed as the total number of accurate predictions on soil class labels divided by the total number of images under test classification. Whereas a loss is more like an "error" that calculates how far apart the output/predicted value of a learning model deviates from the ground truth value. To fit parameters while training any neural network model, with increase in every epoch, loss decreases and accuracy increases. But with validation accuracy and validation loss, when validation loss starts decreasing, validation accuracy should start increasing, which indicates the model that is build has decent learning. The classification results observed during this process of soil classification performed is shown in Table 7.3.

In the crop suggestion system, four different machine learning algorithms like Logistic Regression with 79 percent, SVM with 88 percent, KNN with 91 percent and Random Forest with 95 percent was observed and the results of this process are shown in Table 7.4.

7.5 CONCLUSION AND FUTURE WORK

Our project proposes the idea of classifying soil and suggesting suitable crops using advanced techniques of image processing. The need for food is rising with the rapid growth of the world's population, but the area of cultivable land is reducing due to modernization. Hence, the soil classification and crop suggestion model is built that was trained and tested using VGG16, which provides a better accuracy rate. For the crop suggestion system, four different algorithms like Logistic Regression, SVM, KNN, and Random Forest were tested and observed that Random Forest has given a better accuracy, 95 percent. Implementation of the system can be helpful for

(a) Home Page

(b) Test Image Upload and Soil Classification

(c) Entering Parameter Values

(d) Final Prediction of Crop

Figure 7.2 Illustration of various pages of crop suggester.

farmers as they can upload the soil image and enter mandatory soil and climate parameters through our web interface. In future, the system's accuracy can be enhanced by increasing the training dataset, resolving the closed identity problem, and finding an optimal solution. The crop recommendation system will be further developed to connect with a yield predictor, another subsystem that would also allow the farmer to estimate the production based on the recommended crop. We likewise anticipate carrying out this framework on a portable stage for the farmers.

REFERENCES

[1] Awad, M.M.: Toward precision in crop yield estimation using remote sensing and optimization techniques. *Agriculture* 9(3) (2019) 54.

[2] Ben Ayed, R., Hanana, M.: Artificial intelligence to improve the food and agri-culture sector. *Journal of Food Quality* 2021 (2021) 1–7.

[3] Waikar, V.C., Thorat, S.Y., Ghute, A.A., Rajput, P.P., Shinde, M.S.: Crop predic-tion based on soil classification using machine learning with classifier ensembling. *International Research Journal of Engineering and Technology* 7(5) (2020) 4857–4861.

[4] Zubair, M., Ahmed, S., Dey, A., Das, A., Hasan, M.: An intelligent model to suggest top productive seasonal crops based on user location in the

context of Bangladesh. In: *Smart Systems: Innovations in Computing*. Springer (2022) 289–300.

[5]   Islam, T., Chisty, T.A., Chakrabarty, A.: A deep neural network approach for crop selection and yield prediction in Bangladesh. In: 2018 IEEE Region 10 Humanitarian Technology Conference (R10-HTC), IEEE (2018) 1–6.

[6]   Kedlaya, A., Sana, A., Bhat, B.A., Kumar, S., Bhat, N.: An efficient algorithm for predicting crop using historical data and pattern matching technique. *Global Transitions Proceedings* 2(2) (2021) 294–298.

[7]   Pudumalar, S., Ramanujam, E., Rajashree, R.H., Kavya, C., Kiruthika, T., Nisha, J.: Crop recommendation system for precision agriculture. In: 2016 Eighth Inter-national Conference on Advanced Computing (ICoAC), IEEE (2017) 32–36.

[8]   Khaki, S., Wang, L.: Crop yield prediction using deep neural networks. *Frontiers in plant science* 10 (2019) 621.

[9]   Veenadhari, S., Misra, B., Singh, C.: Machine learning approach for forecasting crop yield based on climatic parameters. In: 2014 International Conference on Computer Communication and Informatics, IEEE (2014) 1–5.

[10] Nevavuori, P., Narra, N., Lipping, T.: Crop yield prediction with deep convolutional neural networks. Computers and electronics in agriculture, 163 (2019) 104859

[11] Kumar, Y.J.N., Spandana, V., Vaishnavi, V., Neha, K., Devi, V.: Supervised machine learning approach for crop yield prediction in agriculture sector. In: 2020 5th International Conference on Communication and Electronics Systems (ICCES), IEEE (2020) 736–741.

[12] Usha Rani, N., Gowthami, G.: Smart crop suggester. In: International Conference On Computational And Bio Engineering, Springer (2019) 401–413.

[13] Simonyan, K., Zisserman, A.: *Very deep convolutional networks for large-scale image recognition*. arXiv preprint arXiv:1409.1556 (2014).

Chapter 8

A machine learning based expeditious Covid-19 prediction model through clinical blood investigations

Bobbinpreet Kaur, Sheenam, and Mamta Arora

8.1 INTRODUCTION

The Covid-19 virus erupted in Wuhan, China, in the last days of 2019, affecting countries around the globe. Covid-19 created tremendous pressure on health care systems throughout all countries because of the very high number of patients. The virus, named as Severe Acute Respiratory Syndrome Corona Virus 2 (SARS CoV-2) is responsible for loss of a huge number of lives [1]. In January 2019 the World Health Organization (WHO) declared this virus outbreak as a Public Health Emergency of International Concern (PHEIC). In 2020 it was named Covid-19 and in March 2020 the WHO declared this outbreak as a pandemic as it touched all corners of the world [2].

The testing for this virus is done either through viral or antibody testing procedures [3]. The testing is available everywhere in the government or private allocated laboratories containing the supporting equipment and procedures. In India, according to the latest guidelines by the Indian Council of Medical Research (ICMR), there should be a testing center within 250 kms distance in the plains and 150 km in the hills. This caused a delay of 5 hours for getting the sample to a lab, followed by a testing procedure lasting approximately 24–48 hours. As of 13 May 2020, 1.85 million samples had been tested as per the report published by ICMR [4]. Internationally, researchers in the medicine worked towards finding the appropriate drug or vaccine and developing rapid and accurate testing procedures [5]. Many rapid testing procedures were failing throughout the globe thereby still leaving the effort for rapid results an open stream for researchers.

In recent years computer-based Machine Intelligence algorithms are replacing the human effort in almost all the fields. Particularly for the medicine, machine learning based systems have proven their capabilities to predict and diagnose various kinds of diseases, including lung cancer, breast cancer detection, fetal heart related diseases, rheumatology, and so forth [6]. Machine learning algorithms maps the knowledge acquired through

DOI: 10.1201/9781003453406-8

the training data into meaningful predictions, thereby supporting medical practitioners in diagnosis.

Some certain parameters that are evaluated rapidly and can be tested in almost all the clinical laboratories for prediction of Covid-19 disease in a individual have been designed. The model can make decisions about confirmed cases [7–9] from the possible suspects and thereby can help restricting the spreading of disease by exercising suitable safety measures. The limitations of delay and approachability of testing laboratories are overcome using this model-based prediction. The model's decision is based only upon the parameters obtained through common tests. The dataset consisting of sample history of both positive and negative test result patients will be passed through suitable parameter selection mechanism, thereby improving the accuracy of the system being modeled.

8.2 LITERATURE SURVEY

Mohammad Saber [17] conducted a study on detection of Covid-19 effects on human lungs and proposed a new hybrid approach based on neural networks, which extracts features from images by using 11 layers and, on that basis, another algorithm was also implemented to choose valuable qualities and hide unrelated qualities. Based on these ideal characteristics, lung x-rays were identified using a Support Vector Machines (SVM) classifier to get better results. This study demonstrated that the accuracy indicator and the quantity of pertinent characteristics extracted outperformed earlier methods using the same data.

Fátima Ozyurt [18]. A fresh, handcrafted feature generator and feature selector were employed. Four main procedures make up the system: preprocessing, fused dynamic scaled exemplars-based pyramid feature generation; Relief; iterative neighborhood component analysis-based feature selection; and deep neural network classifier. In the preparation stage, CT images are shrunk to 256 × 256 pixels and transformed into 2D matrices. The main objective of the suggested framework is to utilize manually created features from the CT scans to achieve a greater accuracy than convolution neural networks (CNN).

Zeynep Gündo Gar [19] reviewed deep learning models and attribute-mining techniques. Utilization of matrix partition in the TMEMPR [19] method provides 99.9 percent data reduction; the Partitioned Tridiagonal Enhanced Multivariance Products Representation (PTMEMPR) method was proposed as a new attribute-mining method. This method is used as a preprocessing method in the Covid-19 diagnosis scheme. The suggested method is compared to the cutting-edge feature extraction techniques, Singular Value Decomposition (SVD), Discrete Wavelet Transform (DWT), and Discrete Cosine Transform, in order to assess its effectiveness (DCT). Additionally, feature extraction and classification using artificial neural networks are presented for the diagnosis of Covid-19, which had a detrimental impact on public health and caused the suspension of social life due to the global trade economy.

Rasha H. Ali [20]. Due to COVID's lack of a good vaccine, 19 spread quickly over the world. As a result, it is crucial to identify those who have the virus early in order to try and contain it by isolating the affected individuals and providing any necessary medical care to prevent further transmission. This research provides a feature selection-based prediction model against virus. The preprocessing stage, the features selection stage, and the classification stage are the three stages that make up this model. For this study's data collection, which includes 8,571 records and 40 attributes for patients from various nations, to choose the best features that have an impact on the proposed model's prediction, two feature selection strategies are used. These are the Extra Tree Classifier (ETC) [20] as embedded feature selection and the Recursive Feature Elimination (RFE) as wrapper feature selection.

Mokhalad Abdul [21]. A study was done using new Caledonian crow learning. In the first stage, the best features related to COVID-19 disease are picked using the crow learning algorithm. The artificial neural network is given a set of COVID-19 patient-related features to work within the proposed method, and only those features that are absolutely necessary for learning are chosen by the crow learning algorithm.

Ahmed T. Sahlol [22]. The dramatic spread of COVID-19, which produced lethal complications in both humans and animals, is something we are still witnessing. Although convolution neural networks (CNNs) are thought to be the most advanced image categorization method currently available, their deployment and training require extremely high computational costs. In order to modify hybrid classification for COVID-19 images, the present author suggests combining the advantages of CNNs and a swarm-based feature selection algorithm to choose the best pertinent characteristics. On two open COVID-19 X-ray datasets, a proposed method that combines high performance and computational complexity reduction was tested.

Ekta Gambhir et al. [23]. From the Johns Hopkins University visual dashboard, a regression analysis was performed to describe the data before feature selection and extraction. The SVM method and Polynomial Regression were then used to create the model and obtain the predictions.

Table 8.1 refers to comparative study of all the Machine learning approach on the basis of accuracy, sensitivity and specificity.

Table 8.1 Summarized ML approaches and results

S.No.	Researcher	Approach Used	Used For	Result
1.	Mohammad Saber et al.[18]	SVM classifier.	Lungs disease	99.43%accuracy, 99.16%sensitivity, and 99.57% specificity
2.	Fátima Ozyurt[19]	Chest X-ray and computed tomography (CT) images.	Chest disease	94.10% and 95.84% classification accuracies
3.	Zeynep Gündo gar[20]	PTMEMPR	Medical Application	accuracy 99.8%.
4.	Rasha H.Aiii [21]	Recursive Feature Elimination, Extra Tree Classifier, Restricted Boltzmann Machine,Naïve Bayesian.	Used for all types of disease from different countries and filter out at every stage.	Accuracy 66.329%, 99.924%
5.	Mokhalad abdul[22]	crow learning algorithm	All types of diseases	Accuracy 94.31%, sensitivity 94.38%, precision 94.27%,
6.	Ahmed T. Sahlol[23]	Fractional order marine predators algorithm	Both Human and animals	98.7% accuracy, 98.2% and 99.6%, 99% F-Score
7.	Ekta Gambhir et. al. [24]	Polynomial Regression Algorith	All types of diseases	Accuracy = 93%
8.	David Goodman Mezza et al. [25]	Ensemble machine learning model	All types of diseases	Area = 0.91, sensitivity 0.93 and specificity 0.64.

With the aid of an ensemble machine learning method, **Goodman-Meza et al.** [24] assessed seven machine learning models for the final categorization of Covid-19 diagnosis if PCR is insufficient.

Yazeed Zoabi et al. [25] established a new gradient boosting machine model that used decision tree to predict the Covid-19 in patients by asking eight specific questions. Mahajan [24–27] proposed a new hybrid technique based on the Aquila optimizer (AO) and the arithmetic optimization algorithm (AOA). Both AO and AOA are recent meta-heuristic optimization techniques. They can be used to solve a variety of problems such as image processing, machine learning, wireless networks, power systems, engineering design, and so on. The impact of various dimensions is a standard test that has been used in prior studies to optimize test functions that indicate the impact of varying dimensions on AO-AOA efficiency.

8.3 METHODOLOGY

8.3.1 Dataset and its preparation

In this chapter we discuss a dataset available through Kaggle, provided by a Brazilian hospital resource for research purposes. A total of 5,644 records (samples) are available in this dataset, with 111 different parameters of clinical test acting out of which 110 will act as predictors and one column (test positive or negative) will act as target class. Out of the total cases, nearly 10 percent account for the positive cases (558). Since the dataset is incomplete in nature, a manual modification of data before passing on to preceding stages is required.

Figure 8.1 shows the basic manual editing operation to be performed on data to make it ready for training a machine learning algorithm. The parameter values must be numeric in nature, and they can be replaced with 0, 1 for two possible outcomes, for example, detected, not detected, or with three numerals such as 0, 1, 2 for three possible outcomes: detected, not detected, not defined. All blank entries need to be replaced by a suitable literal, which differentiates it from rest of the data. The target variable needs to be moved to the last column of the dataset and the positive and negative resultants

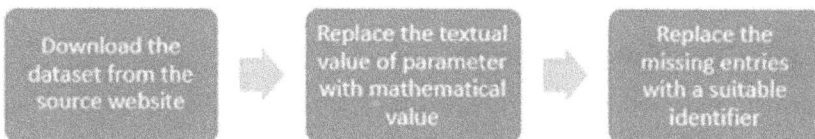

Figure 8.1 Manual editing of data for training.

Algorithm Relief

Input: for each training instance a vector of attribute values and the class value

Output: the vector W of estimations of the qualities of attributes

1.   set all weights $W[A] := 0.0$;
2.   **for** i := 1 **to** m **do begin**
3.   randomly select an instance R_i;
4.   find nearest hit H and nearest miss M;
5.   **for** A := 1 **to** a **do**
6. $W[A] := W[A] - \text{diff}(A, R_i, H)/m + \text{diff}(A, R_i, M)/m$;
7.   **end**;

Figure 8.2 Pseudo code for relief Algorithm.

Source: [11].

can be replaced with a numeric class label 0 or 1 for ease of interpretation. Since the data contains a high number of missing entries, a manual screening for removal of parameters with high missing values is required at this stage. This will result in appropriate reduction of irrelevant data. Subsequent to this is removal of redundant features by a suitable technique. In our work we have applied the Relief algorithm for feature selection to select highly appropriate features for the classification process.

The relief algorithms work as a filter-based method for feature selection. For each and every feature a score value is calculated, and the features are sorted in descending order according to the score value. The score value is generated on the basis of neighbor relationships between the values. The ordered features are then screened to select top features. In our implementation the dataset containing different blood testing based parameters acting as features are first manually screened then selection is done based on relief evaluated feature score. In this way highly appropriate and less redundant features reach the classification model. Thus the redundant features of the data are filtered and can later lead to improving the accuracy of the system model thus developed [10]. Figure 8.2 shows the Pseudo code for the basic Relief algorithm. The estimation of quality for each vector designated as W is calculated, and distance value is evaluated for the features in order to assign the score.

8.3.2 Classification set up

The data thus obtained after the selection process is used for training and testing the prediction model that is shown in Figure 8.3.

In total, 750 [7] samples were filtered for consideration as they comprise nearly complete information of all the variables. The samples are than divided

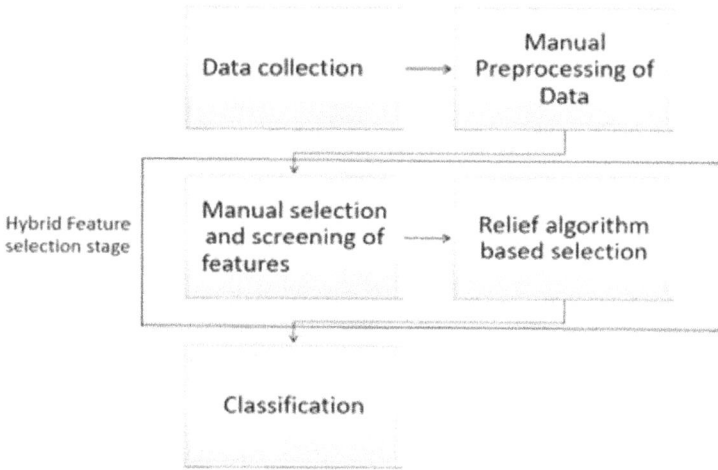

Figure 8.3 Proposed methodology.

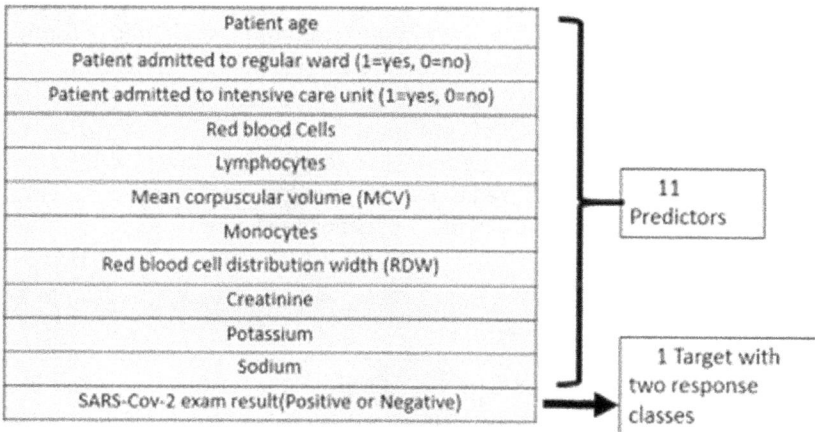

Figure 8.4 Predictors and target for classification.

in an 80:20 ratio for training versus testing. Out of this total, 750 samples, 602 are selected for training purposes and among the training samples 519 samples belong to negative tested patients and 83 of positive tested patients. In our research the various machine learning based classifiers [12–16] are implemented with this training: testing setup and the performed evaluation is performed. The classifiers used are Naïve Bayes (NB), Support Vector Machine (SVM), K Nearest Neighbor (KNN), and Decision Tree (DT) that is shown in Figure 8.4. Their results are implemented and evaluated.

8.3.3 Performance evaluation

For evaluation of developed models we have used a 10 fold cross validation scheme with non-overlapped data. The accuracy of all the machine learning algorithms stated in Section 2 are evaluated, and a comparison is drafted for the same. The confusion matrix is also plotted, and a ratio of 80:20 without any overlapping case between training and testing is ensured. The total training samples number 602, and testing samples are 148.

8.4 RESULTS AND DISCUSSION

The purpose of this work is to develop a machine learning based system capable of expditing the virus prediction based on general blood tests that can be performed in available laboratories. This can help the health care practitioners make decisions about further quarantine and treatment measures for the predicted Covid-19-positive patients.

Since this disease is spreading mainly through human-to-human interaction, timely detection can result in reducing the overall spreading of the disease, thereby saving many lives of those who come in contact with Covid-19 patients. As explained in previous sections, we implemented four state-of-the-art classifiers, and an accuracy comparison of these classifiers for 602 training samples is tabulated and shown in Table 8.2 This can be concluded from the accuracy values that Naïve Bayes Classifier is outperforming the rest of the classifiers.

The comparison elaborates the model is achieving accuracy of 94.2 percent on this dataset, with 11 predictors. The response of the classifier is either negative Covid-19 or positive Covid-19 patient-based on the predictor values.

8.5 CONCLUSION

While medical facilities and government councils try to develop improved rapid-testing procedures for detecting Covid-19 in patients, a delay in a report may result in spreading disease to persons who come in contact with infected patients. The machine learning based system proposed in this chapter can help predict the result with simple clinical lab tests of the blood sample of the suspected patient is shown in Table 8.2. The data availability

Table 8.2 Performance evaluation

Classifier	Training Vs. Testing data	Number or Predictors	Cross Validation	Accuracy
NB	602:148	11	5	94.2%
KNN	602:148	11	5	91.2%
DT	602:148	11	5	93.6%
SVM	602:148	11	5	93.7%

is the major issue for designing such system. Data true and complete in all respects can help in improving system accuracy and, with the future availability of complete data with more relevant parameters, can help develop the system with highly accurate predictions and thereby can lead to reducing or minimizing the number of tests done for the suspected patients.

REFERENCES

1. Pourhomayoun, Mohammad, and Mahdi Shakibi. "Predicting Mortality Risk in Patients with COVID-19 Using Artificial Intelligence to Help Medical Decision-Making." *medRxiv* (2020).
2. www.who.int [Accessed on 2-04-20]
3. www.cdc.gov/coronavirus/2019-ncov/symptoms-testing/testing.html [Accessed on 2-04-20]
4. https://main.icmr.nic.in/content/covid-19 [Accessed on 13-5-20]
5. Burog, A. I. L. D., C. P. R. C. Yacapin, Renee Rose O. Maglente, Anna Angelica Macalalad-Josue, Elenore Judy B. Uy, Antonio L. Dans, and Leonila F. Dans. "Should IgM/IgG rapid test kit be used in the diagnosis of COVID-19?." *Asia Pacific Center for Evidence Based Healthcare* 4 (2020): 1–12.
6. Mlodzinski, Eric, David J. Stone, and Leo A. Celi. "Machine Learning for Pulmonary and Critical Care Medicine: A Narrative Review." *Pulmonary Therapy* (2020): 1–11.
7. www.kaggle.com/einsteindata4u/covid19 [Accessed on 22-3-20]
8. Wu, Jiangpeng, Pengyi Zhang, Liting Zhang, Wenbo Meng, Junfeng Li, Chongxiang Tong, Yonghong Li et al. "Rapid and accurate identification of COVID-19 infection through machine learning based on clinical available blood test results." *medRxiv* (2020).
9. Peker, Musa, Serkan Ballı, and Ensar Arif Sağbaş. "Predicting human actions using a hybrid of Relief feature selection and kernel-based extreme learning machine." In *Cognitive Analytics: Concepts, Methodologies, Tools, and Applications*, pp. 307–325. IGI Global, 2020.
10. Urbanowicz, Ryan J., Melissa Meeker, William La Cava, Randal S. Olson, and Jason H. Moore. "Relief-based feature selection: Introduction and review." *Journal of Biomedical Informatics* 85 (2018): 189–203.
11. Uldry, Laurent and Millan, Jose del R. (2007). Feature Selection Methods on Distributed Linear Inverse Solutions for a Non-Invasive Brain-Machine Interface.
12. Susto, Gian Antonio, Andrea Schirru, Simone Pampuri, Seán McLoone, and Alessandro Beghi. "Machine learning for predictive maintenance: A multiple classifier approach." *IEEE Transactions on Industrial Informatics* 11, no. 3 (2014): 812–820.
13. Lanzi, Pier L. *Learning classifier systems: From foundations to applications.* No. 1813. Springer Science & Business Media, 2000.
14. Kononenko, Igor. "Semi-naive Bayesian classifier." In *European Working Session on Learning*, pp. 206–219. Springer, Berlin, Heidelberg, 1991.
15. Rueping, Stefan. "SVM classifier estimation from group probabilities." (2010).

16. Rish, Irina. "An empirical study of the naive Bayes classifier." In *IJCAI 2001 workshop on empirical methods in artificial intelligence*, vol. 3, no. 22, pp. 41–46. 2001.

17. Iraji, Mohammad Saber, Mohammad-Reza Feizi-Derakhshi, and Jafar Tanha. "COVID-19 detection using deep convolutional neural networks and binary differential algorithm-based feature selection from X-ray images." *Complexity* 2021 (2021).

18. Ozyurt, Fatih, Turker Tuncer, and Abdulhamit Subasi. "An automated COVID-19 detection based on fused dynamic exemplar pyramid feature extraction and hybrid feature selection using deep learning." *Computers in Biology and Medicine* 132 (2021): 104356.

19. Gündoğar, Zeynep, and Furkan Eren. "An adaptive feature extraction method for classification of Covid-19 X-ray images." *Signal, Image and Video Processing* (2022): 1–8.

20. Ali, Rasha H., and Wisal Hashim Abdulsalam. "The Prediction of COVID 19 Disease Using Feature Selection Techniques." In *Journal of Physics: Conference Series*, Vol. 1879, no. 2, p. 022083. IOP Publishing, 2021.

21. Kurnaz, Sefer. "Feature selection for diagnose coronavirus (COVID-19) disease by neural network and Caledonian crow learning algorithm." *Applied Nanoscience* (2022): 1–16.

22. Sahlol, Ahmed T., Dalia Yousri, Ahmed A. Ewees, Mohammed A.A. Al-Qaness, Robertas Damasevicius, and Mohamed Abd Elaziz. "COVID-19 image classification using deep features and fractional-order marine predators algorithm." *Scientific Reports* 10, no. 1 (2020): 1–15.

23. G. Ekta, J. Ritika, G. Alankrit and T. Uma, "Regression analysis of COVID-19 using machine learning algorithms.," in In 2020 International conference on smart electronics and communication (ICOSEC), 2020.

24. Goodman-Meza, A. R. David, N. C. Jeffrey, C. A. Paul, E. Joseph, S. Nancy and B. Patrick, "A machine learning algorithm to increase COVID-19 inpatient diagnostic capacity.," Plos one, vol. 15, no. 9, p. e0239474, 2020

25. Y. Zoabi, D.-R. Shira and S. Noam, "Machine learning-based prediction of COVID-19 diagnosis based on symptoms," *npj digital medicine*, vol. 4, no. 1, pp. 1–5, 2021

26. Mahajan, S., Abualigah, L., Pandit, A.K. et al. Fusion of modern meta-heuristic optimization methods using arithmetic optimization algorithm for global optimization tasks. *Soft Comput* 26, 6749–6763 (2022).

27. Mahajan, S., Abualigah, L., Pandit, A.K. et al. Hybrid Aquila optimizer with arithmetic optimization algorithm for global optimization tasks. *Soft Comput* 26, 4863–4881 (2022).

28. Mahajan, S., Pandit, A.K. Hybrid method to supervise feature selection using signal processing and complex algebra techniques. *Multimed Tools Appl* (2021).

29. Mahajan, S., Abualigah, L. & Pandit, A.K. Hybrid arithmetic optimization algorithm with hunger games search for global optimization. *Multimed Tools Appl* 81, 28755–28778 (2022).

30. Mahajan, S., Abualigah, L., Pandit, A.K. et al. Fusion of modern meta-heuristic optimization methods using arithmetic optimization algorithm for global optimization tasks. *Soft Comput* 26, 6749–6763 (2022).

Chapter 9

Comparison of image based and audio based techniques for bird species identification

Jyoti Lele, Naman Palliwal, Sahil Rajurkar, Vibor Tomar, and Anuradha C. Phadke

9.1 INTRODUCTION

Gathering data about bird species requires immense effort and is very time consuming. Bird species identification finds the specific category a bird species belongs to. There are many methods of identification, such as through image, audio, or video. An audio processing technique captures the audio signals of birds and, similarly, image processing technique identifies the species by capturing the image in various parameters (i.e., distorted, mirror, hd quality) of birds.

The research focuses on identifying the species of birds using audio and images. To predict these bird species, first it requires accurate information about their species, for which it needs to select a large dataset for both images and audio, which will be needful to train Neural Network for bird species information. By catching the sound of different birds, their species can be identified with different audio processing techniques. A system for audio-based bird identification has proven to be particularly useful for monitoring and education. People analyze images more effectively than sounds or recordings but, in some cases, audio is more effective than images, so we opted for the audio approach to classify bird species using images and audio both. Many institutes are working on ecological and societal consequences of biodiversity, many of the bird species are extinction-prone, and a few are functionally extinct. The proposed system aims to help such institutes and ornithologists who study the ecology of birds, identify the key threats and find out the ways of enhancing the survival of species.

9.2 LITERATURE SURVEY

Mahajan Shubham et al. (2021) have proposed multilevel threshold based segmentation. For determining optimal threshold, Fuzzy Entropy Type II technique is combined with Marine Predators Algorithm. Rai, Bipin Kumar (2020) and his group proposes a model explaining working of the project in which the user would be able to capture and upload the image to the

DOI: 10.1201/9781003453406-9

system and can store the image in the database. A. C. Ferreira, L. R. Silva, and F. Renna, et al. (2020) described methods for automation in the process of collection and generation of the training data of individual bird species. They developed a CNN based algorithm for the classification of three small bird species.

Nadimpalli et al. (2006) uses the Viola-Jones algorithm for bird detection with accuracy of 87 percent. In the automated wildlife monitoring system proposed by Hung Nguyen et al. (2017) a deep learning approach is used. Yo-Ping Huang (2021) uses a transfer learning based method using Inception-ResNet-v2 to detect and classify bird species. They claim accuracy of 98 percent for the 29 bird species they have considered. Tejas Khare and Anuradha C. Phadke (2020) have used the You Only Look Once algorithm for classification of animals to build automated surveillance systems using computer vision.

M. T. Lopes et al. (2011) focuses on the automatic identification of bird species from the audio recorded songs of birds. They use 64 features such as means and variances of timbral features, calculated in the intervals, for the spectral centroid, roll off, flux, the time-domain zero crossings, including the 12 initial MFCCs in each case. Kansara et al. (2016), in his paper on speaker identification uses, MFCC in combination with deep neural network. Rawat, Waseem and Wang, and Zenghui (2017) have taken a review of Deep Convolutional Neural Networks for image classification. They have mentioned applications of Inception and improved Inception models in their paper.

Existing techniques deal either with an image based or audio based approach. But adverse environmental conditions may damage the sensors if we rely on any one approach. So to make the system perfect and to cope with the changes in weather and environment, we propose here a framework to identify bird species with both approaches. We can combine both the methodologies to authenticate the results as well, as if one of them fails other can support.

In Section 3 these methodologies are explained; in Section 4 is the system design; while in Section 5 results are discussed.

9.3 METHODOLOGY

Classification of images can be done with different state-of-the art approaches. One suitable approach for bird species' identification is the Vision Transformer (ViT) technique, which works as an alternative to Convolutional Neural Networks. The ViT was originally designed for text-based tasks, which is a visual model based on the architecture of a transformer. In this model an input image is represented as a series of image patches, like the series of word embeddings used while using ViT to text and predict class labels for given images. When enough data are available for training, it gives an extraordinary performance, with one-fourth fewer

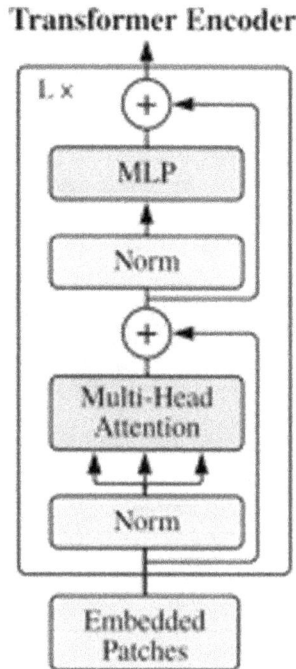

Figure 9.1 VIT architecture.

Source: Y.- P. Huang and H. Basanta (2021).

computational resources of a similar state-of-art CNN. The encoder of the transformer includes Multi-Head Self Attention Layer (MSP), Multi-Layer Perceptrons (MLP), and Layer Norm (LN). The architecture of ViT is shown in Figure 9.1. ViT is used for classification of bird species based on images.

For audio-based bird species identification Mel-frequency cepstral coefficients (MFCCs) derived from an audio signal are stored in the form of images as a spectrogram and are given as input to InceptionNet.

When multiple deep layers of convolutions were used in a model it resulted in the overfitting of the data. The inception V1 model employs the idea of using multiple filters of different sizes on the same level. In the inception models, instead of having deep layers, we have parallel layers, making it wider rather than deeper. Its architecture is shown in Figure 9.2.

9.4 SYSTEM DESIGN

Bird species identification carried out in proposed work is with both image based and audio based techniques. While building datasets the common

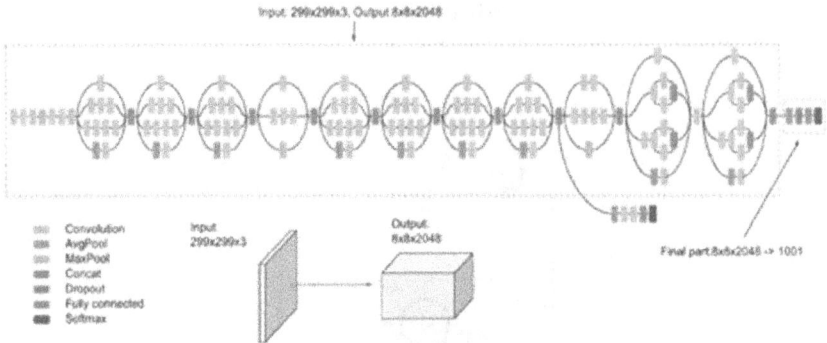

Figure 9.2 Architecture of InceptionNet.

Table 9.1 Image dataset

	Before Augmentation	After Augmentation
Training Images	7,581	1,44,039
Validation Images	250	4,790
Testing Images	250	No No Augmentation

species available in both datasets of image and audio were respectively considered for experimentation and comparison.

9.4.1 Dataset used

Due to small size of dataset (about 100 images for each class/species) the image augmentation method was used to increase Dataset size. Each image is provided with 18 different augmentations that means each image have total of 19 variations. Table 9.1 shows the total number of images used for training with augmentation. Table I contains Image Data sets of 50 different bird species.

Types of Augmentation used:

- Original Image
- Rotation – 4 instances of Rotating image once in each quadrant.
- Flip – Flipping image upside down.
- Mirror – Mirror image of original.
- Illumination – Changing contrast and brightness so that night light (low light) images are made clear (2 Instances).
- Translation – The shifting of an image along the x- and y-axis (4 instances)

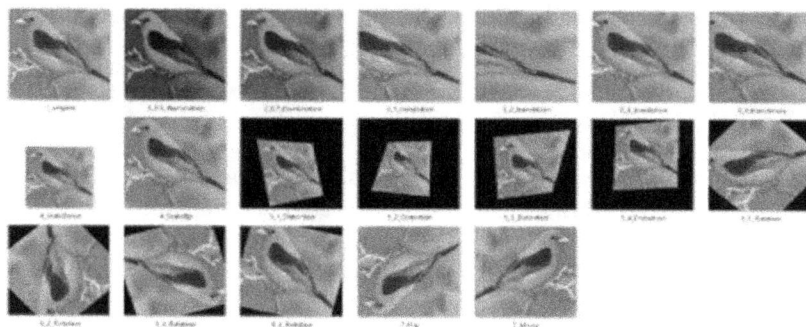

Figure 9.3 Augmented images used in Dataset.

Table 9.2 Audio dataset

	Number of Files
Training Images	3342
Validation Images	409
Testing Images	432

- Scale – Scaling refers to the resizing of a digital image
- Distortion – Distortion causes straight lines to appear curved

Different augmentations used are shown in Figure 9.3. The total number of images increased as shown in Table 9.1 after augmentation.

Dataset used for Audio signal-based bird species identification consists of 50 species which are common to dataset used for image-based techniques as shown in Table 9.2.

9.4.2 Image based techniques

A model based on images as input was trained and validated l on Py-Torch for bird image classification. Vision Transformer was used with a training batch size of 128, while the validation batch size was 24. The results are disused in next section. Testing results were analyzed with various parameters such as precision, accuracy, recall, and F1_score.

9.4.3 Audio based techniques

Different pre-processing techniques as explained below were used to process an audio dataset to get ready for classification. A model based on

InceptionNet was trained and validated on Py-Torch for bird audio classification. with the training batch size as 16 and validation batch size as 8. The results are discussed in next section.

- Mono audio channel

The dataset used was a mixture of files, mono format (audio with one channel) and stereo format (audio with two channels, left and right).

To standardize the dataset, the whole dataset was converted to mono since it sounds better. Mono takes less equipment, less space, and is cheaper. If two or more speaker stereo inputs are used then it gives better experience, but if a single speaker is used then mono input gives louder music than stereo input.

- Standardize Sampling Rate at 44.1kHz

Humans can hear frequencies between 20 Hz and 20 kHz. Therefore, maximum frequency required can be 40 kHz according to Nyquist Theorem. So, for many music applications, the sample rate considered is 44.1 kHz or 48 kHz when creating an audio for video.

- Noise Reduction

Many of the signal processing devices are susceptible to noise. Noise can be introduced by a device's mechanism or signal processing algorithms, which are random in nature with an even frequency distribution (white noise), or frequency-dependent distribution.

In many electronic recording devices, "hiss" created by random electron motion due to thermal agitation at all temperatures above absolute zero is a major source of noise. Detectable noise is generated by agitation of electrons, which rapidly add and subtract from the voltage of the output signal.

Therefore, Noise Reduction is an important part of pre-processing, which is the process of removing noise from a signal.

- Split Audio

Since all audio files are not of the same length, they were split into the same length clips to make dataset consistent. Audio files were split in 10-second clips, and the last clip (of less than 10 seconds) was concatenated with itself until it was 10 seconds.

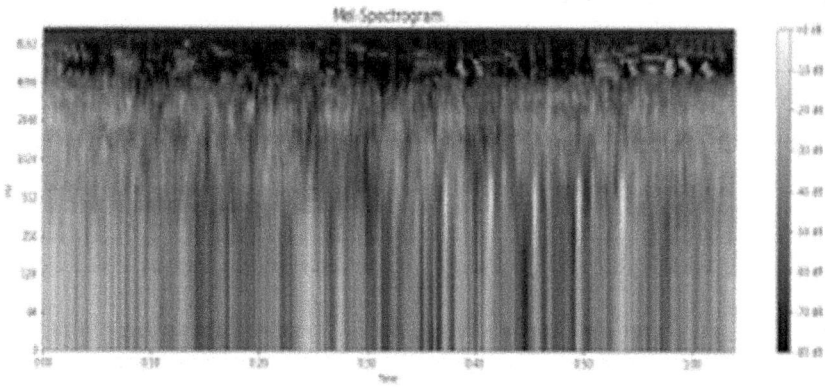

Figure 9.4 Mel-Spectrogram.

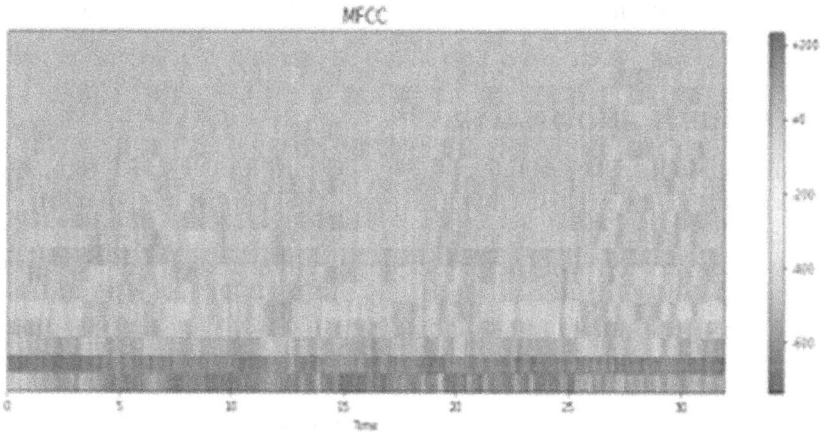

Figure 9.5 MFCC.

- Feature Extraction – MFCC (Mel-Frequency Cepstral Coefficients)

While obtaining MFCCs, the first step was computing the Fourier transformation of audio data, which takes the time domain signal and turns it into a frequency domain signal. This is computed by the fast Fourier transformation, which is an incredibly important algorithm of time.

Then the Mel-spectrogram (power spectrum) was computed, and the Mel-Filter bank was applied to it. The Mel-Frequency scale relates to perceived frequency of an audio signal and relates to pitches judged by listeners. The Mel Spectrogram of the sample from our dataset is shown in Figure 9.4 while a plot of MFCC is shown in Figure 9.5. Instead of converting MFCC

Table 9.3 Training and validation results

	Training Accuracy	Validation Accuracy
Image based Techniques	88.37%	90.48%
Audio based Techniques	51.57%	40.09%

Table 9.4 Testing results

	Accuracy	F1_Score	Precision	Recall
Image based Techniques	98.8%	98.82%	99.1%	98.8%
Audio based Techniques	34.21%	25%	25.66%	29.34%

coefficients into an image that is given directly as input to InceptionNet, MFCC can be given as input to the Artificial Neural Network. But the proposed work image based MFCC are compared with image based techniques.

9.5 RESULTS AND ANALYSIS

Results obtained after training the model for image based technique by Vision, transformer and audio based technique by InceptionNet are presented in Tables 9.3 and 9.4.

The image based techniques give good validation accuracy of 90.48 percent as compared to audio based techniques as shown in Table 9.3. Table 9.4 gives analysis of testing results. It can be seen from Table 9.4 that image based techniques outperform in all the metrics of testing, like accuracy, F1_score, Precision and Recall.

Through an analysis of the image dataset, it was found that the relative number of training samples for each bird species is quite uneven, which seems to lead to favoritism from the model of bird species, and that some bird species are more difficult to classify than others.

Accuracy of the MFCC based approach can be increased using Artificial Neural Network, and the system can be built by combining both image and audio based approaches to make it foolproof.

This work of combining both methods is under process and will be published in the near future. Due to adverse weather conditions if capturing the image of a bird is not possible then audio based approach will support.

9.6 CONCLUSION

From the two approaches implemented for bird species identification, audio based technique has shown poor accuracy whereas image based technique has demonstrated very good accuracy of 98.8 percent. Thus the image based bird identification tool can be used as assistant to bird-watchers. It will also play an important role in ecology studies such as identification of endangered bird species. In many circumstances the birds are not visible, and there is need to identify the bird that is singing. In such cases, the audio based bird identification technique can solve the problem of bird identification. And, hence, there is need to improve an audio based bird detection algorithm using features such as MFCC with classifier instead of converting it to image. Vision transformer is one of the state of the art techniques of machine learning and gives very good performance for the image based approach.

REFERENCES

Alexey Dosovitskiy, Lucas Beyer, Alexander Kolesnikov, Dirk Weissenborn, Xiaohua Zhai, Thomas Unterthiner, Mostafa Dehghani, Matthias Minderer, Georg Heigold, Sylvain Gelly, Jakob, Uszkoreit, Neil Houlsby (2021). An Image is worth 16x16 words: Transformers for Image Recognition at Scale. ICLR 2021.

Balint Pal Toth, Balint Czeba (2016). Convolutional Neural Networks for Large-Scale Bird Song Classification in Noisy Environment. Conference and Labs of the Evaluation forum (CLEF).

Ferreira, A.C., Silva, L.R, Renna F., et al. (2020) Deep learning-based methods for individual recognition in small birds. *Methods in Ecology and Evolution.* 11, 1072–1085. https://doi.org/10.1111/2041-210X.13436

Hung, Nguyen, Sarah J. Maclagan, Tu Dinh Nguyen, Thin Nguyen, Paul Flemons, Kylie Andrews, Euan G. Ritchie, and Dinh Phung (2017). Animal Recognition and Identification with Deep Convolutional Neural Networks for Automated Wildlife Monitoring. International Conference on Data Science and Advanced Analytics (DSAA).

Kansara, Keshvi, Suthar and Anilkumar (2016). Speaker Recognition Using MFCC and Combination of Deep Neural Networks. doi: 10.13140/RG.2.2.29081.16487

M.T. Lopes, A.L. Koerich, C.N. Silla and C.A.A. Kaestner (2011). Feature set comparison for automatic bird species identification. IEEE International Conference on Systems, Man, and Cybernetics, 2011, 965–970, doi: 10.1109/ICSMC.6083794

Mahajan, Shubham and Mittal, Nitin and Pandit, Amit. (2021). Image segmentation using multilevel thresholding based on type II fuzzy entropy and marine predators algorithm. *Multimedia Tools and Applications.* 80. 10.1007/s11042-021-10641-5

Rai, Bipin Kumar (2020). Image Based Bird Species Identification. *International Journal of Research in Engineering, IT and Social Sciences*, Vol. 10, Issue 04, April 2020, pp. 17–24.

Rawat, Waseem and Wang, Zenghui. (2017). Deep Convolutional Neural Networks for Image Classification:A Comprehensive Review. *Neural Computation*, 29, 1–98.

Tejas Khare, Anuradha C. Phadke (2020). Automated Crop Field Surveillance using Computer Vision. 4th IEEE International Conference on Distributed Computing, VLSI, Electrical Circuits and Robotics doi: 10.1109/DISCOVER50404.2020.9278072, ISBN:978-1-7281-9885-9

Uma D. Nadimpalli, R. Price, S. Hall, Pallavi Bomma (2006). A Comparison of Image Processing Techniques for Bird Recognition. Biotechnology Progress 22 (1), 9–13

Y. -P. Huang and H. Basanta. (2021). Recognition of Endemic Bird Species Using Deep Learning Models: IEEE Access, vol. 9, pp. 102975–102984, 2021, doi: 10.1109/ACCESS.2021.3098532

Chapter 10

Detection of Ichthyosis Vulgaris using SVM

Talha Fasih Khan, Pulkit Dubey, and Yukti Upadhyay

10.1 INTRODUCTION

Ichthyosis Vulgaris is described by means of amendments in Full Length Gene (FLG) Genetic linkage inspection of four families planned to the epidermal separation mixed up on the chromosome. More genotyping has, as of late, declared that a shortage of highlight transformations within the FLG quality are the explanation of, the events that are acquired in a semi-predominant manner with 83–96 percent penetrance. Amendments achieve a shortened PRO filaggrin protein, which cannot be controlled into intentional filaggrin subunits. It seems reasonable that amendments in related qualities ought to achieve shortened filaggrin proteins. FLG changes rationale in every Caucasian and Asian population [15, 16] any way they will generally be people with extraordinary and unique, once in a while mutually uncommon changes among those meetings. Indeed, even inside European populaces, there are nearby contrasts. While the R501X and changes represent approximately 80 percent of amendments in northern European relatives, they are significantly less in southern European successors. Heterozygous addition considered, change own toleration benefit over homozygous passive or potentially homozygous predominant genotypes. As there are all the earmarks of there being a scope-based occurrence propensity all through Europe, FLG changes could likewise give better endurance rates. The Chinese Singaporean people's eight unmistakable amendments represent approximately 80 percent. [30] Also, S2554X and 3321delA changes are exceptionally consistent with the Japanese, to whatever degree it is more uncommon in Koreans. The event of IV in darkly pigmented populaces is by all accounts low [10,11], but nevertheless more prominent inspection is expected to attest to those perceptions. The event gauges might need to presumably underrate the real event of in those populaces because FLG changes unique to Europeans had been above all else used to find amendments supplier frequencies in Asians.

DOI: 10.1201/9781003453406-10

10.2 LITERATURE SURVEY

After many investigations, we studied image processing for detecting skin diseases. In this, we give a brief overview of a number of the strategies as suggested in this literature. A device is proposed for the detection of skin disease sicknesses with the usage of color pictures without intervention of a physician. [3] The device includes different stages, the primary one being the use of color photo processing strategies, k-approach clustering, and color gradient strategies to discern the problem in the skin. The second one is the class of the disorder kind, the usage of synthetic neural networks. After preparing the system for detecting the illness, it was tested on six styles of pores and skin sicknesses, where the mean accuracy of the first degree was 95.99 percent and the second degree 94.061 percent. In this approach, the wider the variety of functions extracted from the photo, the higher the accuracy of the device. Melanoma is a disease that can result in death as it leads to skin cancer if it is not diagnosed at an early stage. Numerous segmentation strategies were targeted that might be carried out to discover cancer by using photo processing. [5] The segmentation system is defined that falling at the inflamed spot barriers to extracting greater functions. The paintings proposed the improvement of a Melanoma prognosis device for darkish pores and skin and the usage of a specialized set of rules databases such as pictures from many Melanoma resources. Similarly with the class of pores and skin sicknesses mentioned together with Melanoma, Basal Biliary Carcinoma (BCC), Nevus and Seborrheic Keratosis (SK) through the usage of the method guide vector system, termed the Support Vector Machine (SVM). SVM yields excellent accuracy from several different strategies. For one, the unfolding of persistent pores and skin sicknesses in special areas might also additionally result in extreme consequences. Therefore, we proposed a personal computer gadget that routinely detects eczema and determines its severity. [10] The gadget includes three stages, the primary stage being powerful segmentation through detecting the pores and skin; the second one extracts a fixed series of functions, specifically color, texture, and borders and the 0.33 determines the severity of eczema and the usage of an SVM. A brand-new method has been proposed to discover pore and skin sicknesses, a method that mixes pc imaginative and prescient with system mastering.[15] The position of pc imaginative and prescient is to extract the functions from the photo at the same time as system mastering is used to discover pore and skin sicknesses. The gadget examined six styles of pores and skin sicknesses with an as it should be 95 percent.

10.3 TYPES OF ICHTHYOSIS

10.3.1 Ichthyosis Vulgaris

This normal sort of ichthyosis is acquired as an autosomal prevailing. It is rarely noticed sooner than 90 days after acquisition, and large numbers of victims are impacted during their lives with injuries to their palms and legs. Further, many sufferers enhance as they get older, so that no medical results can be glaring in the summer, and the sufferers appear to be normal. Nonetheless, rehashed clinical assessment will for the most part show that this is not generally the case and to the point that either the mother or father had a couple of indications of the extent of their illness.

10.3.2 Hyperkeratosis

In inclusion to the ichthyosis, characterized with the aid of using pleasant white branny scales, its miles frequently viable to look at tough elevations around the hair follicles (keratosis pilaris), in inclusion to expanded palmar and plantar smudge. A significant part of the face can be impacted. The pores and skin of the flexures and neck are generally typical. At the point that the storage compartment is involved the ichthyosis is substantially less distinguished at the stomach than at the back, and much of the time there is limited hyperkeratosis on the elbows, knees, and lower legs. A considerable number of those victims have at least one of the signs of atopy (asthma, skin inflammation, and roughage fever disorder), and drying of the palms and heels is a typical concern. The cytology might also display a few hyperkeratosis, with a faded or truant granular layer, and its miles more likely are a few discounts within the range of sweat and sebaceous gland.

10.4 SEX-CONNECTED ICHTHYOSIS

Sex-associated ichthyosis may be well thought about under this heading, with Ichthyosis Vulgaris held totally responsible, with some distance more and typical spot polygenic assortment. Synthetically, IV might be strongly associated with Ichthyosiform Erythroderma. This state of ichthyosis is communicated through clinically unaffected females and shows itself most readily in men. It follows that the parents and children of men with sex-associated ichthyosis might have typical skin effects, notwithstanding, their little child can be mandatory heterozygotes ready to communicate what is going on to their children. A child cannot acquire the disease from the father.

10.5 SYMPTOMS

Ichthyosis scales appear dialogue box
Ichthyosis Vulgaris slows the pores' and skin's herbal dropping process. The reason is the extreme development of the protein inside the top layer of the pores and skin (keratin). Side effects incorporate dry, layered pores and tile-like skin; little scales that are white, dim or brown scales, depending on the pores; and a variety of flaky scalp-deep excruciating outbreaks. The scales by and large appear on elbows and lower legs and can be thick and darkish over the shins. Most cases of Ichthyosis Vulgaris are mild, but nonetheless, it sometimes can be extreme. The severity of the signs may also range extensively amongst one' own circle of relative participants becoming affected. Indications generally get worse or are more expressed in cool, parched habitats and tend to enhance or maybe resolve in mild, muggy habitats.

10.6 COMPLICATIONS

Some human beings with ichthyosis may also sense burnout. In uncommon cases, the pores and skin solidify and scales of ichthyosis can intrude with dripping. This can impede cooling. In some people, additional perspiring (hyperhidrosis) can happen. Skin parting and breaking may likewise bring about further contamination. The prognosis or infants could be very poor. Most of the affected newborn infants do not live past the primary week of life. It has been said that survival rates vary from 10 months to twenty-five years with supportive remedies depending on the severity of the condition.

10.7 DIAGNOSIS

There is no remedy for inherited Ichthyosis Vulgaris. The treatment especially decreases the size and dryness of pores and epidermis. Treatment plans require taking showers frequently. Drenching empowers hydration of pores and skin and melts the size of the IV infection. Assuming you have open bruises, your dermatologist may propose using petroleum jelly, or something comparable, on those sbruises before venturing into water. This can decrease the consumption and stinging caused by the water. A few victims say that including ocean salt (or common salt) in the water decreases the consumption and stinging. Adding salt can likewise reduce the tingling. Absorbing water relaxes the infection's size. Your dermatologist may likewise recommend that you decrease the size simultaneously as it is milder, gently scour with a rough wipe, Buff Puff, or pumice stone. Apply lotion to soggy pores and skin within two minutes of washing. The lotion can seal water from a shower or wash it into your pores and skin. Your dermatologist may likewise recommend a lotion that comprises an effervescent component like urea, alpha hydroxyl corrosive, or lactic corrosive. These and other such components also can assist to lessen scale. Apply petroleum jelly to the worst breaks. This can help eliminate them. On the off chance that you develop pore and skin contamination, your dermatologist will suggest a cure that you either take or practice for your pores and skin.

10.8 METHODOLOGY

SVM is a measurement technique in light of factual learning hypothesis, and is reasonable for shopping centre example size order. You can minimize training errors and gain confidence. Analyze a given training set or test set that uses SVMs to identify three things. Skin disease. A number is chosen from the main example number and for preparing extricated highlights (e.g., variety and surface capacity) and utilization of the judicious center capacity of the help vector machine. You can construct a characterization model. In this composition, the three archetypal skin after-effects are herpes, dermatitis, and psoriasis classes I, II, III. The Support Vector Machine 1, achy zone emphasize classifier is coupled with Support Vector Machine 2, along with the element model picked up is coupled with the effects. σ is the value of the radial basis function parameter. The Support Vector Machine can be observed along with the miniature applied in the K-Nearest Neighbor classifier. Presume that enduring a peculiar kitty family for certain elements of canines, so assuming we really bear a miniature that can exactly perceive whether it is a catlike or canine, such a miniature is made by applying the Support Vector Machine appraisal. At the beginning set up our miniature along with stacks of film and of pussycats and doggies so it can get knowledge from the various species of pussycats and doggies, along with subsequently, we try it with this unusual critter. Accordingly, as the assistance vector pursues a call limit among the two data (catlike along with canine) and picks incredible cases (support vectors), it will endure the crazy example of catlike and canine. Given help assist vectors, it with willing request it as a feline, where aj is a Lagrange multiplier, b* is the predisposition, and k(x1, x2) is a part work; x1 alludes to the eigenvector that is acquired from the trademark model; alludes to the outcomes; and σ is boundary esteem in the outspread premise work.

10.9 RESULTS

The images are classified using Support Vector Machine (SVM). We have used a system with Intel Core i7 processor 10 generation 2.60 GHz with 16 GB RAM. We have classified images on basis of skin color also and explained it in methodology section. Support vector is explained through graphs (Figure 10.1). The images are classified as shown in Figure 10.2. In Figure 10.3 There are two types of images circle and rectangle, in 2-D it is difficult to classify both images, as we go to 3-D we can easily identify the

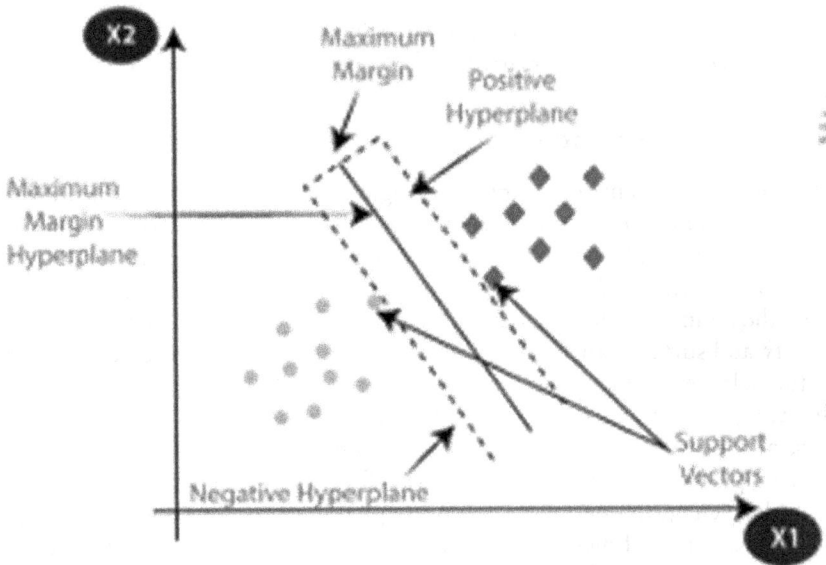

Figure 10.1 Classification using a hyperplane.

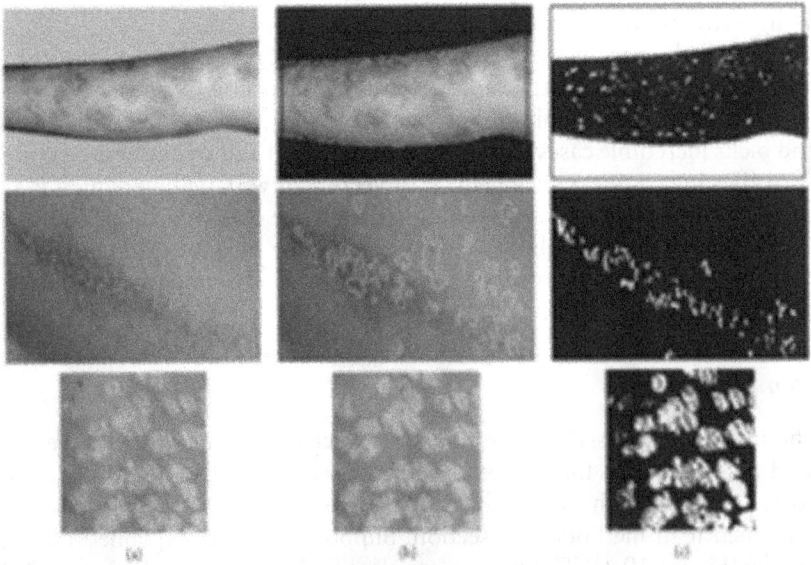

Figure 10.2 Picture division (a) Unique pictures. (b) Marker-controlled watershed division (c) Imprint controlling + grouping.

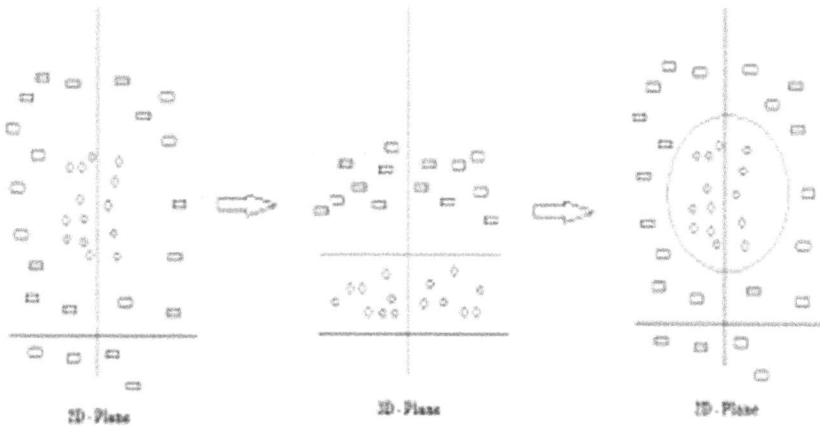

Figure 10.3 SVM Mechanism.

images and then, coming back to the original state, we can now classify the images. Ichthyosis is a non-curable disease but by using SVM methodology we can predict it early stages and can prevent it by taking precautionary measures.

10.10 FUTURE WORK

In future we can classify images on the basis of their color, which will be easy to predict, and classify different skin diseases. We can also develop a website or an application through which anyone can identify the disease. Further we can study other diseases that are curable, but due to late identification becomes serious. Other technologies can also be viewed for detection.

10.11 CONCLUSION

This chapter utilizes the investigation technique for vertical picture division to recognize ichthyosis. A couple of unessential elements can be decreased through picture isolating, picture turn, and Euclidean length change appertained in picture preprocessing. On a contrary line for every tip on the essential center is not altogether settled. What is more, the epithelium can be isolated into ten vertical picture areas. Given the dim position co-event grid embraced to separate the face element, along with the area pixel technique appertained to remove attributes of the achy area. At closing, the aid vector of the appliance is applied toward group data of more than 2 various derma illnesses as per highlights of surface along with the affliction area, negotiating fresh perfection of acknowledgement. In any case, the research focuses on dermatitis, herpes, and psoriasis and does not consider the varied side effects prevalent with analogous kinds of skin sickness. For illustration, Zima, herpes, and rubella all hold a place with analogous series. It will be the focal juncture of the subsequent platform to feel varied skin illnesses of analogous series by exercising an icon handling strategy.

REFERENCES

[1] S. Salimi, M. S. Nobarian, and S. Rajebi. "Skin disease images recognition based on classification methods," *International Journal on Technical and Physical Problems of Engineering.* 22(7), 2015.

[2] M. Ganeshkumar and J. J. B. Vasanthi. "Skin disease identification using image segmentation," *International Journal of Innovative Research in Computer and Communication Engineering.* 5(1): 5–6, 2017.

[3] S. Kolker, D. Kalbande, P. Shimpi, C. Bapat, and J. Jatakia. "Human skin detection using RGB, HSV and YCbCr Color models," *Adv Intell Syst Res.* 137, 2016.

[4] A. L. Kotian and K. Deepa. "Detection and classification of skin diseases by image analysis using MATLAB," *Int J Emerging Res Manage Technol.* 6(5): 779–784, 2017.

[5] Kumar, S. and A. Singh. "Image processing for recognition of skin diseases," *Int J Comp Appl.* 149(3): 37–40, 2016.

[6] Mazereeuw-Hautier, J., Hernandez-Martin, A., O'Toole, E. A., Bygum, A., Amaro, Aldwin, C. "Management of congenital ichthyoses: European guidelines of care," Part Two. *Br J. Dermatol.* 180: 484– 495, 2019.

[7] Wahlquistuist, A., Fischer, J., Törmä, H. "Inherited nonsyndromic ichthyoses: An update on pathophysiology, diagnosis and treatment." *Am J Clin Dermatol.* 19: 51–66, 2018.

[8] Schlipf, N. A., Vahlquist, A., Teigen, N., Virtanen, M., Dragomir, A., Fishman, S. "Whole-exome sequencing identifies novel autosomal recessive DSG1 mutations associated with mild SAM syndrome." *Br J Dermatol* 174: 444–448, 2016.

[9] Kostanay, A. V., Gancheva, P. G., Lepenies, B., Tukhvatulin, A. I., Dzharullaeva, A. S., Polyakov, N. B. "Receptor Mincle promotes skin allergies and is capable of recognizing cholesterol sulfate," *Proc Natl Acad Sci U S A.* 114: E2758–E2765, 2017.

[10] Proksch, E. "pH in nature, humans and skin." *J Dermatol.* 45: 1044–10. 2018.

[11] Bergqvist, C. Abdallah, B.. Hasbani, D. J., Abbas, O., Kibbi, A. G., Hamie, L., Kurban, M., Rubeiz, N. "CHILD syndrome: A modified pathogenesis-targeted therapeutic approach." *Am. J. Med. Genet.* 176: 733–738, 2018.

[12] McAleer, M. A., Pohler, E., Smith, F. J. D., Wilson, N. J., Cole, C., MacGowan, S., Koetsier, J. L., Godsel, L. M., Harmon, R. M., Gruber, R., et al. "Severe dermatitis, multiple allergies, and metabolic wasting syndrome caused by a novel mutation in the N-terminal plakin domain of desmoplakin." *J. Allergy. Clin. Immunol.* 136: 1268–1276, 2015.

[13] Zhang, L., Ferreyros M, Feng W, Hupe M, Crumrine DA, Chen J, et al. "Defects in stratum corneum desquamation are the predominant effect of impaired ABCA12 function in a novel mouse model of harlequin ichthyosis." *PLoS One.* 11: e0161465, 2016.

[14] Chan A., Godoy-Gijon, E., Nuno-Gonzalez, A., Crumrine, D., Hupe, M., Choi, E. H,, et al. "Cellular basis of secondary infections and impaired desquamation in certain inherited ichthyoses." *JAMA Dermatol.* 151: 285–292, 2015.

[15] Zhang, H., Ericsson, M., Weström, S., Vahlquist, A., Virtanen, M., Törmä, H. "Patients with congenital ichthyosis and TGM1 mutations overexpress other ARCI genes in the skin: Part of a barrier repair response?" *Exp Dermatol.* 28: 1164–1171, 2019.

[16] Zhang, H., Ericsson, M., Virtanen, M., Weström, S., Wählby, C., Vahlquist, A., et al. "Quantitative image analysis of protein expression and colocalization in skin sections." *Exp Dermatol.* 27: 196–199, 2018.

[17] Honda, Y., Kitamura, T., Naganuma, T., Abe, T., Ohno, Y., Sassa, T., et al. "Decreased skin barrier lipid acyl ceramide and differentiation- dependent gene expression in ichthyosis gene Nipal4-knockout mice." *J Invest Dermatol.* 138: 741–749, 2018.

Chapter 11

Chest X-Ray diagnosis and report generation

Deep learning approach

Chinmay Dixit and Amit D. Joshi

11.1 INTRODUCTION

Chest radiography is a cost-effective, commonly accessible, and easy to use medical imaging technology and is a form of radiological examination that can be used for diagnosis and screening of lung diseases. A chest X-ray image includes the chest, lung, heart, airways, and blood vessels, and it can be used by trained radiologists for diagnosing several abnormal conditions. X-ray imaging is inexpensive and has a simple generation process. Computer-aided techniques have the potential to use chest X-rays to diagnose thoracic diseases accurately and with accessibility [1]. Computers can be made to learn features that depict the data optimally for certain problems. Increasingly higher-level features can be learned while the input data is being transformed to output data using models [2]. The field of deep learning has seen significant progress in applications like classification of natural and medical images using computer vision approaches over the past few years. Applications of these techniques in modern healthcare services still remains a major challenge. A majority of chest radiographs around the world are analyzed visually, which requires expertise and is time-consuming [3]. The introduction of the ImageNet database has improved the performance of image captioning tasks. In addition to this, improvements in deep Convolutional Neural Networks (CNN) enable them to recognize images effectively.

Recent studies also use Recurrent Neural Networks (RNN), using features from the deep CNNs to generate image captions accurately [4]. These features may also be modified and combined with other observed features from the image to retain important details that can help form a richer and more informative description [5]. Caption generation models must be capable of detecting objects present in an image and also capture and represent relationships between them using natural language. Attention based models work on training on local salient features and ignore redundant noise. Localization and recognition of regions with salient features also allows us to generate richer and diverse captions [6]. A detailed diagnostic medical

DOI: 10.1201/9781003453406-11

image report consists of multiple forms of information. A detailed chest X-ray report contains an impression, findings, and a list of tags. Automatic radiology report generation can be done with the help of computer vision methods and has received particular attention, as it makes the interpretation of images easier for clinicians to understand [7]. However, clinical accuracy is very important in the resulting generated reports. Compared to actual radiology reports, traditional approaches for image captioning produce short and simple pieces of text [8]. Study and comparison of different approaches and methods for such tasks may help in quick and useful diagnosis of medical conditions [9].

This work focuses on providing an accurate approach to detecting thoracic diseases and generating accurate reports for them. The models that this work uses will be evaluated on various metrics and compared to existing models and approaches. As mentioned before, many of the conventional methods used today involve manual analysis of chest radiographs by experts in the field, which proves to be time-consuming, involves human effort, and hinders the rate of diagnosis and subsequent treatment of patients, especially in challenging times like pandemics. The main motivation behind this work is to study and evaluate the performance of approaches in the deep learning domain for the task of chest X-ray diagnosis and report generation. This study also emphasizes understanding different techniques used for implementing existing models. Understanding and implementing these modern approaches will help in future studies and development of different and more efficient methods, while increasing accuracy, reducing time and human effort, and may also facilitate better data collection and availability. This work proposes a detailed comparison study and evaluation of a selection of deep learning methods used for image classification, labelling and report generation using a suitable framework.

11.2 LITERATURE REVIEW

Previous related work indicates that significant research has been done in the domain of both feature extraction and text report generation. This literature review has been divided into two parts: the first part describes previous work related to methods for chest X-ray image analysis, and the second part discusses image captioning and text generation studies.

Hu, Mengjie, et al. proposed an approach for quick, efficient, and automatic diagnosis of chest X-rays, called the multi-kernel depth wise convolution (MD-Conv). Lightweight networks can make use of MD-Conv in place of the depth wise convolution layer. The lightweight MobileNetV2 is used instead of networks like ResNet50 or DenseNet121. The approach aims to provide a foundation for later research and studies related to lightweight networks and the probability of identifying diseases in embedded and mobile devices [10]. Albahli, Saleh, et al. evaluated the effectiveness of different CNN models that use Generative Adversarial Networks (GAN) to generate synthetic data. Data synthesis is required as over-fitting is highly possible because of the imbalanced nature of the training data labels. Out of the four models whose performance was evaluated for the automatic detection of cardiothoracic diseases, the best performance was observed when ResNet152 with image augmentation was used [11]. Existing methods generally use the global image (global features) as input for network learning that introduce a limitation as thoracic diseases occur in small localized areas and also due to the misalignment of images. This limitation is addressed by Guan, Qingji, et al. It proposes a three-branch attention guided CNN (AG-CNN). This approach learns on both local and global branches by first generating an attention heat map using the global branch. This heat map is used for generating a mask, which is later used for cropping a discriminative region to which local attention is applied. Finally, the local and global branches are combined to form the fusion branch. Very high accuracy is achieved using this approach with an average value of Area Under Receiver Operating Characteristic Curve (AUC) as 0.868 using ResNet50 CNN, and 0.871 using DenseNet121 [12].

In addition to using existing data, some studies also proposed collection and analysis of new data. Bustos, Aurelia, et al. proposed a dataset called PadChest, which contains labels mapped onto the standard unified medical language system. Instead of solely relying on automatic annotation tools, trained physicians performed the task of manually labelling the ground truth targets. The remaining reports that were not labeled manually were then tagged using a deep learning neural network classifier and were evaluated using various metrics [13]. Four deep learning models were developed by Majkowska, Anna, et al. to detect four findings on frontal chest radiographs. The study used two datasets, the first one was obtained with approval from a hospital group in India and the second one was the publicly available ChestX-ray14 dataset. A natural language processing system was created for the prediction of image labels by processing original radiology reports. The models performed on par with on-board radiologists. The study performed analysis for population-adjusted performance on ChestX-ray14 dataset images and released the adjudicated labels. It also aims to provide a useful resource for further development of clinically useful approaches for chest radiography [14].

Text generation is an important objective, and previous research indicates that performance is dependent on the data as well as the approach. An RNN Long Short-Term Memory (LSTM) model that takes stories as input, creates and trains a neural network on these input stories, and then produces a new story from the learned data is described by Pawade, D., et al. The model understands the sequence of words and generates a new story. The network learns and generalizes across various input sequences instead of learning individual patterns. Finally, it has also been observed that by adjusting the values of different parameters of the network architecture, the train loss can be minimized [15].

Melas-Kyriazi, et al. proposed a study in which the issue of lack of diversity in the generated sentences in image captioning models is addressed. Paragraph captioning is a relatively new task compared to simple image single-sentence captioning. Training a single-sentence model on the visual genome dataset, which is one of the major paragraph captioning datasets, results in generation of repetitive sentences that are unable to describe the diverse aspects of the image. The probabilities of the words that would result in repeated trigrams are penalized to address the problem. The results observed indicate that self-critical sequence training methods result in lack of diversity. Combining them with a repetition penalty greatly improves the baseline model performance. This improvement is achieved without any architectural changes or adversarial training [16].

Jing et al. proposed a multi-task learning framework that can simultaneously perform the task of tag prediction and generation of text descriptions. This work also introduced a co-attention mechanism. This mechanism was used for localizing regions with abnormalities and to generate descriptions for those regions. A hierarchical LSTM was also used to generate long paragraphs. The results observed in this work were significantly better as compared to previous approaches with a Bilingual Evaluation Understudy (BLEU) score of 0.517 for BLEU-1 and 0.247 for BLEU-4. The performance observed on other metrics was also better than other approaches [17]. Xue, Yuan, et al. tackle the problem of creating paragraphs that describe the medical images in detail. This study proposed a novel generative model which incorporates CNN and LSTM in a recurrent way. The proposed model in this study is a multimodal recurrent model with attention and is capable of generating detailed paragraphs sentence by sentence. Experiments performed on the Indiana University chest x-ray dataset show that this approach achieves significant improvements over other models [18].

A Co-operative Multi-Agent System (CMAS), which consists of three agents – the planner, which is responsible for detecting an abnormality in an examined area, the normality writer, which is used for describing the observed normality, and the abnormality writer, which describes the abnormality detected – is proposed by Jing, et al. CMAS outperforms all other described approaches as evident from the value of many of the metrics used for evaluation. Extensive experiments, both quantitative and qualitative, showed that CMAS could generate reports that were meaningful by describing the detected abnormalities accurately [19].

11.3 PROPOSED METHODOLOGY

This section gives an outline of the deep learning algorithms, terminologies, and information about the data used in this work. It also explains the two proposed methods in detail.

11.3.1 Overview of deep learning algorithms

1. Convolutional Neural Network

CNNs are most commonly used for analyzing images as a part of various computer vision tasks. CNN makes use of convolution kernels or filters as a part of a shared weight architecture. The network generates a feature map for the input image by extracting local features with fewer number of weights as compared to traditional artificial neural networks. The feature map can be passed through pooling or sub-sampling layers to decrease its spatial resolution. Reducing the resolution also reduces the precision with which the position of the feature is represented in the feature map. This is important because similar features can be detected at various positions in different images [20][21].

Deep CNNs have been made possible recently because of improvements in computer hardware. As the network grows deeper, the performance improves as more features are learned. But it also introduces the problem of vanishing gradient where the gradient becomes smaller and smaller with each subsequent layer, making the training process harder for the network.

2. Recurrent Neural Network

Sequence data consists of data that follows a certain order or sequence. An RNN can be used for modelling sequence data. It contains a hidden state h along with an optional output y. The network works on a sequence $x = (x_1, ..., x_t)$, which may vary in length. Equation 11.1 is used for updating the hidden state h_t of the RNN at each time step t.

$$h_t = f\left(h_{\langle t-1 \rangle}, x_t\right) \tag{11.1}$$

where f is a non-linear function [22]. The hidden state contains information which represents previous inputs. While modelling the sequence data, the probability distribution is learned by the RNN by learning to predict the next symbol in a sequence. Each time step produces an output, which is a conditional distribution that depends on the previous outputs.

In standard neural networks, the features learned by the network are not shared between different positions in the text, meaning if the same word occurs in two different positions, it may not have the same parameters. In an RNN, parameters are passed between the layers of the network to maintain relationship between words in the sequence. RNNs face problems when trying to learn long-term dependencies. In conventional gradient-based learning algorithms like back-propagation through time, where the error signal propagates backwards through the network, gradients tend to either vanish or explode. LSTM is a recurrent network architecture that can overcome these problems. LSTM networks are capable of learning long-term dependencies [23]. An LSTM unit consists of three gates, a cell state and a hidden state. Another similar mechanism is Gated Recurrent Unit (GRU), which has fewer parameters and only contains a hidden state that makes it faster to train than LSTM.

3. Transfer Learning and Fine-Tuning

Transfer learning involves using features learned on one problem and using them on a similar problem. The feature-maps learned by CNNs while training on one task can be used on a similar but new task. Transfer learning is usually done when insufficient data is available to train a model from scratch. In actual implementation, transfer learning is accomplished by using several approaches depending on the differences between the two problems. The old layers and weights learned to detect features from the original problem can either be fixed or be allowed to update. The features detected at each layer are different and unique. A certain layer in the model may learn weights that detect corners or edges in the input image, another layer might detect lines and bends present in the network and so on. Since the new problem in transfer learning is similar to the original problem, most of the features still remain relevant while training on the new task.

Fine Tuning is an approach of transfer learning in which either only the output layer is trained on the new data or the entire model is retrained after being initialized with the weights of the pretrained model so that the pretrained features may adapt to the new data. This work focuses on applying transfer learning the pretrained models using the ChestX-ray14 dataset.

4. Pretrained models

DenseNet121 is a CNN architecture that directly connects any two layers with equal feature map sizes. Every layer in the network receives feature maps as inputs from all previous layers. This architecture not only addresses the vanishing gradient problem, but also strengthens qualities like propagation of features, reuses existing features and reduces the number of parameters [24]. Rajpurkar, Pranav, et al. used DenseNet121 to train on the ChestX-ray14 dataset to achieve highly accurate results [25].

MobileNetV2 is a mobile neural network architecture that is specifically designed for environments which are mobile and limited in terms of resources. In comparison with modern widely used networks which require high computational resources, MobileNetV2 tries to focus on improving performance on mobile and embedded applications. It introduces inverted residuals with linear bottleneck in addition to the existing depth wise separable convolution feature [26].

Along with the two models mentioned above, the performance of VGG16, ResNet152V2 and InceptionV3 is also compared on the same task. The models were loaded with weights learned on the ImageNet dataset. The ImageNet dataset contains 1.2 million images and 50,000 images for training and validation respectively labelled with 1,000 different classes. It is a dataset organized according to the WordNet hierarchy and aims to provide an average of 1,000 images to represent each synset present in WordNet.

11.3.2 Data

ChestX-ray14 is a medical imaging dataset containing 112,120 frontal-view X-ray images of 30,805 unique patients; 14 common disease labels were text-mined from the text radiological reports via natural language processing techniques and assigned to each of the images [27]. For training the feature extraction models in this work, a random sample of the ChestX-ray14 dataset (available on Kaggle) consisting of 5 percent of the full dataset was used. The sample consists of 5,606 images and was created for use in kernels. Additionally, the dataset also contains data for disease region bounding boxes for some images that can be used for visualizing the disease region.

There is also a certain amount of data bias in the ChestX-ray14 dataset. A few disease conditions present in the dataset have less prevalence as compared to the number of examples where no abnormality is detected. This bias can affect the performance on unseen test examples. This problem can be resolved by assigning class weights to each class in the dataset. These weights are inversely proportional to the frequency of each class, so that a minority class is assigned a higher weight, and a majority class is assigned a lower weight. Assigning class weights lowers the training performance but ensures that the algorithm is unbiased towards predicting the majority class.

ChestX-ray14 contains images of chest X-rays along with the detected thoracic disease labels. However, it does not contain the original reports from which the labels were extracted. In this work, the Indiana University chest X-ray collection from Open-i dataset was used for the report generation task. The dataset contains 7,470 images containing frontal and lateral chest X-rays with 3,955 reports annotated with key findings, body parts, and diagnoses.

11.3.3 Feature extraction

Feature extraction involves generating feature maps for input images that represent important features in the input. This task can be defined as a sub-task of chest disease detection in the current problem since the features extracted are finally used for predicting labels. A deep CNN can be trained on the dataset and made to learn important features. In this work, five different CNN architectures and pre-trained models were employed to solve a multi-label problem defined on a subset of the ChestX-ray14 dataset. These models were trained on the dataset by making the entire model learn new sets of features specific to this task. In a multi-label problem, a set of classification labels can be assigned to the sample, as opposed to a single label out of multiple possible labels in a multi-class problem.

The default final network layer was removed and a new dense layer with 15 activation units was added. The images in the dataset were resized to 224 × 224 before using them as input to the initial network. Adam optimizer was used with the value of learning rate as 0.001. The value of the learning rate was reduced by a factor of 10 whenever the validation loss stopped decreasing. Training was done with binary cross-entropy loss:

$$\text{Loss} = -\frac{1}{N_c} \sum_{i=1}^{N_c} y_i \log\left(\widehat{y}_i\right) + \left(1 - y_i\right) \log\left(1 - \widehat{y}_i\right) \tag{11.2}$$

where $\widehat{y}_i$ is the i-th predicted scalar value, y_i is the corresponding target value and N_c is the number of output classes or output scalar values. Since the problem is a multi-label problem, each class was determined with a separate binary classifier, and where each output node should be able to predict whether or not the input features correspond to that particular class irrespective of other class outputs. In case of a multi-class problem, categorical cross-entropy loss is used. The new final classification layer used sigmoid activation instead of softmax.

$$\text{Sigmoid}(x) = \frac{1}{1 + e^{-x}} \tag{11.3}$$

Local features of the input image can be directly extracted from the previous layers of the CNN. These features were passed through an average pooling layer while training the CNN to decrease the feature map dimension. The feature map was then fed to the fully connected classification layer. Global features, on the other hand, can be extracted by reducing the spatial resolution of the image with the help of average pooling layers.

A CNN used for feature extraction serves as an encoder for the input data. The data is encoded into a set of features by using various operations like convolution (for detecting features), dimensionality reduction, and so forth. The CNN learns to detect features in the input data automatically, and manual feature definition is not required. Feature extraction can also be used in other tasks like image segmentation and class-activation mapping.

11.3.4 Report generation

Report generation for the input image is the next important task. In this task, the findings part of a detailed report, which lists the observations for each examined area, was generated. The CNN encoder can be used to infer the image and output a set of features as discussed above. The sequence-to-sequence model accepts these features as input. This model uses the input features to output a sequence of words that form the report and is an example of feature extraction where the encoder is simply used to predict the features instead of learning them for the input images.

As mentioned above, the Open-i dataset was used for this task. Instead of using both frontal and lateral images, only frontal images with the corresponding findings from the report were used, as these images are used as input for the encoder, which was originally trained on only frontal images of chest x-rays. While the encoder is still capable of detecting features in lateral images, these features will not accurately represent the condition observed in the chest x-ray.

The findings for each example were pre-processed and tokenized to remove redundant characters and words. Each tokenized word was mapped to a 300 dimensional vector defined in the pre-trained GloVe (Global Vectors for Word Representation) model. These vectors were then made part of an embedding matrix that was later used to represent each word, using a 300 dimensional vector.

Two existing methods were adopted and implemented for this task:

1. CNN-RNN Encoder-Decoder architecture

In the encoder-decoder architecture, the CNN serves as an encoder used for encoding the input image and the RNN as the decoder. One of the implementations described by Vinyals, Oriol, et al. involves feeding the output of the encoder (CNN) to an LSTM network as the first input to inform the LSTM network about the contents of the image and generate states [28]. In this work, in the CNN-RNN architecture, the encoded image features were first input to the LSTM network and then the updated internal state was used as an initial state for the embedded sequence.

The model takes images as inputs and the next word in the sequence is generated by using partial sequences. The image input from the CNN encoder was first passed through a fully connected layer and then normalized before being passed on to the LSTM layer. The sequence input was passed through an embedding layer after which it was fed to an LSTM layer (with the initial state updated), which was followed by another LSTM layer with

50 percent dropout. Another dense layer followed by dropout was used. Finally, a dense layer with softmax activation units for each word in the vocabulary was used to get a probability distribution:

$$\text{softmax}(x_i) = \frac{\exp(x_i)}{\sum_j \exp(x_j)} \tag{11.4}$$

The categorical cross-entropy loss function was used to train the model:

$$\text{Loss} = -\sum_{i=1}^{N} y_i \cdot \log \hat{y}_i \tag{11.5}$$

where $\hat{y}_i$ is the i-th predicted scalar value, y_i is the corresponding target value and N is the output size.

The final report generation model can be evaluated by using methods like greedy search and beam search. Greedy search chooses the word which maximizes the conditional probability for the current generated sequence of words. On the other hand, beam search chooses the N most likely words for the current sequence, where N is the beam width.

2. Attention-based architecture

An increase in the length of the sequence negatively affects the model performance of the first method. The Attention-based method focuses on important parts of the sequence. Xu, Kelvin, et al. describe the use of CNN as an encoder in an attention-based encoder decoder architecture. The features extracted from the image by the CNN are also known as annotation vectors. In order to obtain a correspondence between portions of the input image and the encoded features, the local features were extracted from a lower convolutional layer without pooling the outputs to get a higher dimensional output [29].

The decoder, as introduced by Bahdanau, et al. conditions the probability on a context vector c_i for each target word y_i, where c_i is dependent on the sequence of annotation vectors. These annotation vectors are generated by the encoder as stated above. Annotation vector a_i contains information about the input, with strong focus on the parts surrounding the i-th feature extracted at a certain image location. A weighted sum of annotations $\{a_1,...,a_L\}$ is used for computing the context vector c_i.

$$c_i = \sum_{j=1}^{L} \alpha_{ij} a_j \tag{11.6}$$

For each annotation a_j, weight α_{ij} is computed using

$$\alpha_{ij} = \frac{\exp(e_{ij})}{\sum_{k=1}^{L} \exp(e_{ik})} \tag{11.7}$$

where

$$e_{ij} = f_{att}(h_{i-1}, a_j) \tag{11.8}$$

where f_{att} is an attention model implemented using a multilayer neural network where the previous hidden state h_{i-1} was used to condition the network [30]. The input report is initially pre-processed, as mentioned above, and then processed by an embedding layer that converts the words to 300-dimensional vectors. The concatenation of the output of the embedding layer is done with the context vectors generated to form the input for the RNN layer. The target word is passed to the decoder as its next input while training the RNN. This technique is known as Teacher Forcing. An RNN layer that can consist of either LSTM or GRU units was used to output the current hidden state and the output. This output is fed to a fully connected layer with number of units equal to the size of the vocabulary. In this way, every time the model generates a word, it searches for a set of positions where the most relevant information is concentrated in the input.

11.3.5 Evaluation metrics

This work reports the AUC for the feature extraction task. The AUC is computed for each of the labels individually and then averaged across those labels. A Receiver Operating Characteristic curve is a graph that plots the classification performance for a model for all classification thresholds. Two parameters are plotted in this curve:

1. True Positive Rate (TPR) or Recall

$$TPR = \frac{count(True\,positives)}{count(True\,positives) + count(False\,negatives)} \tag{11.9}$$

2. False Positive Rate (FPR)

$$FPR = \frac{count(False\,positives)}{count(False\,positives) + count(True\,negatives)} \tag{11.10}$$

A threshold can be defined as a value used for deciding whether another value represents a positive example of a class or not. A threshold value of 0.5 will mean that all values above 0.5 will be classified as positive class examples and all those below will be classified as negative examples. More examples are classified as positive if the classification threshold is lowered. Unlike metrics such as accuracy, precision and recall, which only consider a single threshold value, all classification thresholds are considered to provide an aggregate measure of performance by AUC. A higher value of AUC indicates better performance.

In the case of report generation, the BLEU score is reported as a measure of performance for our models. BLEU scores can be calculated for varying lengths of n-grams. BLEU-1, BLEU-2, BLEU-3, and BLEU-4 scores are reported in this work. A higher value of BLEU score signifies better performance [31].

This work also uses Metric for Evaluation of Translation with Explicit Ordering (METEOR) score for evaluating the predicted reports. METEOR was designed to address the weaknesses observed in BLEU [32]. The average value of all sentence-level METEOR values for each individual report is reported in this work.

11.4 RESULTS AND DISCUSSIONS

This section explains the observed results from the two objectives discussed in this work. The main criteria of evaluation is model performance as observed from both the implementations.

11.4.1 Feature extraction

Section 3.2 states that a smaller (5%) sample of the dataset was used for training the models for feature extraction. The results indicate that despite the small size of training data, the models generalize moderately well on unseen data. The AUC values for the five models that were trained have been compared in Figure 11.1. As mentioned in section 3.5 AUC is a superior metric for multi-label problems compared to accuracy. The highest AUC with a value of 0.7165 is achieved by the pre-trained DenseNet-121 model. The performance of the model in the case of MobileNetV2 indicates overfitting according to the training and validation AUC values observed as the performance on the training set was significantly better as compared to the test/validation set.

Section 3.2 also discusses the data imbalance problem faced in ChestX-ray14 dataset. The best performing model, DenseNet-121 was also trained on the dataset after assigning class weights. However, the performance declines, which is justified by the actual frequency of classes. VGG-16 was the worst performing model of the five with the lowest training AUC of

Comparison of AUC values for feature extraction models

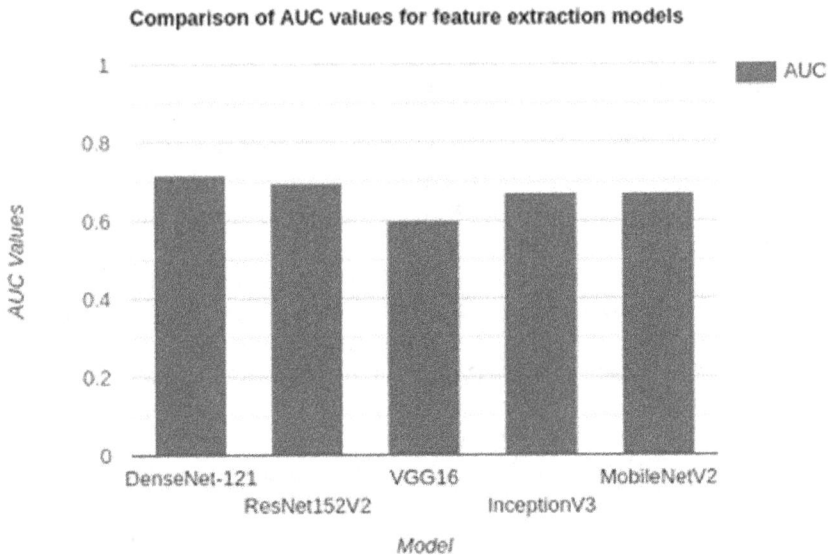

Figure 11.1 Comparison of AUC values observed for different models.

0.6012. ResNet152V2 (0.6956), InceptionV3 (0.6707) and MobileNetV2 (0.6697) performed moderately well, with ResNet152V2 getting the closest to the top performing model. The primary aim behind this objective was to evaluate the performance and the ability of widely used pretrained CNN models to train on a smaller sample of the dataset accurately. While related work and other popular models use the entire dataset, training on a smaller subset of the dataset is much less resource intensive and time expensive and provides an insight into the performance of these models on similar data.

11.4.2 Report generation

Table 11.1 contains the various BLEU scores and METEOR score calculated for various models that were trained. These scores were evaluated for individual reports and represent how accurately the model predicts a report for an input chest x-ray image. The two approaches described in section 3.4 are compared in Table 11.1. Observations indicate that the attention-based model performed better than the CNN-RNN model when greedy search was used. Beam search with beam width of 5 was used for generating the inference for the encoder-decoder model and performed significantly better than greedy search. However, beam search used for the attention-based model yielded poor results as compared to greedy search. As mentioned in section 3.4 only frontal chest x-ray images from the Open-i dataset were

Table 11.1 Comparison of models

Method	CNN	BLEU-1	BLEU-2	BLEU-3	BLEU-4	METEOR
Encoder-	DenseNet-121	0.319	0.189	0.123	0.0796	0.278
Decoder	DenseNet-121 (Beam Search)	0.35	0.21	0.143	0.095	0.312
	MobileNetV2	0.227	0.148	0.108	0.081	0.208
	ResNet152V2	0.229	0.134	0.092	0.061	0.208
Attention-	DenseNet-121	0.354	0.208	0.134	0.084	0.294
based	MobileNetV2	0.289	0.177	0.119	0.084	0.276
	InceptionV3	0.271	0.162	0.106	0.08	0.273
	ResNet152V2	0.268	0.155	0.103	0.071	0.267

Table 11.2 Report generation example

Image	Original vs Generated Report
	Actual report (preprocessed): the cardiomediastinal silhouette and pulmonary vasculature are within normal limits size . there patchy airspace disease the right lower lobe . the lungs are otherwise grossly clear . there no pneumothora pleural effusion . **Predicted report: (encoder-decoder model, beam search):** the heart size and pulmonary vascularity appear within normal limits . the lungs are free focal airspace disease . no pleural effusion pneumothora seen . no acute bony abnormalities .

used for training. This reduces the number of features extracted by half, as lateral images are not considered. The original and predicted report for a random image is shown in Table 11.2.

11.5 CONCLUSION

In the medical field, the amount of openly available task-specific data is limited. Data collection is difficult as precision and expertise are required when performing tasks such as data labelling. This work touches upon the use of deep learning techniques in solving medical problems, specifically the problem of automated chest x-ray classification and report generation. In this work, two important objectives were discussed along with methods used for achieving them. The methods include comparison of various pretrained CNN models for feature extraction and implementation of two CNN-RNN encoder-decoder models, one with attention and the other without attention. An overview of the major deep learning algorithms and techniques used for image captioning was also given. This work was able to achieve comparable results with related works for report generation with limited resources. The two objectives can be considered as two completely different tasks and can be optimized separately to produce better results on their respective tasks.

REFERENCES

1. Dong Y, Pan Y, Zhang J, Xu W (2017) Learning to read chest X-ray images from 16000+ examples using CNN. In: 2017 IEEE/ACM International Conference on Connected Health: Applications, Systems and Engineering Technologies (CHASE), pp. 51–57.
2. Litjens G, Kooi T, Bejnordi BE, Setio AAA, Ciompi F, Ghafoorian M, Van Der Laak JA, Van Ginneken B, Sánchez CI (2017) A survey on deep learning in medical image analysis. Medical image analysis, 42, pp. 60–88.
3. Wang H, Xia Y: Chestnet (2018) A deep neural network for classification of thoracic diseases on chest radiography. *arXiv preprint* arXiv:1807.03058.
4. Shin HC, Roberts K, Lu L, Demner-Fushman D, Yao J, Summers RM (2016) Learning to read chest x-rays: Recurrent neural cascade model for automated image annotation. In: Proceedings of the IEEE conference on computer vision and pattern recognition, pp. 2497–2506.
5. Ma S, Han Y. (2016) Describing images by feeding LSTM with structural words. In: 2016 IEEE International Conference on Multimedia and Expo (ICME), pp. 1–6,
6. Johnson J, Karpathy A, Fei-Fei L. (2016) Densecap: Fully convolutional localization networks for dense captioning. In: Proceedings of the IEEE conference on computer vision and pattern recognition, pp. 4565–4574,
7. Boag W, Hsu TMH, McDermott M, Berner G, Alesentzer E, Szolovits P. (2020) Baselines for chest x-ray report generation. In: *Machine Learning for Health Workshop*, pp. 126–140.
8. Liu G, Hsu TMH, McDermott M, Boag W, Weng WH, Szolovits P, Ghassemi M. (2019) Clinically accurate chest x-ray report generation. In: Machine Learning for Healthcare Conference, pp. 249–269.
9. Rahman T, Chowdhury ME, Khandakar A, Islam KR, Islam KF, Mahbub ZB, Kadir MA, Kashem S (2020) Transfer learning with deep convolutional neural network (CNN) for pneumonia detection using chest X-ray. *Applied Sciences*, 10(9): p. 3233.
10. Hu M, Lin H, Fan Z, Gao W, Yang L, Liu C, Song Q (2020) Learning to recognize chest-Xray images faster and more efficiently based on multi-kernel depthwise convolution. IEEE Access, 8, pp. 37265–37274.
11. Albahli S, Rauf HT, Arif M, Nafis MT, Algosaibi A (2021) Identification of thoracic diseases by exploiting deep neural networks. *Neural Networks*, 5: p. 6.
12. Guan Q, Huang Y, Zhong Z, Zheng Z, Zheng L, Yang Y (2018) Diagnose like a radiologist: Attention guided convolutional neural network for thorax disease classification. arXiv preprint arXiv:1801.09927.

13. Bustos A, Pertusa A, Salinas JM, de la Iglesia-Vayá M (2020) Padchest: A large chest x-ray image dataset with multi-label annotated reports. *Medical Image Analysis*, 66:101797.

14. Majkowska A, Mittal S, Steiner DF, Reicher JJ, McKinney SM, Duggan GE, Eswaran K, Cameron Chen PH, Liu Y, Kalidindi SR, Ding A (2020) Chest radiograph interpretation with deep learning models: assessment with radiologist-adjudicated reference standards and population-adjusted evaluation. *Radiology*, 294(2), pp. 421–431.

15. Pawade D, Sakhapara A, Jain M, Jain N, Gada K (2018) Story scrambler-automatic text generation using word level RNN-LSTM. *International Journal of Information Technology and Computer Science* (IJITCS), 10(6), pp. 44–53.

16. Melas-Kyriazi L, Rush AM, Han G (2018) Training for diversity in image paragraph captioning. In: Proceedings of the 2018 Conference on Empirical Methods in Natural Language Processing, pp. 757–761.

17. Jing B, Xie P, Xing E (2017) On the automatic generation of medical imaging reports. arXiv preprint arXiv:1711.08195.

18. Xue Y, Xu T, Long LR, Xue Z, Antani S, Thoma GR, Huang X (2018) Multimodal recurrent model with attention for automated radiology report generation. In: International Conference on Medical Image Computing and Computer-Assisted Intervention (pp. 457–466). Springer, Cham.

19. Jing B, Wang Z, Xing E (2020) Show, describe and conclude: On exploiting the structure information of chest x-ray reports. arXiv preprint arXiv: 2004.12274.

20. LeCun Y, Bottou L, Bengio Y, Haffner P (1998) Gradient-based learning applied to document recognition. Proceedings of the IEEE, 86(11), pp. 2278–2324.

21. LeCun Y, Bengio Y (1995) Convolutional networks for images, speech, and time series. *The Handbook of Brain Theory and Neural Networks*, 3361(10):1995.

22. Cho K, Van Merriënboer B, Gulcehre C, Bahdanau D, Bougares F, Schwenk H, Bengio Y (2014) Learning phrase representations using RNN encoder-decoder for statistical machine translation. arXiv preprint arXiv:1406.1078.

23. Hochreiter S, Schmidhuber J (1997) Long short-term memory. *Neural Computation*, 9(8), pp. 1735–1780.

24. Huang G, Liu Z, Van Der Maaten L, Weinberger KQ (2017) Densely connected convolutional networks. In: Proceedings of the IEEE conference on computer vision and pattern recognition, pp. 4700–4708.

25. Rajpurkar P, Irvin J, Zhu K, Yang B, Mehta H, Duan T, Ding D, Bagul A, Langlotz C, Shpanskaya K, Lungren MP (2017) Chexnet: Radiologist-level pneumonia detection on chest x-rays with deep learning. arXiv preprint arXiv:1711.05225.

26. Sandler M, Howard A, Zhu M, Zhmoginov A, Chen LC (2018) Mobilenetv2: Inverted residuals and linear bottlenecks. In: Proceedings of the IEEE conference on computer vision and pattern recognition, pp. 4510–4520.

27. Wang X, Peng Y, Lu L, Lu Z, Bagheri M, Summers RM (2017) Chestx-ray8: Hospital-scale chest x-ray database and benchmarks on

weakly-supervised classification and localization of common thorax diseases. In: Proceedings of the IEEE conference on computer vision and pattern recognition, pp. 2097–2106.

28. Vinyals O, Toshev A, Bengio S, Erhan D (2015) Show and tell: A neural image caption generator. In: Proceedings of the IEEE conference on computer vision and pattern recognition, pp. 3156–3164.

29. Xu K, Ba J, Kiros R, Cho K, Courville A, Salakhudinov R, Zemel R, Bengio Y (2015) Show, attend and tell: Neural image caption generation with visual attention. In: International conference on machine learning, pp. 2048–2057.

30. Bahdanau D, Cho K, Bengio Y (2014) Neural machine translation by jointly learning to align and translate. arXiv preprint arXiv:1409.0473.

31. Papineni K, Roukos S, Ward T, Zhu WJ (2002) Bleu: a method for automatic evaluation of machine translation. In: Proceedings of the 40th annual meeting of the Association for Computational Linguistics, pp. 311–318.

32. Banerjee S, Lavie A (2005) METEOR: An automatic metric for MT evaluation with improved correlation with human judgments. In: Proceedings of the acl workshop on intrinsic and extrinsic evaluation measures for machine translation and/or summarization, pp. 65–72.

Chapter 12

Deep learning based automatic image caption generation for visually impaired people

Pranesh Gupta and Nitish Katal

12.1 INTRODUCTION

Image captioning is a term to describe a given task by generating text, also known as *caption for the image*. It is easy for humans to write short, meaningful sentences for an image by understanding the constituents and activities in the given image. When the same work has to be performed automatically by a machine, it is termed as image captioning. It is a worthy challenge for researchers to solve with the potential applications in real life. In recent years a lot of research has been focused on object detection mechanisms. But the automated image captioning generation is a far more challenging task than object recognition, because of the additional task of detecting the actions in the image and then converting them into a meaningful sentence based on the extracted features of the image. As long as machines do not talk, behave like humans, natural image caption generation will remain a challenge to be solved.

Andrej Karapathy, the director of AI at Tesla, worked on the image captioning problem as part of his PhD thesis at Stanford. The problem involves the use of computer vision (CV) and natural language processing (NLP) to extract features from images by understanding the environment and then generate the captions by sequence learning. Figure 12.1, shows an example to understand the problem. A human can easily visualize an image while machines cannot easily do it. Different people can give different captions for an image. In Figure 12.1 following are the various sentences that can be used to understand the image, that is:

- One person can say *"A dog catching Frisbee,"*
- Some other person can say *"A white dog is leaping in the air with a green object in its mouth."*

Certainly, both of the above descriptions are applicable for Figure 12.1. But the argument is this: it is easy for humans to engender various descriptions

DOI: 10.1201/9781003453406-12

Figure 12.1 Example image for image caption generation.

for an assumed image, but it is a difficult task to train a machine to have so much accuracy.

The problem of image caption generation can help us to solve many other real-life problems. The advancement of this task over object recognition opens up many enormous opportunities in real life applications. Image captioning can be used in self-driving cars to caption the scene around the car. Self-driving cars are one of the prime challenges facing researchers. Solving this challenge can give a boost to automatic driving systems. Image description can be used as a mode to describe the essential information in an image. Moreover, scene description can be used for human and robot interaction and can open new horizons for the development of humanized robotics.

As image captioning is a challenging problem, so have researchers expressed their interest in this problem and have advanced applications to large classification datasets to extract more features accurately. Some have worked on this problem by developing the combination of convolution neural networks (CNNs) to obtain the feature vector and long short-term memory (LSTM) unit to decode those feature vectors into a sequence. Some of these researchers have used the concept of attention mechanism to find the salient features. It is important when there are a lot of objects relating to each other in the given image. Researchers have also worked on two variants, that is, hard and soft attention mechanism in respect to the image captioning problem. Some tried to solve this problem with ensemble learning approaches.

The image captioning problem deals with both deep learning techniques and sequence learning, but this problem suffers from the high variance problem. So, ensemble learning has been used in image captioning with multiple models and at last these models are combined to offer better results. But there is a drawback also: if the number of models increase in ensemble learning then its computational cost also increases. So, it is good to use with a limited number of models in ensemble learning.

In the proposed work, deep learning is used for the generation of descriptions for images. The image captioning problem has been solved in three phases: (a) image feature extractor; (b) sequence processing; and (c) decoding. For the task of image feature extraction, a 16-layer pre-trained VGG model and Xception model are used on the ImageNet dataset. The model is based on an CNN architecture. The extracted feature vector, after reducing the dimension, is fed into a LSTM network. The decoder phase combines the output of other two.

This chapter has been divided into following sections. Section 2 presents the literature review. Section 3 contains datasets, different architectures, and proposed work. Section 4 contains some of the results. Section 5 gives a brief dialogue and future prospects. And at last Section 6 completed it with the conclusion of the proposed work.

12.2 RELATED WORK

The previous approaches in visual recognition have been focused on image classification problems. It is now easy to provide labels for a certain number of categories of objects in the image. During the last few years, researchers have had good success in image classification tasks with the use of deep learning techniques. But the image classification using deep learning provides us only with a limited statistics regarding the items that are there in the scene. Image captioning is a much more multifaceted undertaking and requires a lot more knowledge to find the relations amongst various objects by combining the information with their characteristics and happenings. After processing is complete, the objects have been identified and their various features have been combined, the last task being to express this gathered information as a conversational message in the form of a caption.

Earlier, Caltech 101 was one of the first datasets for multiclass classification that comprehends 9,146 images and 101 classes. Later Caltech 256 increased the number of different object classes to 257 (256 object classes, 1 clutter class). It consists of 30,607 real world images. For classification, ImageNet dataset is given and the dataset is larger both in terms of scale and diversity when compared to Caltech101 and Caltech 256.

Several algorithms and techniques were anticipated by the researchers to solve the image captioning problem. The latest approaches to this project follow deep learning-based architectures. Visual features can be extracted using convolution neural networks and, for describing the features, LSTM can be used to generate some meaningful and conversational sentences. Krizhevsky et al. [4] has implemented the task with non-saturated neurons for the neural networks and proposed the use of dropout for the regularization to decrease overfitting. The proposed network contains convolution layers followed by Maxpooling layers and used softmax function for the next predicted word. Mao at el. proposed combining both the architectures of CNN and RNN. This model extracts the features by using CNN and then generates words sequentially by predicting prospects for every next word and generates a sentence by combining the predicted words. Xu et al. [5] proposed a model by using attention mechanism with the LSTM network. This focuses on salient features on the image and increases the probability of the words based on these salient features by neglecting other less important features. The model was learned by optimizing a vector lower bound using normal backpropagation techniques. The model learnt to identify the entity's border although still generating an effective descriptive statement. Yang et al. [6] suggested a method for automatically generating a natural language explanation of a picture that will help to understand the image. Pan et. al [7] investigated broadly with various network designs on massive datasets containing a variety of subject formats and introduced an inimitable model that outperformed previously proposed models in terms of captioning accuracy. X. Zeng et al. [11] has worked on region detection of ultrasound images for medical diagnoses and worked on gray scale noisy medical images with low resolution. The authors used a pretrained model VGG16 on ImageNet dataset to isolate the attributes from the imageries and for sequence generation. LSTM network has been used and an alternate training method is used for the detection model, evaluating the performance using BLEU, METEOR, ROUGE-L and CIDEr. The image captioning problem was well researched by Karpathy & Feifei [8] in their PhD work and proposed a multimodal RNN that fixes the co-linear arrangements of features in their model. They used the dataset of images to learn the interaction between objects in an input image.

Another different approach model, used by Harshitha Katpally [9], was based on ensemble learning methods to resolve the unruly low variance during training of deep networks. The proposed ensemble learning approaches was different from conventional approaches and analyze the previous ones to find the best ensemble learning approach and has compared all the results with different approaches. Deng et al. [10] launched ImageNet database, a wide-ranging array of pictures based on the structure of WordNet. The various classes are organized by the ImageNet database in a semantic hierarchy that is heavily inhabited. Vinyals et al. [12] proposed deep recurrent neural network architecture to generate descriptions for the pictures, while ensuring that the produced sentence accurately describes the target image with the highest likelihood. Christopher Elamri [13] proposed a CNN-RNN model and used the LSTM network to create an explanation of the picture. The proposed work is based on the VGG-16 network and also used the PCA to decrease the extent of the vocabulary and also expanded the work on MS COCO dataset and used the performance metrices of BLEU, METEOR, and CIDEr scores.

12.3 METHODS AND MATERIALS

12.3.1 Data set

In the present study, for the image caption generation, the Flickr_8k dataset has been used. It is easy to work on this dataset as it is relatively small and realistic. In this dataset there are a total of 8,000 images, which have been predefined as 6,000 images for training, 1,000 images for development and another 1,000 images are for testing. In the dataset each image has 5 diverse descriptions that provide a strong understanding of the prominent objects and actions in the image. Each image has a predefined different label name / identity so it is easy to find the related captions from the flick8k_text dataset. Other larger datasets that can also be used are Flick_30k dataset, MS COCO dataset. But these datasets can take days just to train the network and have only a marginal accuracy than has been reported in the literature.

12.3.2 Deep neural network architectures

The image caption generation model uses deep learning techniques that work on a combination of CNN and RNN based architecture. CNN extracts the characteristics from the image and the RNN is used for sequence learning to create words for the caption. For extracting features, two pretrained models, namely VGG16 and Xception net, have been used. The use of these models for feature extraction reduces the training time. These models are exercised on the ImageNet dataset. This dataset is used for classification of images. Other architectures like Inception-v3 and ResNet50 have also been explored in the literature for classification of images on the ImageNet dataset. These architectures can also be used for mining the attributes from the pictures and then fed to an LSTM to generate captions.

12.3.2.1 Convolution Neural Networks (CNNs)

Currently, CNN is being widely used for solving visual recognition problems [21, 23]. The pictures can be characterized as 2D matrix and the network takes the input image as a 2D matrix. Mostly, CNN are deployed for image classification problems and identify different categories of objects. CNNs

20	16	0	18
15	12	10	5
5	35	40	48
100	120	80	36

2 x 2 Max-Pooling →

20	18
120	80

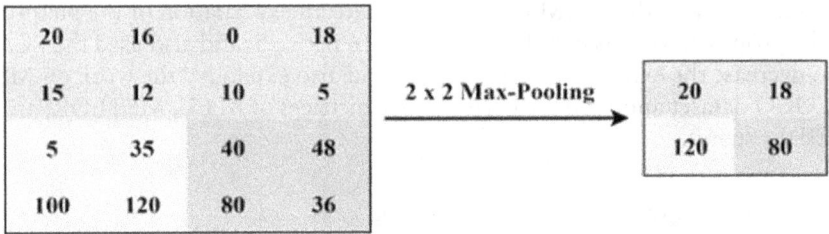

Figure 12.2 Max pooling layer convolution operation in CNN.

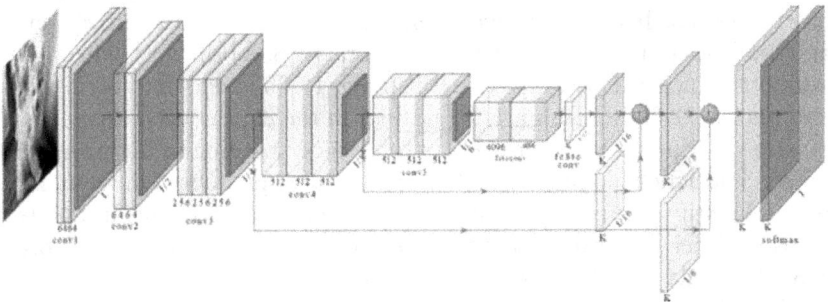

Figure 12.3 CNN architecture for classification.

Source: [1].

take an input picture, process it, allocate rank to things within the image, and then identify and classify the different objects in the images. Figure 12.2 shows the convolution operation in the maxpooling layer in CNNs [22, 24].

The CNN architecture is composed of following layers (a) a convolution layer (CL); (b) a pooling layer (PL); and (c) a fully connected (FC) layer. Figure 12.3 shows the CNN architecture used for the classification. The CL is used to generate unruffled depictions of the input picture. Matrix multiplication is executed amid the filter with every single slice of the picture matrix with equal dimensions as that of the kernel. The PL also have a similar purpose as that of the CL, but the PL has extra ability to find more dominant features in picture. PL has two categories, Average pooling and Max pooling, these layers reduce the dimension of the input matrix. Max pooling has been well thought-out to be the better pooling method as it contributes a de-noising component and removes the noisy features from the image. The last step is to compress all the characteristics we have obtained from the last layers and feed them into an Neural Network (NN) for classification. In the FC layer, all the non-linear relationships amongst the significant characteristics are learned. This compressed vector is fed into the NN using backpropagation and categorizes all the objects via softmax function.

CNNs extract features from the images by glancing over the picture from left to right and top to bottom, followed by combination of all the features that are detected. In the preset work, the feature vector having a dimension of 2048 has been extracted from the last third layer of the pretrained VGG-16 network.

12.3.2.2 Long Short-term Memory (LSTM)

LSTM is primarily operated for sequence learning and contains a memory cell. After extracting the features from the CNN, it reduces the proportions of the picture feature vector that have been reduced using principal component analysis (PCA). Now, the obtained characteristic vector is provided to the LSTM cell and this LSTM network generates the description by sequence learning of words.

The architecture of the LSTM network is analogous to RNNs (recurrent neural networks) but the LSTM stores the previous cell output in a hidden state. But conventional RNN are negatively affected by the short-term memory complications, that is, they tend to forget important information when they are working on long sequences. This problem is due to the vanishing gradient problem in RNNs. This occurs when the network tries to change the important weights and biases and shrinks due to backpropagation over time. This makes the gradient so small, that they do not contribute in network learning.

In the LSTM network, the information is carried out all the way from the start state to the present status and not only on the preceding states. Due to this ability, the LSTM networks are free from short-term memory problems. The LSTM network architecture has two apparatuses, that is, cell state (CS) and different gates (DG). A CS can transfer information to present sequencing state working as memory of the network. In LSTM, some neural network units govern the flow of the information, that is, the information has to be forwarded to the current state or not. In an LSTM cell, three types of gates are there: (a) forget gate (FG); (b) input gate (IG); and (c) output gate (OG). FG resolves whether the data from the preceding state is to be reserved or has to be overlooked. To do so, the information is passed through the sigmoid function, which decides whether the information should be kept or not. IG is used to regulate the values from the previous states of the cell and the OG is used to combine the outputs for new concealed state that forwards it to next unit of LSTM. A LSTM model is shown in Figure 12.4.

12.3.3 Proposed model

The proposed model has been divided into two parts. The first part uses a CNN for the attribute abstraction from images and followed by LSTM. The LSTM network uses the data from CNN to assist generation of sentences of the image. So, in the proposed work, both these networks have been

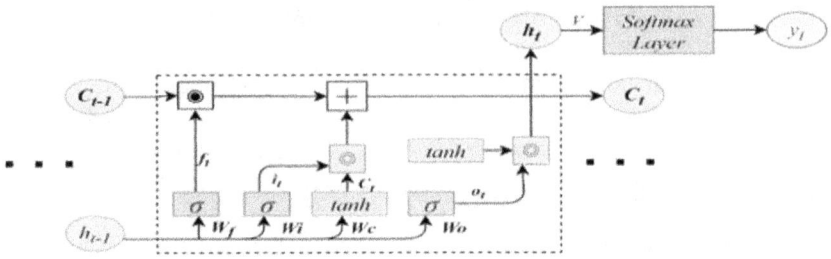

Figure 12.4 Architecture of LSTM cell.

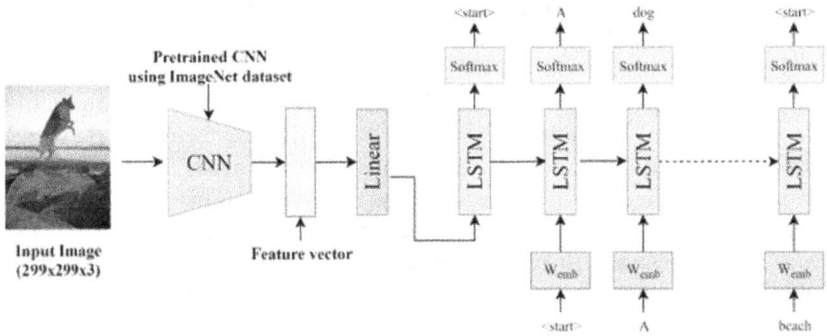

Figure 12.5 Model of Image caption generator.

merged to design a hybrid CNN-RNN representation. The schematics of the anticipated representation is shown in Figure 12.5.

12.3.3.1 Feature extraction models

These two CNN architectures, namely VGG16 and Xception networks, are exercised on a large ImageNet dataset. This dataset contains 14 million images and classifies the objects in 1,000 classes. The details of these models are discussed as below:

1. The VGG16 model is proposed for categorization of objects on ImageNet dataset. Input image should be of dimensions 224 × 224 RGB. The architecture of the model is the combination of convolution layers followed by max pooling layers and at the last, fully connected layers are present. VGG16 CNN network architecture has been proven to achieve 92.7 percent test accuracy on the ImageNet dataset. The features can be mined using this pretrained VGG16 model by removing the last softmax classification layer. The VGG16 network

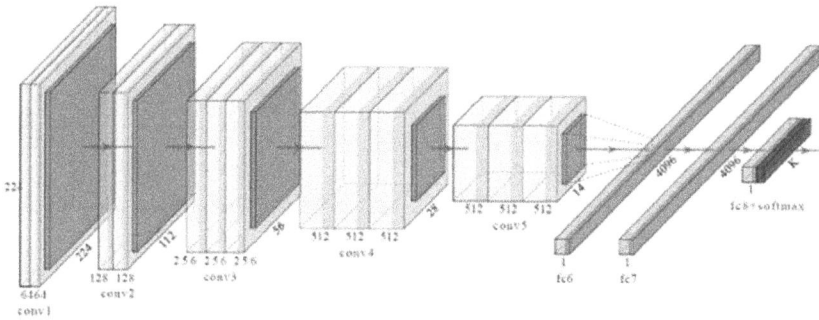

Figure 12.6 Architecture of VGG16 network.

Source: [1].

is slow in training. Figure 12.6 shows the architecture of VGG16 network.

2. Xception Network: The Xception model was anticipated by Francois Chollet and is an extension of the Inception network and also has been trained on ImageNet dataset. Xception reportedly gives 94.5 percent test accuracy on the ImageNet dataset, and the accuracy is in the top-5 ranking. For this network the input image is provided with the dimensions 299 × 299 × 3. Xception architecture contains 36 convolution layers forming a convolution base, and these layers are followed by a logistic regression layer. The architecture of this model is a linear stack of convolution layer that are separable by depth. Now, to excerpt the characteristics from network, the final classification layer from the network has been removed to acquire the attribute vector. The attribute vector has a size of 2048. These features are dumped into a pickle file, so they can again be sent as input to the CNN-RNN model.

The image captioning merge model can be described in three steps:

i. *Feature extraction process:* To extract the features from the training dataset, a pretrained model on the ImageNet dataset is used. Then these extracted characteristics are benefitted as the response for the proposed CNN-RNN model. Figure 12.7 shows the proposed hybrid architecture based on CNN-RNN architecture.

ii. *Sequence processor:* In this step, an embedding layer will take the textual input, followed by LSTM network.

iii. *Decoder:* Here the output of the above two layers and the output layers are administered by a dense layer to get the ultimate predictions. The

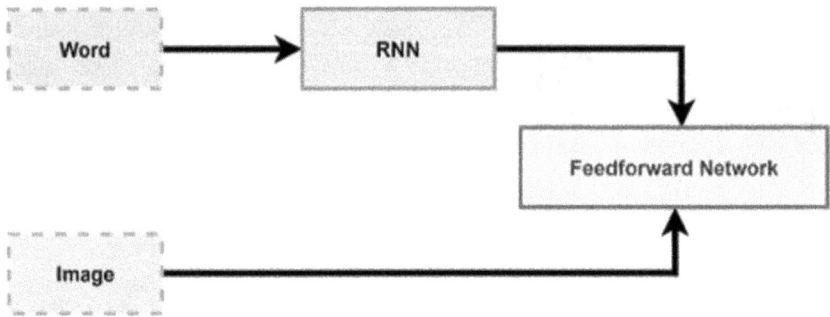

Figure 12.7 The proposed hybrid CNN-RNN model considered for image caption generation.

last layer has a size equivalent to that of the vocabulary size. This is done to get the probability for each word in the sequence learning process.

12.3.3.2 Workflow for image caption generation

The work flow includes the subsequent phases:

i. *Extracting the characteristic vector:* The abstraction of features from images is done by transfer learning approach by using two different pre-trained models. At first, preprocessing of the dataset has to be done to extract the feature vector. Pretrained CNN models of VGG16 and Xception network have been used to extract the features by removing the last two layers of the networks. This is done in order to get a feature vector for all the images. Then these features are dumped into a pickle file, so that the information can use again to train the model.

ii. *Cleaning descriptions:* This step involves preparing the vocabulary from the set of descriptions by using preprocessing steps. To clean the data and work with textual data, the test dataset has been modified by making the text in small case letters, removing punctuation marks and also removing the alphanumeric numbers from the texts. Now the descriptions have been tokenized to generate the vocabulary. Indexes are given to every word in the terminology. Keras has been used for the tokenizer function. The tokenizer will create tokens from the vocabulary.

iii. *Defining the model:* The assembly of the model is defined for caption generation as a series of arguments that, once merged form a sentence, and the generated sentence explains about the picture that was used as input. An RNN network performs a sequence learning process to forecast the subsequent term grounded on the features and also using the information from the previous output. Then the output is provided to the next cell as input to forecast the subsequent word. The input of the RNN network is word embedding layer which was generated for the words of the vocabulary.

iv. *Evaluation of model:* Testing is done on 1000 images of the Flicker8k dataset. For evaluation, BLEU metric is used. BLEU metric matches the similarity between the generated caption and reference sentence. So, we get different BLEU scores every time. Figure 12.8 shows the architecture of the proposed model.

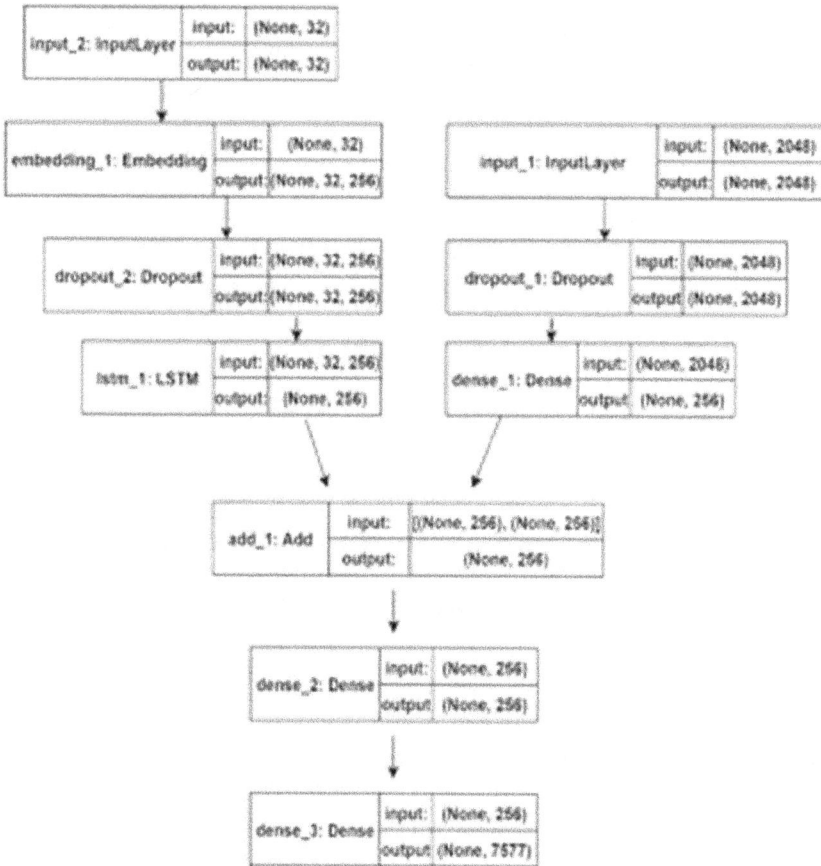

Figure 12.8 Model architecture.

12.4 RESULTS AND DISCUSSION

12.4.1 Evaluation metrics

a. Accuracy metrices

The precision of model is tested proceeding the test data. Each image in the test dataset is fed as a record and an output caption is engendered. The closeness of this caption with the captions of the dataset for the same image will give the accuracy of the model. The evaluation of generated captions can be done using the metrics: BLEU (Bilingual Evaluation Understudy), CIDEr (Consensus-based Image description evaluation), METEOR (Metric for Evaluation of Translation with Explicit ordering). Each metric gives the grade that established how adjacent evaluated text is to references text. In present work, BLEU metrics has been used for the performance evaluation. BLEU scores are used to evaluate the translated text against the one or more reference translation sentences. For the flicker8k dataset there are 5 references given for each input image. So, the BLEU score is calculated against all of the reference sentences for the input image and the BLEU scores are calculated for 1 gram, 2 gram, 3 gram and 4 cumulative n-grams. The BLEU score will be in the range from 0 to 1 and a score near to 1.0 is considered as improved results.

b. Calculation of BLEU score for 1, 2, 3 and 4 n-grams

The generated caption is compared against all of the reference sentences for an image. In the present work, BLEU score for 1, 2, 3, and 4 cumulative n-grams has been evaluated. The weights for the amassed and discrete 1-gram BLEU are the same, for example (1,0,0,0). Each of the 1-gram and 2-gram scores receives 50 percent of the 2-gram weighting factor, for example (0.5,0.5,0,0), while individually 1, 2 and 3-gram scores receives 33 percent of the 3-gram weights, for example (0.33,0.33,0.33,0). In the present work, the BLEU score for 1, 2, 3 and 4 grams have been evaluated for the VGG16 model and are given in Table 12.1.

Table 12.1 Calculated BLEU score for VGG16 model

No. of Grams	BLEU Score
BLEU – 1	0.556201
BLEU – 2	0.288290
BLEU – 3	0.191689
BLEU – 4	0.083565

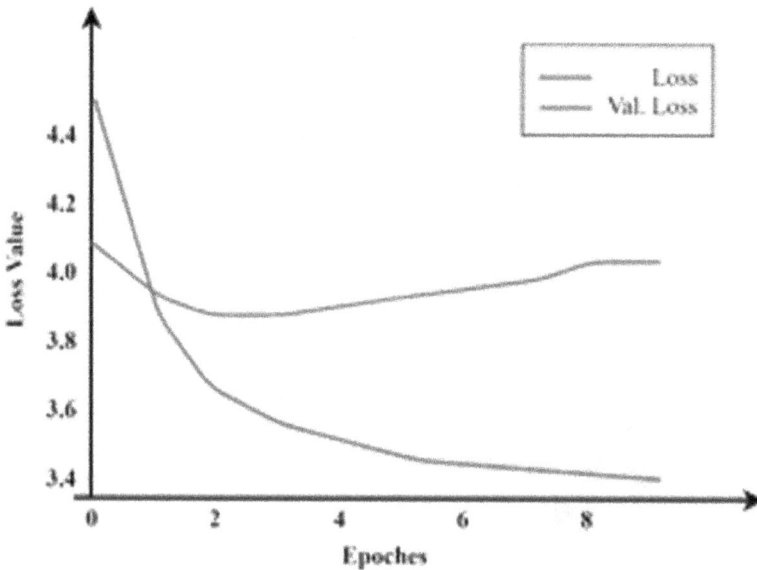

Figure 12.9 Plot for loss and validation loss curve.

12.4.2 Analysis of results

The model is trained for 20 epochs, and training loss of 3.3690 and a validation loss of 4.1698 have been obtained for the VGG16 model. The plot for loss and validation loss is given in Figure 12.9.

From the obtained results, we can deduce that the Xception model gives better results as compared to the VGG16 model. For improving even further, a good and large vocabulary that can be obtained by using MS COCO dataset and also a more accurate model that can identify objects in the image more accurately and based on the activities, will be able to generate a meaningful and conversational sentence.

12.4.3 Examples

Table 12.2 Examples

S. No	Original Image	BLEU Score	
1		**Original:**	"A boy in his blue swim shorts at the beach."
		Xception:	"Man in red shorts is walking along the beach."
		VGG16:	"Man in red shirt is standing on the beach."
2		**Original:**	"A black dog chases a boy wear red and blue."
		Xception:	"Black dog is running through field."
		VGG16:	"Two men are playing in the grass."
3		**Original:**	"A man ride a motorcycle with the number 5 on it."
		Xception:	"Man in red shirt rides bike on the street"
		VGG16:	"Man in red shirt is riding bike on the street."
4		**Original:**	"A brown dog is running through the field."
		Xception:	"Brown dog is running through the grass."
		VGG16:	"Two dogs are running through the grass."
5		**Original:**	"A man and woman stand on the sidewalk of a busy street."
		Xception:	"Man in black shirt and black shirt is standing in front of an urban archway."
		VGG16:	"Man in red shirt is standing on the street."

12.5 DISCUSSION AND FUTURE WORK

Several researchers have proposed many models to generate meaningful captions for the images, but these existing models also have certain constraints. Improving accuracy is the main objective of describing actions and objects in images. By improving accuracy, we can further aid people with audio of the generated captions by converting the captions into voice by using neural machine translation. A higher BLEU score can be achieved by using large datasets like MS COCO as it will increase the extent of the vocabulary and numeral of unique words in the descriptions. So, the key point is that the model totally depends on the dataset size. A smaller dataset cannot give so much accuracy due to the smaller size of the vocabulary and cannot predict words that are not in the vocabulary.

In future several improvements in the model can be made, such as:

i.   A bigger dataset can be used for the purpose.
ii.   Model architecture can be modified to improve the accuracy (like dropout layers can be used for regularization, adding batch normalization layer, etc.).
iii. Better hyperparameter tuning of the model can be done in future to get better performance.
iv.   Creating API for this model using FLASK that can be deployable also.
v.   It can be used to give a detailed overview of the objects' actions by searching the generated caption on Google.
vi.   This work can be used to give information to visually impaired persons in their particular regional language at the time of converting text into voice by using neural machine translation techniques.
vii. It can be used for human–robot interaction that can increase more humanized functioning of robots. It can be done by generating captions from an image and these generating captions will give direction to the robot to work.
viii. In the future, instead of creating captions for pictures, it might be used to translate videos directly into sentences

12.6 CONCLUSIONS

In the present work, deep learning techniques have been used to generate captions automatically from images. This will aid visually compromised individuals for improved perception of their environments. The proposed model draws it inspiration from a hybrid CNN-RNN model. The CNN has been employed to excerpt the characteristics from pictures. The obtained characteristics are fed as input to an LSTM network through a dense layer. Based on the existing research, the proposed model achieves analogous to state-of-art work reported in the literature. It engenders meaningful and expressive sentences with the limited vocabulary size. So, the proposed model can benefit the visually impaired people to perceive their environments.

REFERENCES

1. Theckedath, Dhananjay and R. R. Sedamkar. "Detecting affect states using VGG16, ResNet50 and SE-ResNet50 networks." *SN Computer Science* 1 (2020): 1–7.
2. Vinyals, Oriol, Alexander Toshev, Samy Bengio, and Dumitru Erhan. "Show and tell: A neural image caption generator." In *Proceedings of the IEEE conference on computer vision and pattern recognition*, pp. 3156–3164. 2015.
3. Tan, Ying Hua and Chee Seng Chan. "Phrase-based image caption generator with hierarchical LSTM network." *Neurocomputing* 333 (2019): 86–100.
4. Krizhevsky, Alex, Ilya Sutskever, and Geoffrey E. Hinton. "Imagenet classification with deep convolutional neural networks." Advances in neural information processing systems 25 (2012): 1097–1105.
5. Xu, Kelvin, Jimmy Ba, Ryan Kiros, Kyunghyun Cho, Aaron Courville, Ruslan Salakhudinov, Rich Zemel, and Yoshua Bengio. "Show, attend and tell: Neural image caption generation with visual attention." In International conference on machine learning, pp. 2048–2057. PMLR, 2015.
6. Yang, Zhongliang, Yu-Jin Zhang, Sadaqat ur Rehman, and Yongfeng Huang. "Image captioning with object detection and localization." In International Conference on Image and Graphics, pp. 109–118. Springer, Cham, 2017.
7. Pan, Jia-Yu, Hyung-Jeong Yang, Pinar Duygulu, and Christos Faloutsos. "Automatic image captioning." In 2004 IEEE International Conference on Multimedia and Expo (ICME) vol. 3, pp. 1987–1990. IEEE, 2004.
8. Karpathy, Andrej, and Li Fei-Fei. "Deep visual-semantic alignments for generating image descriptions." In Proceedings of the IEEE conference on computer vision and pattern recognition, pp. 3128–3137. 2015.
9. Katpally, Harshitha. "Ensemble Learning on Deep Neural Networks for Image Caption Generation." PhD diss., Arizona State University, 2019.
10. Deng, Jia, Wei Dong, Richard Socher, Li-Jia Li, Kai Li, and Li Fei-Fei. "Imagenet: A large- scale hierarchical image database." In 2009 IEEE conference on computer vision and pattern recognition, pp. 248–255. Ieee, 2009.
11. Zeng, Xianhua, Li Wen, Banggui Liu, and Xiaojun Qi. "Deep learning for ultrasound image caption generation based on object detection." *Neurocomputing* 392 (2020): 132–141.
12. Vinyals, Oriol, Alexander Toshev, Samy Bengio, and Dumitru Erhan. "Show and tell: A neural image caption generator." In Proceedings of the IEEE conference on computer vision and pattern recognition, pp. 3156–3164. 2015.
13. Christopher Elamri, Teun de Planque, "Automated Neural Image caption generator for visually impaired people" Department of Computer science, Stanford University.

14. Tanti, Marc, Albert Gatt, and Kenneth P. Camilleri. "What is the role of recurrent neural networks (rnns) in an image caption generator?." arXiv preprint arXiv:1708.02043 (2017).

15. Maru, Harsh, Tss Chandana, and Dinesh Naik. "Comparitive study of GRU and LSTM cells based Video Captioning Models." In *2021 12th International Conference on Computing Communication and Networking Technologies (ICCCNT)*, pp. 1–5. IEEE, 2021.

16. Aneja, Jyoti, Aditya Deshpande, and Alexander G. Schwing. "Convolutional image captioning." In *Proceedings of the IEEE conference on computer vision and pattern recognition*, pp. 5561–5570. 2018.

17. Bahdanau, Dzmitry, Kyunghyun Cho, and Yoshua Bengio. "Neural machine translation by jointly learning to align and translate." arXiv preprint arXiv:1409.0473 (2014).

18. Karpathy, Andrej and Li Fei-Fei. "Deep visual-semantic alignments for generating image descriptions." In *Proceedings of the IEEE conference on computer vision and pattern recognition*, pp. 3128–3137. 2015.

19. Jonas, Jost B., Rupert RA Bourne, Richard A. White, Seth R. Flaxman, Jill Keeffe, Janet Leasher, Kovin Naidoo et al. "Visual impairment and blindness due to macular diseases globally: a systematic review and meta-analysis." *American Journal of Ophthalmology* 158, no. 4 (2014): 808–815.

20. Russakovsky, Olga, Jia Deng, Hao Su, Jonathan Krause, Sanjeev Satheesh, Sean Ma, Zhiheng Huang et al. "Imagenet large scale visual recognition challenge." *International Journal of Computer Vision* 115, no. 3 (2015): 211–252.

21. Abouhawwash, M., & Alessio, A. M. (2021). Multi-objective evolutionary algorithm for pet image reconstruction: Concept. IEEE Transactions on Medical Imaging, 40(8), 2142–2151.

22. Balan, H., Alrasheedi, A. F., Askar, S. S., & Abouhawwash, M. (2022). An Intelligent Human Age and Gender Forecasting Framework Using Deep Learning Algorithms. *Applied Artificial Intelligence*, 36(1), 2073724.

23. Abdel-Basset, M., Mohamed, R., Elkomy, O. M., & Abouhawwash, M. (2022). Recent metaheuristic algorithms with genetic operators for high-dimensional knapsack instances: A comparative study. *Computers & Industrial Engineering*, 166, 107974.

24. Abdel-Basset, M., Mohamed, R., & Abouhawwash, M. (2022). Hybrid marine predators algorithm for image segmentation: analysis and validations. *Artificial Intelligence Review*, 55(4), 3315–3367.

Chapter 13

Empirical analysis of machine learning techniques under class imbalance and incomplete datasets

Arjun Puri and Manoj Kumar Gupta

13.1 INTRODUCTION

Problems in datasets may lead to loss or to misinterpreting the data instances. Nowadays, machine learning methods are becoming so efficient that they can handle these problems in datasets. The problem in real-world datasets may be due to distribution disparity among the class or may be due to missing values. The distribution disparity is also called class imbalance, which may arise due to an imbalance in the distribution of data instances among classes. A class imbalance problem may create an impact only when there is a need to study the behavior of minority class instances. Class imbalance problems are seen in many applications, such as fault detection [1], bankrupt prediction [2], natural language processing[3], credit scores [4], twitter spam detection [5] and so forth. A class imbalance problem with missing value problems becomes more complicated. The missing value itself is a huge problem that creates problems in classification. There are various types of missing values, such as MCAR (missing completely at random), MAR (missing at random), NMAR (not missing at random) [6], and so forth.

Real-time datasets contain missing values as well as class imbalance problems. So far, techniques developed for these problems are operated in different phases, which may lead to a problem in classification. Class imbalanced datasets are usually not so complicated, but with other intrinsic difficulties, they become more complex like small disjunct, class overlapping, missing values, and so forth. Many types of techniques are developed to deal with class imbalance problems such as data level class imbalance handling techniques, algorithmic level, and hybrid level. Data level imbalanced handling techniques use two different techniques, like oversampling and undersampling. Oversampling will generate synthetic instances, either by repeating examples (ROS) [7] or by using some random value methods like linear interpolation methods (SMOTE) [8]. Undersampling is used to remove the majority of class instances and create balance in the

DOI: 10.1201/9781003453406-13

dataset by losing information [9]. Algorithmic level handling methods are also used to mitigate the effect of class imbalance. Some of the techniques, like ensembles (Boosting, Bagging) are used as a multi-classifier to solve the class imbalance problem. The hybrid level uses the data level and algorithm level in combination to deal with class imbalance problem [10] (RUSBoost, SMOTEBoost, SMOTEBagging, ROSBagging).

A missing value problem creates a hugely misleading situation when classifying the datasets. Missing value can be treated with two different methods: one is to remove the missing value from the datasets, and they perform analysis on the rest of the datasets available; the other is to impute the missing value and perform analysis on the whole of the datasets with assigned missing value. Many researchers work in the direction to impute the missing value so that these values may contribute in classifying the instance belonging to a particular class. These methods are classified into two different categories: one is to impute by using traditional statistical methods, and the other is to impute by using soft computing with a machine learning algorithm. Conventional methods are based on statistics, some of which are hot deck imputation, regression substitution, Expectation Maximization (EM), Multivariate imputation by chained equations (MICE), Mean, Median, and so forth, whereas in case of soft computing we use k-nearest neighbor technique (KNN), support vector regression (SVR), Bayesian Network, and so forth.

These techniques developed in a separate context and deal with individual problems like missing value or class imbalance. None of them deal with both issues together. Although, in [11] the authors suggested that missing value in class imbalance creates a colossal impact. If it follows ignorance strategy, then there may arise a huge class imbalance ratio. Most of the time, missing values are treated separately. But in the case of class imbalance, the missing value can handle along with class imbalance by using data level class imbalance handling techniques [12–15]. The main focus of our study is to analysis the behavior of techniques used for both class imbalance and missing value handling (i.e., fuzzy information decomposition (FID) technique) with other missing values along with class imbalance handling techniques, under 0 percent, 5 percent, 10 percent, 15 percent and 20 percent missing value in the datasets.

The main organization of this chapter is to state: section 2 covers related work, section 3 covers methodology and results, and finally section 4 covers conclusion.

13.2 RELATED WORK

This section, we present a review on class imbalance, missing value in detail, and also deal with missing values within class imbalance.

13.2.1 Class imbalance

Over the last decade, a lot of research has been reported on tackling the problem of class imbalance. This research is mainly divided into three categories: data sampling, algorithm level, and hybrid level. In data sampling technique, which is also known as rebalancing technique, there is a further division into two primary strategies named undersampling and oversampling methods. Undersampling technologies mainly used to remove the majority instances, and oversampling methods are used to induce minority instances. At the algorithmic level, there is need of modification in the algorithm to deal with class imbalance, some of the methods, cost-sensitive SVM, meta cost SVM, AdaBoost [16], and boosting [17] and bagging [4] ensemble learner. Finally, hybrid technologies are used to deal with class imbalance; some of these techniques are ROSBoost, RUSBoost, ROSBagging, RUSBoost, SMOTEBoost, and SMOTEBagging.

The simple undersampling technique is a random undersampling that removes a random subset of samples from the majority instances (RUS) [18]. But the problem with the random undersampling method is to remove the majority instances so that classifier may include loss of potential information. Later work on undersampling may provide a solution to this problem by using the One Side selection method, which removes redundant or noisy instances from the datasets and makes datasets balance. Later, Cluster-Based Undersampling techniques [9, 19] are used to overcome the problem of underfitting in which majority instances are clustered first, depending upon the number of minority instances, and then take a centroid of each cluster as an example of majority instance, and then classify them with minority instances and predict the result. Moreover, some other authors also develop undersampling techniques to deal with class imbalance like diversified sensitivity based undersampling [20] technique, which selects a sample from both classes.

The basic oversampling technique developed is Random Oversampling technique [21], which generates repeated instances and leads to the problem of overfitting, which is defined as the problem in which training instances show less error but testing datasets show more error. When instances generated by using repeated instances in the training set may lead to classifying only training instances and not consider the position of test instances, this may lead to the problem of overfitting. To tackle the problem of overfitting new method called Synthetic oversampling technique [8] (SMOTE) is developed, which is used to generate instances by using linear interpolation. The problem with SMOTE is that instances are taken from the minority are at random, and there is no restriction on this. To provide a restriction on choosing examples for generating instance many researchers work in this context and develop different variants of SMOTE like Borderline-SMOTE [22] which generates borderline instances of the minority class are calculated by using k nearest neighbor and induce those instances which are present in the danger zone. Another researcher is trying to create instances present in the safe zone by using SMOTE, which may lead to overfitting and the process called Safe-SMOTE [23]. To remove the problem raised by the SMOTE – overgeneralization of various authors who are trying and develop multiple techniques like in [24] authors are working to provide oversampling smote with the direction of generating instances and not consider outliers present in the datasets with the minority class. In [25], authors developed radial based oversampling technique to handling class imbalance and tried to overcome the problem of overfitting, overgeneralization. Also, to perform oversampling in noisy imbalanced data classification.

13.2.2 Missing values

Missing value is one of the most prompt problems seen in the dataset. It may occur due to any defect in data collection. A lot of research has been done in this area. Based on the research is done, we categorize the study into two different parts: one is ignoring missing values and then use classifier and other things*-+ to impute missing value using missing value imputation techniques.

Further, based on the literature survey, missing value imputation technique are divided into two parts: statistical and soft computing. Statistical methods include mean, median, MICE, EM, and so forth, where soft computing methods include KNN. In statistical methods, in case of mean and median, we calculate the mean and median of each feature having complete datasets and impute missing value using mean and media, respectively. But in the case of EM known as Expectation Maximization [26] use mean and covariance for estimation for the parameter of interest. Whereas MICE knows as Multiple imputations by chaining equation [27], which uses mean as simple imputation method and then uses the regression model to device the relationship between simple imputation and observed by forming an equation.

On the other side, soft computing methods are developed to deal with a missing value. In [26], define the K-nearest neighbor (KNN) technique, which uses the concept of distance among the nearest points. Here the distance is measured by using Euclidean distance and consider those similar instances that have the least distance among them.

13.2.3 Missing value in class imbalance datasets

In [11], the authors presented a comparative study of methods to deal with missing values in class imbalance datasets. In that paper, the first identification of missing values was imputed based on relative attribute characteristics.

After imputation of values finally for classification algorithms (K-means, C4.5, artificial neural network, and naïve Bayesian) were used to find the classification performance. Their experimental results show that the presence of a missing value in class imbalance creates a colossal problem. In [12], develop the feature projection K-NN classifier model for imbalanced and incomplete medical data. In this technique, they generate missing values by using K-NN in every feature projection subspace and weighting strategy to deal with class imbalance.

Further, in [14], develop a machine learning technique to deal with the class imbalance and missing value together in case of medical data. In this, the authors use tree augmented naïve Bayesian algorithm (TAN) to generate the missing values and also develop algorithm 2 (TAN for incomplete data and imbalance data) and perform analysis on the patient suffering from myocardial infarction from the third world, Syria and the Czech Republic, and shows that algorithm 2 (TAN for incomplete data and imbalance data) perform well in comparison with TAN_SMOTE and Normal TAN.

Fuzzy information decomposition technique developed [13] to handle missing value with class imbalance. It first identifies the number of instances to be induced in the dataset and then at each feature it identifies the number of missing values and then will perform steps in sequence: first it takes original datasets with missing values, then it will consider each synthetic sample in L and missed value of each feature as p and calculate t by using t = p+L and then consider only instances without missed in each feature and obtain upper and lower bound values in each feature, then calculate the h =(upper-lower bound)/t and divide information I = (lower, upper) into t sub bounds by using I_i = [lower + h, lower + 2*h],....,[lower+(t-1)*h, lower +t*h] then it will calculate these information center points denoted by U = ut then calculate the contribution weights of each observed element $\mu(x_i, u_s)=$

$1 - \|xi - us\|, if \|x - u\| \le h \{ h \quad i \quad s$

after that compute information decomposition from X_i to I_s by using formula

$0, otherwise$

m is = $\mu(x_i, u_s)^* x_i$. Then finally recover missing value by using m_s = {

$mean (x) if \, {}^{\Sigma_i} \mu(x_i, u_s) = 0$

$\underline{\Sigma_j \, mis}$, otherwise

$\Sigma_i \, \mu(x_i, u_s)$

In this way, this method provides the required missing value as well as incomplete datasets.

13.3 METHODOLOGY

In this section, we discuss the effect of class imbalance and missing value imputation techniques on datasets having class imbalance as well as missing values and discuss techniques developed for tackling class imbalance and missing value together like fuzzy information decomposition technique (FID). For this we experiment with different algorithms and experiment section and results. In the experiment section, we discuss datasets description and experiments.

Datasets

In order to perform the analysis, we collect 18 class imbalance datasets from different repositories like MC2, kc2, pc1, pc3, pc4 taken from promise repository; datasets belong to software defaults whose instances range from 194 to 1077 and a number of attributes are ranges from 21 to 39 imbalance ratio (IR) ranges from 1.88 to 10.56. Whereas, heart, Bupa, Mammographic, ILPD datasets are taken from the UCI repository and the rest of the datasets are taken from KEEL-dataset repository [13, 23]. Table 13.1 lists the datasets used in this chapter, where the first column describes the name of the dataset, the second tells us the size (total number of instances), the third tells the number of attributes (attributes), and the last tells us imbalance ratio (IR). The IR (majority to minority ratio) range of datasets varies from 1.05 to 22.75; size varies from 125 to 1484 and attributes fields from 5 to 39. We group the datasets according to class imbalance ratio (i.e., 1–5, 5–10, 10–22) for performance evaluation.

To perform analysis, we need to create missing values randomly with different percentages, that is, 5 percent, 10 percent, 15 percent, 20 percent of each attribute excluding target attribute.

Experimental setup

In this study, we need to analyze the effects of missing values under class imbalance environment. To perform this, we use SMOTE [8], ROS [8] as oversampling technique, CNN [24] as undersampling technique and SMOTE-Tomek-link [25] as combined technique (both oversampling and undersampling) for resampling under missing value and class imbalance environment. For missing value imputation, we use Expectation Maximization (EM), MICE, K- nearest- neighbor (KNN), Mean, and Median. Original (ORI) datasets results are also calculated along with these techniques without using missing value imputation technique and resampling technique. Finally, perform a comparison with Fuzzy Information Decomposition technique (FID) used for missing value imputation and oversampling in case of

Table 13.1 Description of data sets

Datasets	Size	Attributes	IR
Mammographic	833	5	1.05
Heart	270	13	1.25
Bupa	345	6	1.38
MC2	125	39	1.84
Yeast1	1484	8	2.46
ILPD	583	9	2.49
Vehicle2	846	18	2.88
Vehicle1	846	18	2.90
Vehicle3	846	18	2.99
Glass0123vs456	214	9	3.20
Kc2	522	21	3.88
Kc3	194	39	4.39
Pc3	1077	37	7.04
Pc1	705	37	10.56
Glass4	214	9	15.47
Pageblocks13vs4	472	10	15.86
Glass016vs5	184	9	19.44
Glass5	214	9	22.75

Table 13.2 Parameter setting

Algorithm	Parameter Setting
KNN	K=3
SMOTE	K=5
CNN	Default
SMOTE-TOMEK-Link	Default

class imbalance with missing values. In Table 13.2, parameters setting of different methods used for experimental techniques are manifested, where the first column shows the type of algorithms, and the second column shows a parameter setting.

The detail description of 31 algorithms with different missing values imputation and resampling technique are as shown in Table 13.3. Where the algorithm shows sets of the possible model on the given datasets and abbreviation shows the model short form. In algorithms, we use some missing values imputation techniques (EM, MICE, KNN, etc.) directly without using resampling technique in order to compare the effects of missing values techniques on class imbalance handling techniques and in some techniques we use resampling techniques by removing missing values. Moreover, the rest use the model except original datasets follow in Figure 13.1.

For performance evaluation, accuracy is not sufficient for a class imbalance problem. So we take the AUC score [17], F1-measures (F1-Score),

Table 13.3 Models with abbreviation and type

Algorithm	Abbreviation	Type
original datasets	ORI	None
MICE	MICE	Imputation
EM	EM	Imputation
Mean	Mean	Imputation
Median	Median	Imputation
KNN	KNN	Imputation
SMOTE	SM	Resampling
ROS	ROS	Resampling
CNN	CNN	Resampling
SMOTE-TOMEK-LINK	SMT	Resampling
EM+SMOTE	EM_SM	Both
MICE+SMOTE	MICE_SM	Both
KNN+SMOTE	Knn_SM	Both
Mean+SMOTE	Mean_SM	Both
Median+SMOTE	Median_SM	Both
EM+ROS	em_ros	Both
MICE+ROS	Mice_ros	Both
KNN+ROS	Knn_ros	Both
Mean+ROS	Mean_ros	Both
Median+ROS	Median_ros	Both
EM+CNN	Em_cnn	Both
MICE+CNN	Mice_cnn	Both
KNN+CNN	KNN_cnn	Both
Mean+CNN	Mean_cnn	Both
Median+CNN	Median_cnn	Both
EM+SMOTE-Tomek-link	Em_SMT	Both
MICE+ SMOTE-Tomek-link	Mice_SMT	Both
KNN+ SMOTE-Tomek-link	Knn_SMT	Both
Mean+ SMOTE-Tomek-link	Mean_SMT	Both
Median+ SMOTE-Tomek-link	Median_SMT	Both
Fuzzy Information Decomposition	FID	Both

Geometric-mean (GM) [28], and Matthews correlation coefficient (MCC) [28]. F1-score is called a harmonic mean. GM is used to perform the geometric mean of both classes. AUC is defined as the area under the ROC curve ranges from 0 to 1.

For our experiment, we use five cross-validation technique and C4.5 as base classifier because it is sensitive towards class imbalance problem and perform better results. Figure 13.1 shows the working of the model, wherein the first step we load the datasets and if any are missing value then, in the second step, impute the missing value with missing value imputation techniques such as EM, MICE, KNN, Mean and Median. In the third step, we apply cross-validation; under this we split the datasets n splits where one part under n split is considered as a test part, and the remaining n-1 part is

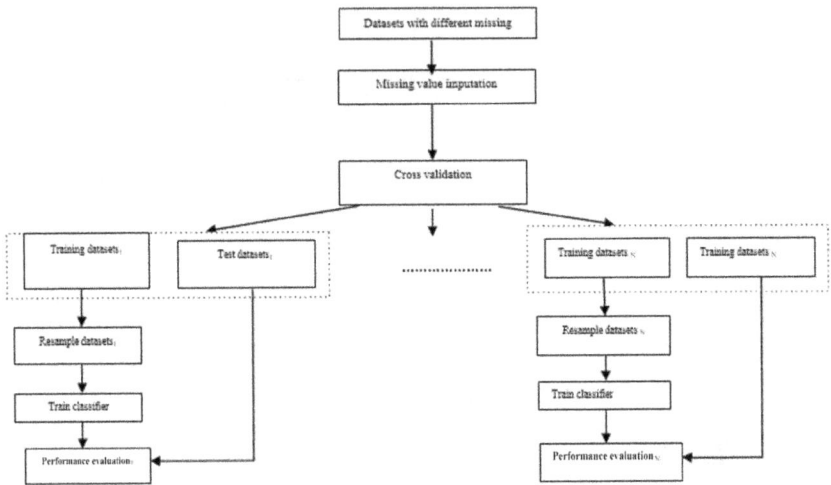

Figure 13.1 General model for missing value with class imbalance handling techniques.

considered as the training part and performs resampling over the training part and train classifier and test on the classifier. The procedure is repeated n number of times, and we take an average of all performance.s

13.4 RESULTS

Our result section is divided into two main sections: Overall performance section and the relationship between imbalance class and missing values.

13.4.1 Overall performance

In this subsection, Figure 13.2 shows overall performance of resampling techniques using G-Mean (Geometric Mean) under 0 percent, 5 percent, 10 percent, 15 percent and 20 percent missing values and depicts that performance of resampling techniques and ORI technique decreases with increase in missing values whereas the performance of FID (Fuzzy Information decomposition) techniques perform consistent with increase in missing values and shows better results at 20 percent missing values as compared with other resampling techniques.

In Figure 13.2, although the overall performance of the FID technique is not as good as compared with SM, SMT, CNN like a resampling algorithm. At 0 percent missing value, condense based nearest neighbors (CNN) perform better than another algorithm. However, as our missing values shifted from 0 percent to 20 percent, resampling techniques and ORI technique

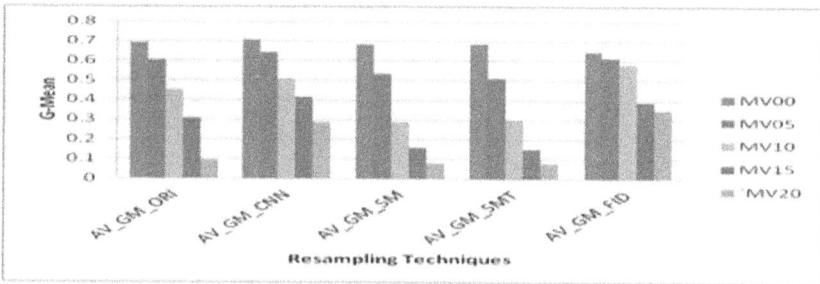

Figure 13.2 Overall performance of resampling techniques in comparison with FID Technique.

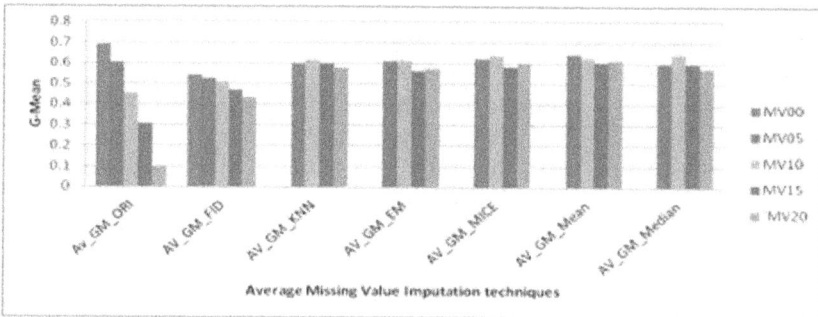

Figure 13.3 Overall performance of missing values imputation techniques compared with FID.

performance degraded. However, FID performance improves with an increase in missing values. In the case of 20 percent, missing value, FID shows 0.348 G-Mean, which is higher among other algorithms.

In Figure 13.3, shows the performance evaluation of FID and ORI techniques with missing value imputation techniques, where AV_GM_ORI, AV_GM_FID, AV_GM_KNN, AV_GM_EM, AV_GM_MICE, AV_GM_ Mean, and AV_GM_Median represents average G-mean original, average G-mean FID, average G-mean KNN, average G- mean EM, average G-mean MICE, average G-mean Mean, and average G-mean Median respectively. Only ORI and FID techniques perform well in the case of 0 percent missing values and rest of techniques will not work under these circumstances. However, with the increase in class, missing values ranges from 5–20 percent Mean technique perform well in totality. However, on comparing, performance evaluation by the missing value, we found that in the case of 0 percent missing value, Average-ORI technique (AV_GM_ORI) with 0.6913 G-Mean perform well in comparison with other techniques.

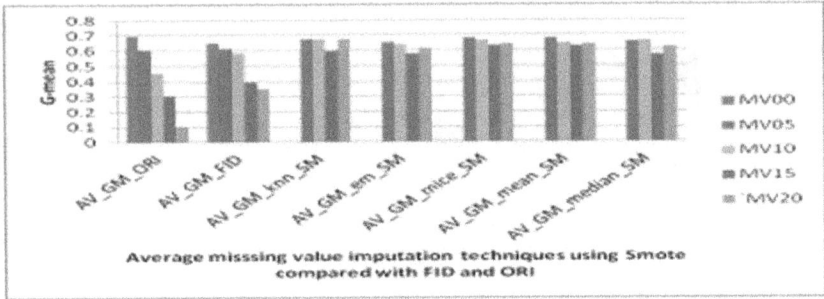

Figure 13.4 Comparison between FID and ORI and missing value imputation techniques using smote technique.

With a 5 percent increase in missing value, Mean technique with 0.64 G-mean performs well in comparison with other techniques, whereas, at 10 percent missing value MICE technique outperforms the rest of the techniques. Again at, 15 percent KNN, Mean, Median techniques perform well. At 20 percent missing value, the performance of Mean technique better than other techniques. On comparing, FID with missing value imputation techniques in case of missing value ranges from 5 percent-20 percent FID performance is not up to mark as compared with missing value imputation techniques.

Figure 13.4, shows the comparative analysis of FID and ORI techniques with missing values imputation using Smote as class imbalance handling technique with C4.5 classifier as the base classifier. Where X-axis shows an average of missing value imputation techniques and the y-axis shows G-Mean with MV00, MV05, MV10, MV15, and MV20 represents 0 percent, 5 percent, 10 percent, 15 percent, 20 percent missing values respectively. On comparing at 0 percent missing value, ORI technique performs well compared with other techniques.

At 5 percent missing value, Mean technique with 0.6828 G-Mean value perform well as compared with other techniques. At 10 percent missing value, KNN imputation technique with 0.6723 G-mean performs better as compared with other techniques. At 15 percent missing value technique, MICE technique with Smote has 0.629 G-Mean value and perform better than other techniques. At 20 percent missing value, knn technique with smote outperform than rest of techniques. In Figure 13.4, we concluded that FID perform better than ORI technique but on comparing with missing values with increase in missing values ranges from 5 to 20 percent we observe that FID does not perform well.

On comparing FID and ORI with missing value imputation technique using CNN (condense Nearest Neighbors) in Figure 13.5, where X-axis

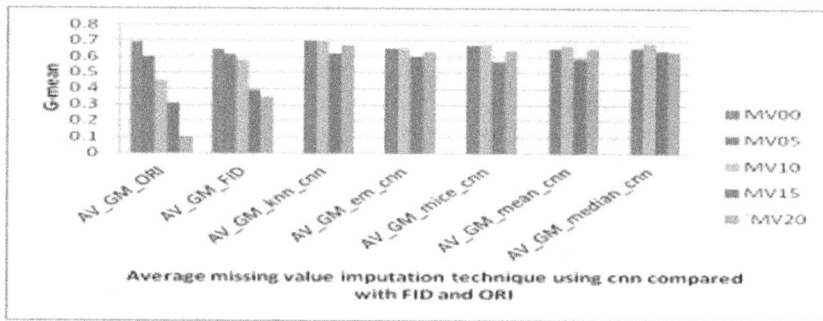

Figure 13.5 Comparison between FID and ORI and missing value imputation techniques using CNN technique.

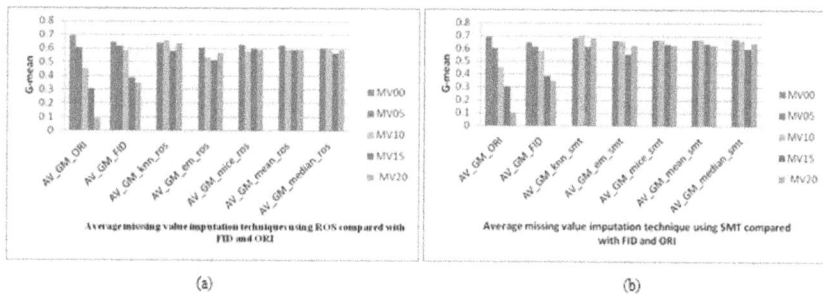

(a)   (b)

Figure 13.6(a & b) Comparison of FID and ORI with missing value imputation techniques using ROS and SMT.

shows average missing value imputation techniques using CNN and FID and ORI techniques and Y-axis shows G-Mean.

At 0 percent missing value, only two techniques, ORI and FID, perform well. However, with an increase in missing value ORI technique shows poor result as compared to other missing value imputation techniques using CNN. The missing value ranges from 5–20 percent, KNN technique performs well in comparison with other missing value and class imbalance handling techniques.

Likely wise in Figure 13.6 (a & b) also show the comparison between FID technique and different missing value and class imbalance handling techniques over 18 different datasets with 0 percent, 5 percent, 10 percent, 15 percent, 20 percent missing value. The comparison is made using G-mean and, finally, we observe that FID performs well in case of missing value 0 percent, 5 percent, and 10 percent but with an increase in missing value from 15–20 percent FID and ORI shows poor performance in comparison with missing value imputation technique using class imbalance handling techniques.

The overall comparison shows that FID performance decreases with an increase in missing value percentage, whereas, missing value imputation techniques along with class imbalance handling technique shows consistently good results as even with a higher percentage of missing value.

Further, we compare FID and ORI technique with missing value imputation technique with metrics like F1 score (harmonic mean) under 0 percent, 5 percent, 10 percent, 15 percent, 20 percent missing value in the datasets shows in Table 13.4. We observe that the FID technique does not show significant results in comparison with the rest of techniques. In the case of 0 percent missing value ORI technique performs well with an increase in missing value at 5 percent MICE_ROS technique performs well. When missing value ranges from 10 to 20 percent, we observe that Knn_ROS technique outperforms as compared with the rest of the techniques.

13.4.2 Effect of class imbalance and missing values

In order to note the effect of class imbalance and missing value on classification. First, we note the behavior of increased missing values percentage in the dataset on classification with an increase in class imbalance and vice versa. In order to perform this, we use AUC as metrics for measurement and made the following observations based on figures given below.

In Figure 13.7, we compare the FID, ORI, and missing value imputation technique under different class imbalance ratio datasets.

On comparing in Figure 13.7, FID with other missing value imputation without addressing class imbalance issues we observe that in case of class imbalance ranges from 1–5 IR and with 5 percent missing value, FID technique performs well and with increase in percentage of missing value from 10–20 percent, we observe that FID deteriorate as compared with missing value imputation technique. However, this process lasts only for class imbalance ratio 1–5. However, with an increase in class imbalance from 5–10 and 10–22, we observe that the FID technique perform well as compared missing value imputation technique because missing value imputation technique does not use class imbalance handling techniques.

Further, we need to performing analysis on FID and ORI technique in comparison with missing value imputation technique like MICE, EM, Mean, Median using class imbalance handling technique like SMOTE, ROS, CNN, SMOTE-TOMEK-LINK (SMT). Figure 13.8 shows a comparison between FID, ORI technique with missing value imputation techniques using SMOTE as class imbalance handling technique in different class imbalance ratio.

Figure 13.8 shows that at 0 percent missing value only FID and ORI technique perform. However, with an increase in missing value FID and ORI shows poor results in comparison with missing value imputation technique

Table 13.4 Average F1 score of each technique

Metrics	Missing values	Techniques											
		ORI	FID	SM	ROS	CNN	SMT	KNN	MICE	EM	Mean	Media n	Knn_SM
F1 Score	0%	0.8624	0.70603	0.853	0.860	0.630	0.851	....	....	....	....	....	....
	5%	0.8212	0.66951	0.702	0.8129	0.592	0.7035	0.80496	0.8373	0.83395	0.8368	0.8324	0.826
	10%	0.6827	0.65720	0.356	0.6869	0.483	0.3583	0.80707	0.8388	0.83915	0.8419	0.8360	0.835
	15%	0.4080	0.38121	0.167	0.3623	0.314	0.1605	0.80495	0.8324	0.83373	0.8385	0.8366	0.822
	20%	0.1222	0.33858	0.097	0.1194	0.192	0.0960	0.79747	0.8305	0.82354	0.8324	0.8236	0.829

Metrics	Missing values	Techniques											
		MICE_SM	EM_SM	Mean_SM	Media n_SM	Knn_ROS	MICE_ROS	EM_R OS	Mean_ROS	Median_ROS	Knn_CNN	Mice_CNN	EM_CNN
F1 Score	0%	....	....	....	....	....	....	....	....	....	....	....	....
	5%	0.8271	0.81947	0.832	0.823	0.834	0.8387	0.82288	0.8386	0.82837	0.7970	0.7740	0.779
	10%	0.8368	0.83000	0.827	0.835	0.857	0.8433	0.82606	0.8382	0.84604	0.8018	0.7949	0.792
	15%	0.8311	0.81861	0.828	0.812	0.842	0.8355	0.82109	0.8349	0.83005	0.7679	0.7303	0.770
	20%	0.8351	0.82211	0.828	0.828	0.844	0.8394	0.83273	0.8420	0.83603	0.7914	0.7648	0.766

Metrics	Missing values	Techniques						
		Mean_CNN	Median_CNN	Knn_SMT	MICE_SMT	EM_SMT	Mean_SMT	Median_SMT
F1 Score	0%	....	....	....	....	....	....	....
	5%	0.7782	0.77343	0.831	0.8289	0.817	0.8273	0.82692
	10%	0.7907	0.77924	0.840	0.8409	0.832	0.8291	0.83405
	15%	0.7439	0.76311	0.819	0.8295	0.809	0.8309	0.81891
	20%	0.7676	0.76081	0.830	0.8265	0.819	0.8276	0.83330

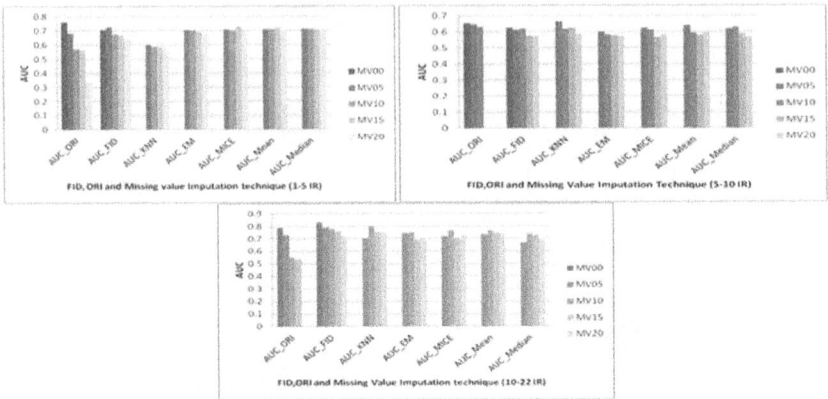

Figure 13.7 Comparison of FID and ORI technique with missing value imputation techniques under different class imbalance ratio.

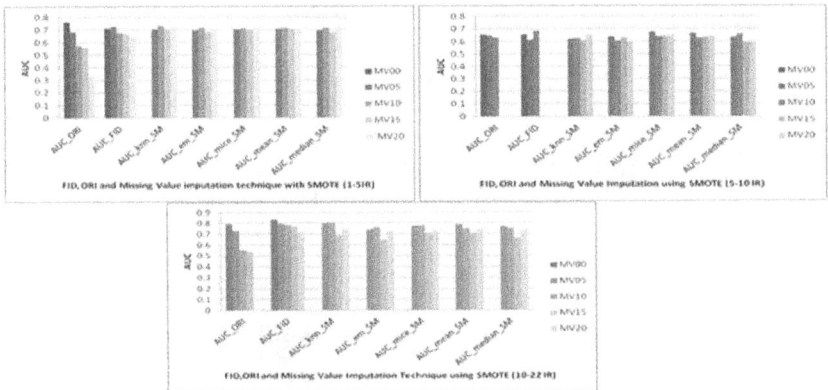

Figure 13.8 Comparison of FID and ORI technique with missing value imputation techniques using SMOTE under different class imbalance ratio.

using class imbalance handling techniques. In the case of class imbalance ratio ranges from 1–5 with missing value 5 percent, FID performs well in comparison with the rest state of art. However, in case of missing value 10 percent and class imbalance range from 1–5, FID is not performing well as compared with the rest of missing value imputation techniques using SMOTE as a class imbalance. Knn_SM method performs well in the case of 10 percent and 15 percent missing value with class imbalance ranges 1–5. However, with an increase in class imbalance ratio from 1–5 to 10–22, we

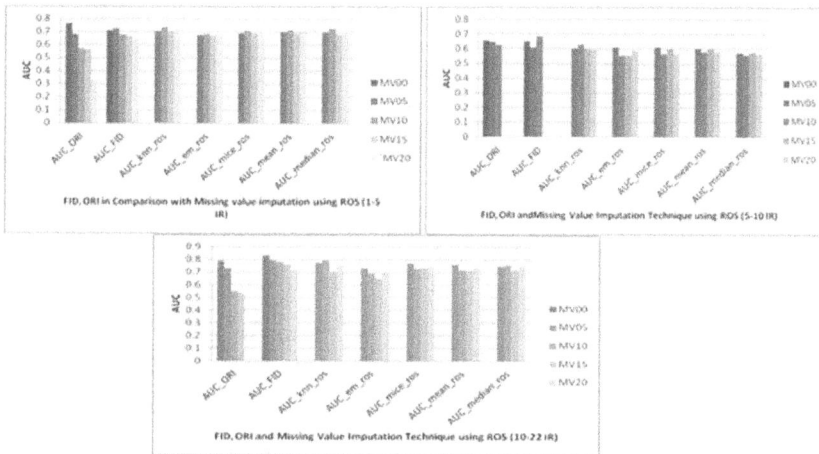

Figure 13.9 Comparison of FID and ORI technique with missing value imputation techniques using ROS under different class imbalance ratio.

observe that FID shows poor performance as compared with the rest of the missing value imputation technique using SMOTE. Although, in class imbalance 10–22 and missing value 20 percent, we observe that FID is not performed well as compared with the rest of the missing value imputation technique using SMOTE as class imbalance handling technique.

Moreover, Figure 13.9 shows that comparison between FID, ORI, and missing value imputation technique using ROS as class imbalance handling technique. Using ROS as class imbalance handling technique use to duplicate the instances of minority instances for handling imbalancing in datasets.

In Figure 13.9, we depict three graphs having techniques for handling missing value using class imbalance with AUC as metrics. From these graphs, we concluded that FID performs better in comparison with ORI, but FID performance degrade with an increase in class imbalance ratio and missing value percentage. At 20 percent missing value with different class imbalance ratio, we observe that FID shows poor performance as compared with other missing value imputation technique using ROS as class imbalance handling technique.

On comparing FID and Missing value imputation techniques using CNN as undersampling technique for class imbalance handling technique as shown in Figure 13.10. We observe that in all class imbalance ratios with a missing value up to 5 percent FID perform better and when missing value ratio increasing from 10–20 percent FID technique shows unsatisfactory results in comparison with missing value imputation techniques using CNN

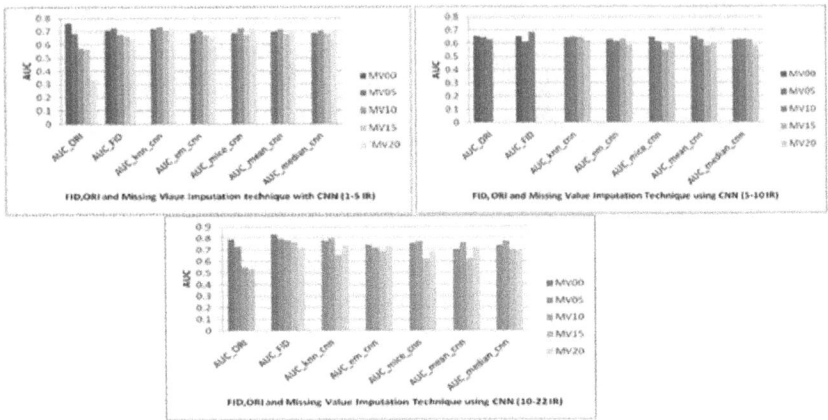

Figure 13.10 Comparison of FID, ORI and missing value imputation techniques using CNN with different class imbalance ratio.

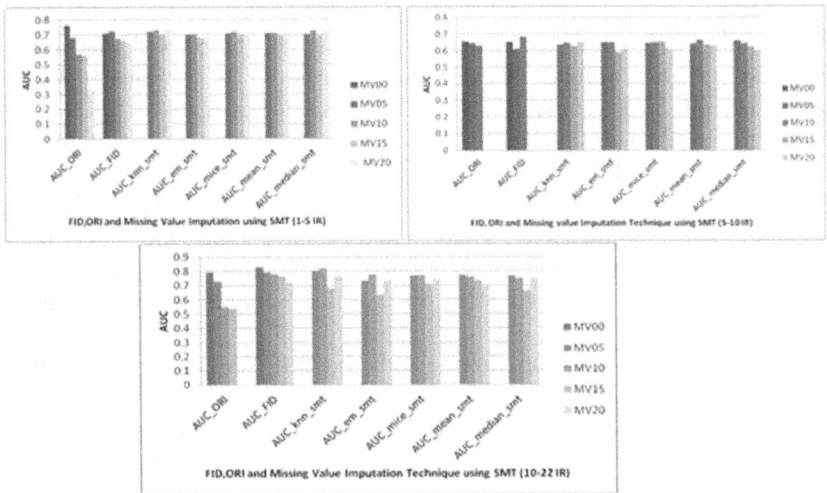

Figure 13.11 Comparison of FID, ORI, and missing values imputation techniques using SMT as class imbalance handling technique.

as an undersampling technique for handling class imbalance. With missing value ranges from 10–20 percent and class imbalance ratio from 1–5 to 10–22, Knn_cnn technique performs well as compared with FID, ORI, and other missing value imputation techniques.

Figure 13.11, shows the comparison FID, ORI, and missing value imputation technique using SMT used as both undersampling and oversampling technique for handling class imbalance. We observed in case of missing value with 5 percent FID technique perform well as compared with the rest of techniques in different class imbalance ratio. However, with an increase in missing value from 10–20 percent, FID shows poor results in comparison with the rest of the missing value imputation technique using SMT as class imbalance handling technique. In class imbalance ratio 1–5 graph shows us that Knn_smt and Median_smt perform well in all missing value percentage whereas, in class imbalance ratio from 5–10, mean_smt perform well. Whereas, in case of class imbalance ratio ranges from 10–22, again Knn_smt perform well. In Figure 13.11, we also observe that FID and ORI go on decreasing with increase in missing value percentage but other missing value imputation techniques using SMT as class imbalance handling technique, especially at missing value 20 percent we observe that AUC increases.

Remarks: In this chapter, we compared FID with other incomplete value imputation techniques using class imbalance handling technique in overall and in the relationship between class imbalance and incomplete value. We observed that considering G-Mean as metrics, overall performance of Fuzzy Information Decomposition technique perform well compared with missing value and class imbalance handling techniques. When we use the combined techniques (missing value with class imbalance handling techniques) to compare with FID, we observe that with an increase in percentage of incomplete value, FID technique shows poor result in comparison with combined techniques. Moreover, FID also shows results in the case of class 0 percent incomplete value, but all combine techniques fails in that case.

On comparing FID with rest of techniques which handles class imbalance and missing values. We observe that the FID technique performs well under missing value percentage in datasets ranges from 0–10 percent in comparison with missing value imputation techniques without using class imbalance handling techniques. However, with an increase in missing value from 15–20 percent, we observe that FID shows degraded results as compared with missing value imputation technique. Further, we observe that with an increase in class imbalance ratio missing value imputation techniques shows poor results. We extend our analysis from missing value imputation to combined technique having both the missing value handling technique and class imbalance handling technique, and further results show that with an increase in class imbalance and missing value FID technique degraded on comparison with combined techniques.

13.5 CONCLUSION

In this study, analysis of 18 publicly available datasets with a class imbalance problem and randomly generated incomplete values with varying percentages (such as 0 percent, 5 percent, 10 percent, 15 percent, and

20 percent) was performed using FID. The remaining techniques for handling class imbalance (SMOTE, ROS, CNN, SMT), missing value imputation (KNN, EM, MICE, Mean, Median), and combined techniques (used for class imbalance and missing value imputation techniques) were also used. We conclude that the FID approach performs effectively when the fraction of missing values is lower. However, we see that the combined techniques work well overall as the percentage of incomplete values rises (15–20 percent).

REFERENCES

[1]   L. Chen, B. Fang, Z. Shang, and Y. Tang, "Tackling class overlap and imbalance problems in software defect prediction," *Software Quality Journal,* vol. 26, pp. 97–125, 2018.

[2]   V. García, A. I. Marqués, and J. S. Sánchez, "Exploring the synergetic effects of sample types on the performance of ensembles for credit risk and corporate bankruptcy prediction," *Information Fusion,* vol. 47, pp. 88–101, 2019.

[3]   S. Maldonado, J. López, and C. Vairetti, "An alternative SMOTE oversampling strategy for high- dimensional datasets," *Applied Soft Computing,* vol. 76, pp. 380–389, 2019.

[4]   C. Luo, "A comparison analysis for credit scoring using bagging ensembles," *Expert Systems,* p. e12297, 2018.

[5]   C. Li and S. Liu, "A comparative study of the class imbalance problem in Twitter spam detection," *Concurrency and Computation: Practice and Experience,* vol. 30, p. e4281, 2018.

[6]   A. Puri and M. Gupta, "Review on Missing Value Imputation Techniques in Data Mining," 2017.

[7]   C. Seiffert, T. M. Khoshgoftaar, J. Van Hulse, and A. Napolitano, "RUSBoost: A hybrid approach to alleviating class imbalance," *IEEE Transactions on Systems, Man, and Cybernetics-Part A: Systems and Humans,* vol. 40, pp. 185–197, 2010.

[8]   N. V. Chawla, K. W. Bowyer, L. O. Hall, and W. P. Kegelmeyer, "SMOTE: synthetic minority over- sampling technique," *Journal of Artificial Intelligence Research,* vol. 16, pp. 321–357, 2002.

[9]   W.-C. Lin, C.-F. Tsai, Y.-H. Hu, and J.-S. Jhang, "Clustering-based undersampling in class-imbalanced data," *Information Sciences,* vol. 409, pp. 17–26, 2017.

[10]   M. Galar, A. Fernández, E. Barrenechea, H. Bustince, and F. Herrera, "An overview of ensemble methods for binary classifiers in multi-class problems: Experimental study on one-vs-one and one-vs-all schemes," *Pattern Recognition,* vol. 44, pp. 1761–1776, 2011.

[11]   L. E. Zarate, B. M. Nogueira, T. R. Santos, and M. A. Song, "Techniques for missing value recovering in imbalanced databases: Application in a marketing database with massive missing data," in *2006 IEEE International Conference on Systems, Man and Cybernetics,* 2006, pp. 2658–2664.

[12] P. Porwik, T. Orczyk, M. Lewandowski, and M. Cholewa, "Feature projection k-NN classifier model for imbalanced and incomplete medical data," *Biocybernetics and Biomedical Engineering,* vol. 36, pp. 644–656, 2016.

[13] S. Liu, J. Zhang, Y. Xiang, and W. Zhou, "Fuzzy-based information decomposition for incomplete and imbalanced data learning," *IEEE Transactions on Fuzzy Systems,* vol. 25, pp. 1476–1490, 2017.

[14] I. Salman and V. Jiří, "A machine learning method for incomplete and imbalanced medical data," in *Proceedings of the 20th Czech-Japan Seminar on Data Analysis and Decision Making under Uncertainty.* Czech–Japan Seminar, 2017, pp. 188–195.

[15] G. Liu, Y. Yang, and B. Li, "Fuzzy rule-based oversampling technique for imbalanced and incomplete data learning," *Knowledge-Based Systems,* vol. 158, pp. 154–174, 2018.

[16] W. Lee, C.-H. Jun, and J.-S. Lee, "Instance categorization by support vector machines to adjust weights in AdaBoost for imbalanced data classification," *Information Sciences,* vol. 381, pp. 92–103, 2017.

[17] M. Galar, A. Fernandez, E. Barrenechea, H. Bustince, and F. Herrera, "A review on ensembles for the class imbalance problem: bagging-, boosting-, and hybrid-based approaches," *IEEE Transactions on Systems, Man, and Cybernetics, Part C (Applications and Reviews),* vol. 42, pp. 463–484, 2012.

[18] J. Li, S. Fong, S. Hu, R. K. Wong, and S. Mohammed, "Similarity majority under-sampling technique for easing imbalanced classification problem," in *Australasian Conference on Data Mining,* 2017, pp. 3–23.

[19] C.-F. Tsai, W.-C. Lin, Y.-H. Hu, and G.-T. Yao, "Under-sampling class imbalanced datasets by combining clustering analysis and instance selection," *Information Sciences,* vol. 477, pp. 47–54, 2019.

[20] W. W. Ng, J. Hu, D. S. Yeung, S. Yin, and F. Roli, "Diversified sensitivity-based undersampling for imbalance classification problems," *IEEE transactions on cybernetics,* vol. 45, pp. 2402–2412, 2015.

[21] A. Liu, J. Ghosh, and C. E. Martin, "Generative Oversampling for Mining Imbalanced Datasets," in *DMIN,* 2007, pp. 66–72.

[22] H. Han, W.-Y. Wang, and B.-H. Mao, "Borderline-SMOTE: a new oversampling method in imbalanced data sets learning," in *International Conference on Intelligent Computing,* 2005, pp. 878–887.

[23] C. Bunkhumpornpat, K. Sinapiromsaran, and C. Lursinsap, "Safe-level-smote: Safe-level-synthetic minority over-sampling technique for handling the class imbalanced problem," in *Pacific-Asia conference on knowledge discovery and data mining,* 2009, pp. 475–482.

[24] T. Zhu, Y. Lin, and Y. Liu, "Synthetic minority oversampling technique for multiclass imbalance problems," *Pattern Recognition,* vol. 72, pp. 327–340, 2017.

[25] M. Koziarski, B. Krawczyk, and M. Woźniak, "Radial-Based Oversampling for Noisy Imbalanced Data Classification," *Neurocomputing,* 2019.

[26] T. Aljuaid and S. Sasi, "Intelligent imputation technique for missing values," in *Advances in Computing, Communications and Informatics (ICACCI), 2016 International Conference on,* 2016, pp. 2441–2445.

[27] M. J. Azur, E. A. Stuart, C. Frangakis, and P. J. Leaf, "Multiple imputation by chained equations: what is it and how does it work?," *International Journal of Methods in Psychiatric Research*, vol. 20, pp. 40–49, 2011.

[28] M. Bekkar, H. K. Djemaa, and T. A. Alitouche, "Evaluation measures for models assessment over imbalanced data sets," *J Inf Eng Appl*, vol. 3, 2013.

Chapter 14

Gabor filter as feature extractor in anomaly detection from radiology images

T. S. Saleena, P. Muhamed Ilyas, and K. Sheril Kareem

14.1 INTRODUCTION

The term *medical imaging* comprises a vast area that includes x-rays, computed tomography (CT), ultrasound scan, nuclear medicines (PET scan), and MRI. They are the key sources used as the input for the diagnosis and detection of any kind of image, especially in the preliminary stage. But image analysis is quite a time-consuming and labor-intensive process. It also suffers from inter-observer variability. So, for the last few decades, the whole process is becoming automated using Artificial Intelligence (AI) techniques, which have touched almost all of these fields as part of digitalization and automation processes. AI has a good source of algorithms that can quickly and accurately identify the abnormalities in medical images. As the number of radiologists is small compared to the number of patients, the introduction of AI techniques in medical imaging diagnosis helps the entire healthcare system in terms of time and money.

Colangelo et al. (2019) have made a study on the impact of using AI in healthcare, and some of the analytics they made are as follows. In Nigeria, the ratio of radiologists to the number of patients is 60:190 million people; in Japan it is 36:1 million; and some African countries do not have any radiologists at all. The researchers have concluded that most often the AI algorithms diagnose and predict the disease more efficiently and effectively than do human experts. The main hurdle in machine-learning techniques is the feature extraction process from the input images. Several filtering methods are available for doing this task, includinhg Gaussian filter, mean filter, histogram equalization, median filter, Laplacian filter (Papageorgiou et al. 2000), wavelet transform, and so forth.

Gabor filter (GF) is a linear filter for the intent of texture analysis, edge detection, and feature extraction. It was named after Dennis Gabor, a Hungarian-British electrical engineer and physicist. It has proved its excellence in various real-life applications like facial recognition, lung cancer detection, vehicle detection, Iris recognition (Zhang et al. 2019), finger-vein

DOI: 10.1201/9781003453406-14

recognition in bio-metric devices, digital circuit implementation, texture segmentation, and so on.

This filter is a combination of a Gaussian filter and a sinusoidal wave. The illustration of this combination is shown in Figure 14.1 (Anuj Shah 2018). The Gaussian filter is normally used for smoothening and blurring effects in images. It is a distribution that is, in turn, a function of standard deviation and mean. If we combine a sinusoidal wave with a particular 2D Gaussian filter, we will get a Gabor filter. This resultant filter can determine the maximum peak that can be allowed in a distribution. It acts as a band pass filter that allows only the waves in the peaks of sinusoidal waves, and all others are banned by the filter. By definition, a band pass filter is a device that can filter out certain frequencies within a particular range.

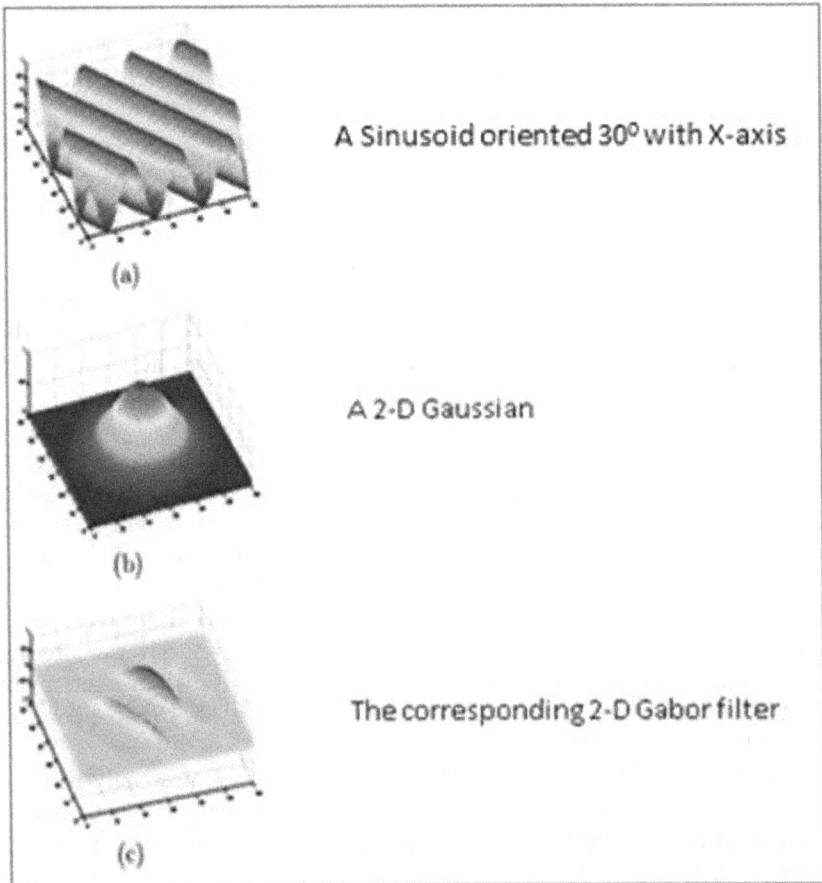

A Sinusoid oriented 30° with X-axis

(a)

A 2-D Gaussian

(b)

The corresponding 2-D Gabor filter

(c)

Figure 14.1 A sinusoidal wave (a) has been combined with a 2D Gaussian filter (b) that results in a Gabor filter(c).

In this chapter, we have used a dataset of MRI brain images that can be used to train a machine-learning model that predicts whether tumor is present in an image or not. In this work, we are only extracting the features of images using the Gabor filter – features that can be later used in classification or segmentation models. This study also depicts the supremacy of such a filter on other filtering and feature extraction methods.

14.2 LITERATURE REVIEW

Gabor filter is one of the finest filters for feature extraction that can be used in both machine learning and deep learning architectures. Gabor filter helps us to extract the features from the image in different directions and angles based on the content of the image and user requirement.

The history of AI-based automatic analysis of medical images for the purpose of assessing health and predicting risk begins at the University of Hawai'i Cancer Center, the world's first AI Precision Health Institute (Colangelo et al. 2019). An optimized moving-vehicle detection model has been proposed by Sun et al. (2005), where an Evolutionary Gabor Filter Optimization (EGFO) approach has been used to do the feature extraction task. Genetic Algorithm (GA) has integrated with the above approach along with incremental clustering methods so that the system can filter the features that are specific for vehicle detection. The features obtained through this method have been used as input for SVM for further processing. A stochastic computation-based Gabor filter has been used in the digital circuit implementation, achieving 78 percent area reduction as compared to the conventional Gabor filter (Onizawa et al. 2015). Due to this stochastic computation, data can be considered as streams of random bits. In order to achieve high speed, the area required for the hardware implantation of digital circuits is usually very large. But this proposed method has area-efficient implementation as this is a combination of Gaussian and sine functions.

The Gabor filter has been used in biometric devices for human identification using Iris recognition (Minhas et al. 2009). As the iris contains several discriminating features to uniquely identify the human being, it is a good choice for authenticating a person. Two methods have been used here, one for collecting global features from the entire image using a 2D Gabor filter and a multi-channel Gabor filter applied locally on different patches of the image. The feature vector obtained through this has shown 99.16 percent accuracy, and it has a good correlation with output obtained through hamming distance, which is a metric to find the similarity between two strings of same length. Biometric systems also make use of the finger vein for the authentication process. Souad Khellat-kihel et al. (2014) have presented an SVM-based classification model where the feature extraction part has been performed by the Gabor filter. Here, two types of pre-processing have been done, one with median filter plus histogram equalization and the second with Gabor filter. Among them, the second approach performs well. This work outperformed the method presented by Kuan-Quan Wang et al. (2012), in which classification is done by the same SVM, but they used the Gaussian filter and Local Binary Pattern Variance for the feature extraction. A combination of GF and Watershed segmentation algorithm has been used for lung cancer diagnosis (Avinash et al. 2016). It helps to make the detection and early-stage diagnosis of the lung nodules easier.

A face recognition system based on GF and Sparse Auto-encoder has been introduced by Rabah Hammouche et al. on seven existing face databases that outperform the other existing systems (Hammouche et al. 2022). The feature extraction has been enriched here using GF, and this has been utilized by the auto-encoder and the system has been tested with different publically available databases, namely JAFFE, AT&T, Yale, Georgia Tech, CASIA, Extended Yale, and Essex (Zhang et al. 2021). The Gabor filter has been used in destriping of hyperspectral images, where the strips have been created due to the error of push-broom imaging devices. There will be vertical, horizontal, and oblique stripes and it affects the quality of the image (Barshooi et al. 2022). This filter has become an important aspect in the diagnosis and classification of the most pandemic disease of this century, Covid19. Here the authors have used chest x-ray dataset for the processing and data scarcity has resolved using combing data augmentation technique with GAN and deeper feature extraction has performed using Gabor filter, Sobel, and Laplacian of Gaussian, where the first one shows better accuracy. This filter has also been used along with genetic algorithm for facial expression recognition (Boughida et al. 2022). The bottleneck of input data size in deep learning methods has been reduced by the use of the 3D-Gabor filter as feature extractor by Hassan Ghassemian et al (2021).

14.3 RESEARCH METHODOLOGY

14.3.1 Data set

We have selected a dataset of brain MRI images from a public repository. The captured MRI images will be in 3D format. But this dataset contains images that are converted into *.jpg* format and thus in 2D format. It has two types of images, one with tumors and another with normal brain images. This dataset was actually meant for the classification problems, but we are limiting our study only to the feature extraction part.

14.3.2 Gabor filter

Gabor filter is a linear filter mainly for texture analysis, and it is a Gaussian kernel function modulated by a sinusoidal plane wave (Wikipedia 2017).

The way this filter does image analysis has many similarities with the human visual system. Some authors have described its resemblance with cells in the visual cortex of mammalian brains (Marčelja et al. 1980).

In this particular case we are considering only the 2D Gabor filter. A 2D Gabor filter is a function $g(x,y, \sigma, \Theta, \lambda, \Upsilon, \varphi)$ and can be expressed as such (Wikipedia 2017).

Explanation of the terms:

x and **y** are the dimensions of kernel

σ or sigma is the standard deviation of Gaussian filter

Θ or theta is the angle of Gaussian filter

λ or lambda is the wavelength of sinusoidal wave

Υ or gamma is the spatial aspect ratio. If gamma value is 1, kernel is circular; if it is zero means the kernel is an ellipse or almost straight line

φ or phi is the phase offset of the sinusoidal wave

GF captures specific frequency content from the image in a particular direction and phase. We can create so many filters – namely, filter bank – by changing the above-mentioned parameters, so that we can capture maximum features from the image in all directions. Figure 14.2 is an illustration of how the Gabor kernel is extracting the features from an image.

For selecting different Gabor filters we can use the function "getGaborKernel()" provided by python library openCV. As the filter is a function of $g(x,y, \sigma, \Theta, \lambda, \Upsilon, \varphi)$, we have given different values to the parameters and have selected the best values of the parameters in a trial-and-error method. We can create a filter bank of any size based on our image. When the input image passes through the filter bank, we will get features in different orientations, and our model will become more reliable.

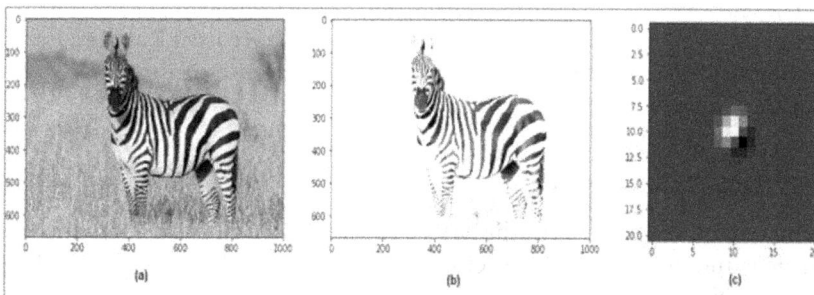

Figure 14.2 (a) Represents the original image; (b) is the image obtained after applying Gabor filter; and (c) is the kernel or filter applied on the image. The combination of values for this case is x,y = 20, sigma = 1, theta = 1*np.pi/4, lamda = 1*np.pi/4, gamma=1, and phi = 0.9.

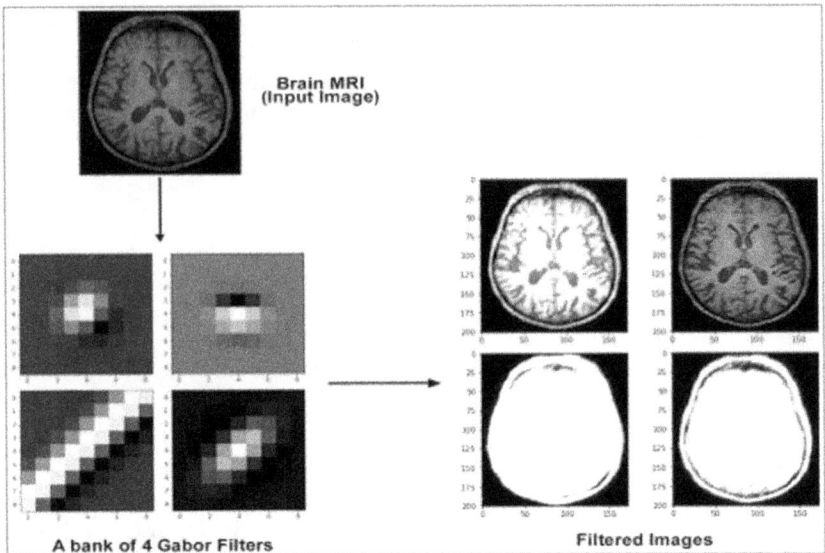

Figure 14.3 Workflow of the proposed work is narrated here. We have created a filter bank with 4 kernels and image passed to them has been filtered and shown as Filtered images.

14.4 RESULTS

In our study, we have collected an MRI dataset of brain images from a public repository and applied different filters and compared the results. The images from the dataset have been tested with different filters, namely Gaussian, median, Sobel, and Gabor, for comparing the efficiency. The outputs obtained from them are displayed in Figures 14.3 and 14.4.

The features extracted from these filters have been used to predict whether or not the brain tumor is present in the image. We have used Random Forest classifier for the classification task. The model that uses Gabor filter as the feature extractor shows better accuracy than anything else.

14.5 DISCUSSION

In our study, we have taken the MRI images of brain to predict whether the image is tumorous or not. For classification purposes, we have selected the Random Forest classifier, which has proven to be good for medical image classification. As we did the classification without applying filters, it shows the accuracy of only 73 percent. But when we have applied the filters to the images it shows better results. We have tested the dataset with different filtering algorithms like Gaussian, Median, Gabor and Sobel. Among them,

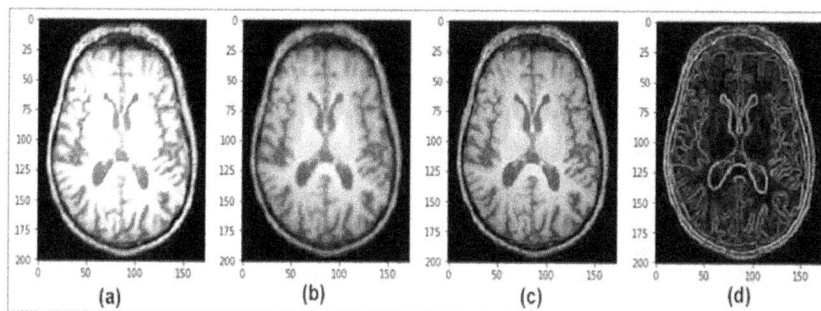

Figure 14.4 Output of (a) Gabor (b) Gaussian with sigma=1 (c) median filter (d) Sobel edge detection algorithm.

Table 14.1 Accuracy of classification using different filters

Filter used	Accuracy for classification
Without fitler	73%
Gabor	83.33%
Gaussian	81%
Sobel	73%
Median	75%

Table 14.2 Parameter values of Gabor function that we have used in this model

Parameter	Value range
σ	1 to 3
Θ	¼*pi, ½*pi
λ	pi/4
Υ	1
Φ	0
Kernel size	8

Gabor performs well with an accuracy of 83.33 percent, as shown in Tables 14.1 and 14.2 and from this we can conclude that they are very much suitable for medical image processing.

14.6 CONCLUSION

Feature extraction is the most important phase in any machine learning or deep learning technique. We have made a comparison study on different filters – Gaussian, median, Sobel and Gabor – to identify the best one and we found that the Gabor filter performs best in our dataset. As every deep learning works, brain tumor classification also suffers from data scarcity and the need for heavy computation power. We have developed Random Forest models using different filters to classify the MRI brain images, and Gabor won to extract the features in the best way. We can do this work with a sounder dataset and also can be extended with other healthcare classification and segmentation tasks. The filter bank that we have created in this work can be customized based on the application where we are using these features.

REFERENCES

Anuj shah, Through The Eyes of Gabor Filter, https://medium.com/@anuj_shah/through- the-eyes-of-gabor-filter-17d1fdb3ac97, Accessed by Jun 17, 2018.

Avinash, S., Manjunath, K., & Kumar, S. S. (2016, August). An improved image processing analysis for the detection of lung cancer using Gabor filters and watershed segmentation technique. In *2016 International Conference on Inventive Computation Technologies (ICICT)* (Vol. 3, pp. 1–6). IEEE.

Barshooi, A. H., & Amirkhani, A. (2022). A novel data augmentation based on Gabor filter and convolutional deep learning for improving the classification of COVID-19 chest X-Ray images. *Biomedical Signal Processing and Control*, 72, 103326.

Boughida, A., Kouahla, M. N., & Lafifi, Y. (2022). A novel approach for facial expression recognition based on Gabor filters and genetic algorithm. *Evolving Systems*, 13(2), 331–345.

Ghassemi, M., Ghassemian, H., & Imani, M. (2021). Hyperspectral image classification by optimizing convolutional neural networks based on information theory and 3D-Gabor filters. *International Journal of Remote Sensing*, 42(11), 4380–4410.

Hammouche, R., Attia, A., Akhrouf, S., & Akhtar, Z. (2022). Gabor filter bank with deep auto encoder based face recognition system. *Expert Systems with Applications*, 116743.

Khellat-Kihel, S., Cardoso, N., Monteiro, J., & Benyettou, M. (2014, November). Finger vein recognition using Gabor filter and support vector machine. In *International image processing, applications and systems conference* (pp. 1–6). IEEE.

Kuan-Quan, W., S. Krisa, Xiang-Qian Wu, and Qui-Sm Zhao. "Finger vein recognition using LBP variance with global matching." Proceedings of the International Conference on Wavelet Analysis and Pattern Recognition, Xian, 15–17 July, 2012.

Marčelja, S. (1980). "Mathematical description of the responses of simple cortical cells". *Journal of the Optical Society of America*. 70 (11): 1297–1300.

Margaretta Colangelo & Dmitry Kaminskiy (2019) AI in medical imaging may make the biggest impact in healthcare. *Health Management*. Vol. 19 – Issue 2, 2019.

Minhas, S., & Javed, M. Y. (2009, October). Iris feature extraction using Gabor filter. In *2009 International Conference on Emerging Technologies* (pp. 252–255). IEEE.

Onizawa, N., Katagiri, D., Matsumiya, K., Gross, W. J., & Hanyu, T. (2015). Gabor filter based on stochastic computation. *IEEE Signal Processing Letters*, 22(9), 1224–1228.

Papageorgiou, C. & T. Poggio, "A trainable system for object detection," *Int. J. Comput. Vis.*, vol. 38, no. 1, pp. 15–33, 2000.

Sun, Z., Bebis, G., & Miller, R. (2005). On-road vehicle detection using evolutionary Gabor filter optimization. *IEEE Transactions on Intelligent Transportation Systems*, 6(2), 125–137.

Wikipedia, the Free Encyclopedia, https://en.wikipedia.org/wiki/Gabor_filter, Accessed February 2017

Zhang, B., Aziz, Y., Wang, Z., Zhuang, L., Ng, M. K., & Gao, L. (2021). Hyperspectral Image Stripe Detection and Correction Using Gabor Filters and Subspace Representation. *IEEE Geoscience and Remote Sensing Letters*, 19, 1–5.

Zhang, Y., Li, W., Zhang, L., Ning, X., Sun, L., & Lu, Y. (2019). Adaptive learning Gabor filter for finger-vein recognition. *IEEE Access*, 7, 159821–159830.

Chapter 15

Discriminative features selection from Zernike moments for shape based image retrieval system

Pooja Sharma

15.1 INTRODUCTION

Images are effective and an efficient medium for presenting visual data. In the present technology the major part of information is images. With the rapid development in multimedia information such as audio visual data, it becomes mandatory to organize it in some efficient manner so that it can be obtained effortlessly and quickly. Image indexing, searching, and retrieval – in other words, content based image retrieval (CBIR) becomes an active research area [1, 2, 3] in both industry and academia. Accurate image retrieval is achieved by classifying image features appropriately. In this chapter we propose a classifier based on a statistical approach to be applied on features extracted by Zernike Moments for selecting appropriate features of images. Image indexing is usually done by low level visual features such as texture, color, and shape. Texture may consist of some basic primitives that describe structural arrangement of a region and its relationship with surrounding regions [4]. However, texture features hardly provide semantic information. Color is another low level feature for describing images, which is invariant to image size and orientation. Color histograms, color correlograms, and dominant color descriptors are used in CBIR. Among them color histograms are most commonly used. However color feature does not include spatial information [5]. Another visual feature shape is related to a specific object in an image. Therefore, shape feature provides more semantic information than color and texture [6], and shape based image retrieval of similar images is extensively studied [6, 7, 8, 9, 10]. In this chapter we pursue the shape based image retrieval system.

Zernike Moments (ZMs) were introduced by Teague [11] and are used as a shape descriptor for similarly based image retrieval applications. ZMs are excellent in image reconstruction [12, 13], feature representation [13, 14] and low noise sensitivity [15]. ZMs are widely used as shape descriptors in various image retrieval applications [17, 18, 19, 20, 21, 22, 23]. ZMs provide appropriate feature extraction from images,

 DOI: 10.1201/9781003453406-15

however the need for suitable classifier emerges by which only effective features can be selected and non effective features can be discarded from the features set. There have been considerable researches done on pattern classification using neural networks [24, 25, 26, 27, 28] and support vector machines [29, 30]. These techniques have their various properties, advantages, and disadvantages. In order to detect and classify images from large and complex databases, we need to select only significant features rather than incorporating all the extracted attributes. In this respect we propose a data dependent classification technique that opts for only those features that describe image more precisely while eradicating less significant features. It is a statistical approach that analyzes all the images of database, emphasizing only on ZMs coefficients with more discriminative power. The discriminate coefficients (DC) have small within class variability and large between class variability. To evaluate the performance of our approach we performed experiments on most frequently used MPEG 7 CE shape 1 part B image database for subject test and rotation test, and observed that ZMs with proposed classifier outperforms by increasing the retrieval and recognition rate more than 3 percent effectively than the traditional approach. The rest of the chapter (described as Section 2) elaborates Zernike Moments shape descriptor; Section 3 describes the proposed discriminative feature selection classifier; Section 4 provides similarity measurement. Detailed experiments are given in Section 5, and Section 6 contains discussions and conclusions.

15.2 ZERNIKE MOMENTS DESCRIPTOR (ZMD)

15.2.1 Zernike Moments (ZMs)

The set of orthogonal *ZMs* for an image intensity function $f(r, \theta)$ with order p and repetition q are defined as [11]:

$$Z_{pq} = \frac{p+1}{\pi} \int_0^{2\pi} \int_0^1 f(r, \theta) V_{pq}^*(r, \theta) r dr d\theta \tag{15.1}$$

where $p \geq 0, \ 0 \leq |q| \leq p, p - |q| = even, \ j = \sqrt{-1}, \ and \ \theta = \tan^{-1}(y/x)$ and $V_{pq}^*(r, \theta)$ is the complex conjugate of the Zernike polynomials $V_{pq}(r, \theta)$, given as

$$V_{pq}(r, \theta) = R_{pq}(r)e^{jq\theta} \tag{15.2}$$

$R_{pq}(r)$ are radial polynomials defined by

$$R_{pq}(r) = \sum_{k=0}^{(p-|q|)/2} (-1)^k \frac{(p-k)!}{k!\left(\dfrac{p+|q|}{2}-k\right)!\left(\dfrac{p-|q|}{2}-k\right)!} r^{p-2k} \tag{15.3}$$

To make ZMs translation and scale invariant the discrete image function is mapped on to unit disc. The set of Zernike polynomials need to be approximated by sampling at fixed intervals when it is applied to discrete image space [31, 32, 33]. For an $N \times N$ discrete space image the Cartesian equivalent to Equation 15.1 is given as

$$Z_{pq} = \frac{p+1}{\pi} \sum_{i=0}^{N-1} \sum_{j=0}^{N-1} f(x_i, y_j) V_{pq}^*(x_i, y_j) \Delta x_i \Delta y_i, \quad x_i^2 + y_j^2 \leq 1 \tag{15.4}$$

The coordinates (x_i, y_j) in a unit disc are given by:

$$x_i = \frac{2i+1-N}{D}, y_j = \frac{2j+1-N}{D}, i, j = 0, 1, 2, \ldots, N-1 \tag{15.5}$$

where D is the digital diameter of the inner or outer circle.

$$D = \begin{cases} N & \text{for inner disc contained in the square image} \\ N\sqrt{2} & \text{for outer disc containing the whole square image} \end{cases} \tag{15.6}$$

$$\Delta x_i = \Delta y_j = \frac{2}{D} \tag{15.7}$$

15.2.2 Orthogonality

ZMs are orthogonal and their orthogonal property makes image reconstruction or inverse transform process easier due to the individual contribution of each order moment to the reconstruction process. The orthogonal properties of Zernike polynomials and radial polynomials are given by Equations 15.8 and 15.9 respectively.

$$\int_0^{2\pi} \int_0^1 V_{pq}(r, \theta) V_{p'q'}^*(r, \theta) r dr d\theta = \frac{\pi}{p+1} \delta_{pp'} \delta_{qq'} \tag{15.8}$$

$$\int_0^1 R_{pq}(r)R_{p'q}(r)r\,dr = \frac{1}{2(p+1)}\delta_{pp'} \tag{15.9}$$

where δ_{ij} is Kronecker delta.

15.2.3 Rotation invariance

The set of *ZMs* inherently possess a rotation invariance property. The magnitude values of *ZMs* remain similar before and after rotation. Therefore the magnitude values of *ZMs* are rotation invariant. *ZMs* of an image rotated by an angle φ are defined as

$$Z'_{pq} = Z_{pq}e^{-jq\varphi} \tag{15.10}$$

where Z'_{pq} are *ZMs* of rotated image and Z_{pq} are *ZMs* of original image. The rotation invariant *ZMs* are extracted by considering only magnitude values as

$$\left|Z'_{pq}\right| = \left|Z_{pq}e^{-jq\varphi}\right| \tag{15.11}$$

$$\left|e^{-jq\varphi}\right| = \left|\cos(q\varphi) + j\sin(q\varphi)\right| = 1 \tag{15.12}$$

Substituting Equation 15.12 in Equation 15.11

$$\left|Z'_{pq}\right| = \left|Z_{pq}\right| \tag{15.13}$$

As $Z^*_{pq} = Z_{p,-q}$ and $\left|Z_{pq}\right| = \left|Z_{p,-q}\right|$, therefore only magnitudes of *ZMs* with $q \geq 0$ are considered [33].

15.2.4 Features selection

One of the important tasks of CBIR is to select the appropriate number of features that can efficiently describe an image. $Z_{0,0}$ and $Z_{1,1}$ moment features are excluded from the features set as $Z_{0,0}$ signifies average gray value of image, and $Z_{1,1}$ is the first order moment which is zero if the centroid of the image falls on the center of the disc. Numbers of features from $p_{max} = 2$ through $p_{max} = 15$ are presented in Table 15.1. For obtaining the optimal solution to our approach we have experimented at $p_{max} = 10, 12, 13, 14$ and 15. By applying the proposed classifier we select 34 features (equivalent to

Table 15.1 Number of Zernike moments from $p_{max} = 2$ through $p_{max} = 15$.

p_{max}	Moments	Features	Features Set
2	$Z_{2,0}\ Z_{2,2}$	2	2
3	$Z_{3,1}\ Z_{3,3}$	2	4
4	$Z_{4,0}\ Z_{4,2}\ Z_{4,4}$	3	7
5	$Z_{5,1}\ Z_{5,3}\ Z_{5,5}$	3	10
6	$Z_{6,0}\ Z_{6,2}\ Z_{6,4}\ Z_{6,6}$	4	14
7	$Z_{7,1}\ Z_{7,3}\ Z_{7,5}\ Z_{7,7}$	4	18
8	$Z_{8,0}\ Z_{8,2}\ Z_{8,4}\ Z_{8,6}\ Z_{8,8}$	5	23
9	$Z_{9,1}\ Z_{9,3}\ Z_{9,5}\ Z_{9,7}\ Z_{9,9}$	5	28
10	$Z_{10,0}\ Z_{10,2}\ Z_{10,4}\ Z_{10,6}\ Z_{10,8}\ Z_{10,10}$	6	34
11	$Z_{11,1}\ Z_{11,3}\ Z_{11,5}\ Z_{11,7}\ Z_{11,9}\ Z_{11,11}$	6	40
12	$Z_{12,0}\ Z_{12,2}\ Z_{12,4}\ Z_{12,6}\ Z_{12,8}\ Z_{12,10}\ Z_{12,12}$	7	47
13	$Z_{13,1}\ Z_{13,3}\ Z_{13,5}\ Z_{13,7}\ Z_{13,9}\ Z_{13,11}\ Z_{13,13}$	7	54
14	$Z_{14,0}\ Z_{14,2}\ Z_{14,4}\ Z_{14,6}\ Z_{14,8}\ Z_{14,10}\ Z_{14,12}\ Z_{14,14}$	8	62
15	$Z_{15,1}\ Z_{15,3}\ Z_{15,5}\ Z_{15,7}\ Z_{15,9}\ Z_{15,11}\ Z_{15,13}\ Z_{15,15}$	8	70

$p_{max} = 10$) with higher discrimination power for each mentioned p_{max} and analyzed that retrieval accuracy increases up to $p_{max} = 12$ then subsequently diminishes. The recognition rate at various moments' orders is depicted in Figure 15.1. However, as the moment's order increases, time complexity also increases. To acquire the best possible solution to our approach we choose $p_{max} = 12$ and by applying the proposed classifier we select merely 34 features with higher discrimination power and without tormenting system speed.

15.3 DISCRIMINATIVE FEATURES SELECTION

Discriminate coefficients (DCs) are computed by dividing the between class variances to the within class variances, as large variation occurs between classes and small variation occurs within classes. The complete procedure for applying the discriminating classification is as follows:

1. Compute Zernike Moments up to a specified maximum order for all the images in the database, for example $p_{max} = 12$.
2. Create the train set matrix T_{pq} by choosing ZMs coefficients of order p and repetition q for all classes C and all samples S as follows:

$$T_{pq} = \begin{bmatrix} Z_{pq}(1,1) & Z_{pq}(1,2) & \cdots & Z_{pq}(1,C) \\ Z_{pq}(2,1) & Z_{pq}(2,2) & \cdots & Z_{pq}(2,C) \\ \vdots & \vdots & \vdots & \vdots \\ Z_{pq}(S,1) & Z_{pq}(S,1) & \cdots & Z_{pq}(S,C) \end{bmatrix}_{S \times C}$$
$$(15.14)$$

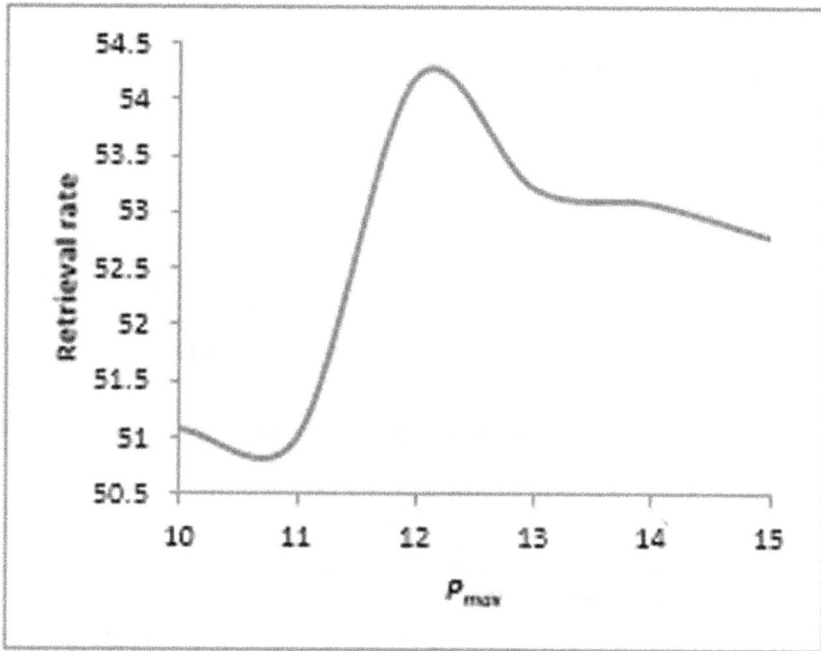

Figure 15.1 Retrieval rate of proposed classifier at p_{max} = 10,11,12,13,14 and 15.

3. Compute mean value of each class

$$M_{pq}^c = \frac{1}{S}\sum_{s=1}^{S} Z_{pq}(s,c), \quad c = 1,2,...,C \tag{15.15}$$

4. Compute variance of each class

$$Var_{pq}^c = \sum_{s=1}^{S}\left(Z_{pq}(s,c) - M_{pq}^c\right)^2, \quad c = 1,2,...C \tag{15.16}$$

5. Average the variance of all the classes

$$Var_{pq}^w = \frac{1}{C}\sum_{c=1}^{C} Var_{pq}^c \tag{15.17}$$

6. Compute the mean of all training samples

$$M_{pq} = \frac{1}{S \times C} \sum_{c=1}^{C} \sum_{s=1}^{S} Z_{pq}(s,c) \qquad (15.18)$$

7. Compute the variance of all the training samples

$$Var_{pq}^{B} = \sum_{c=1}^{C} \sum_{s=1}^{S} \left(Z_{pq}(s,c) - M_{pq} \right)^2 \qquad (15.19)$$

8. Determine the discriminate coefficients (DC) for location (p,q)

$$D(p,q) = \frac{Var_{pq}^{B}}{V_{pq}^{w}}, p \geq 0, 0 \leq |q| \leq p, p - |q| = even \qquad (15.20)$$

9. Arrange $D(p,q)$ values in ascending order. Discriminate coefficients with high values represent more discrimination power and are appropriate for selection.
10. Select the top 34 DC out of 47 and mark their corresponding moment locations. Create a mask array by setting the top 34 DC as value 1 and set remaining DC to 0 values.
11. Multiply the mask array with the ZM coefficients for preserving coefficients with higher discrimination power while discarding less significant coefficients.

15.4 SIMILARITY MEASURE

To compute the similarity of test image to the archived images a suitable similarity metric is required. The database image with the smallest distance to the test image is termed as the most similar image. We apply Euclidean distance similarity measure to evaluate the resemblance of test and training images, given as:

$$d(T,D) = \sqrt{\sum_{k-0}^{M-1} (f_k(T) - f_k(D))^2} \qquad (15.21)$$

Where $f_k(D)$ and $f_k(D)$ represent the kth feature of test image and database image respectively. M represents the total number of features to be

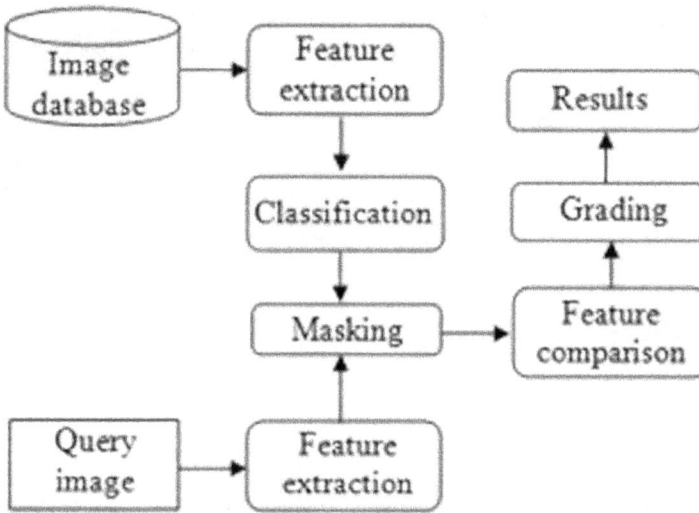

Figure 15.2 Overall process of proposed image retrieval system.

compared. The complete process of proposed image retrieval system is presented in Figure 15.2.

15.5 EXPERIMENTAL STUDY

A detailed experimental analysis is performed to evaluate the performance of the proposed approach to the image retrieval system. The comparison is carried out among three techniques: the proposed approach, traditional Zernike Moments descriptor without classification, and wavelet moments (WM) [35]. Experiments are executed on an Intel Pentium core 2 duo 2.10 GHz processor with 3 GB RAM. Algorithms are implemented in VC++ 9.0.

15.5.1 Experiment setup

(a) *Subject database*: MPEG-7 CE Shape 1 Part-B is a standard image set which contains 1,400 images of 70 classes with 20 samples in each class. Two images from each class are arbitrarily chosen for a test set and the rest of the images are located in the train set. Therefore 140 images are used as queries. All images are resized to 96×96 pixels. Figure 15.3 refers to sample images from MPEG-7 subject database.

(b) *Rotation database:* 70 images from each class of MPEG-7 CE Shape 1 Part-B database are selected and each of them is rotated at angles

Figure 15.3 Sample image from subject database.

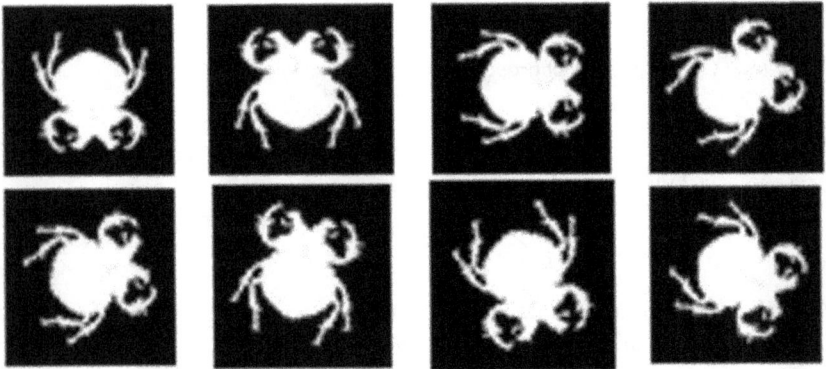

Figure 15.4 Sample images from rotation database.

of $0°, 15°, 30°, 45°, 60°, 75°, 90°, 105°, 105°, 120°, 135°, 150°, 1650°$ and $180°$ thereby creating 13 samples of each class. The database contains 910 images, 70 classes of 13 samples each. Sample images from rotation database are depicted in Figure 15.4. Two images from each class are chosen irrationally as query images. Thus the test set contains 140 query images.

15.5.2 Performance measurement

Precision (P) and recall (R) are used to estimate the image retrieval performance of system. Precision and recall are inversely proportional to each other as precision augments recall reduces. Precision measures the retrieval accuracy and recall measures the ability to retrieve relevant images from the database. Mathematically precision and recall are expressed as

$$p = \frac{n_q}{N_q} \times 100, R = \frac{n_q}{D_q} \times 100 \tag{15.22}$$

where n_q represents the number of similar images retrieved from the database. Nq represents total number of images retrieved. D_q represents number of images in database similar to query image q.

15.5.3 Experiment results

Initially we have analyzed the performance of proposed classifier on subject database by querying 140 images from the test set at $p_{max} = 12$ and selecting 34 higher discriminating coefficients. Table 15.2 presents selected coefficients of ZMs with higher discriminating power. The traditional Zernike Moments descriptor at $p_{max} = 10$ (ZMD_{10}) has 34 features without classification, Wavelet moments with 126 features are used. Their performance is measured through $P - R$ graph as shown in Figure 15.5 which signifies that the retrieval rate of proposed ZM is higher than that of traditional (ZMD_{10}),followed by wavelet moments. The retrieval accuracy of Proposed ZM, ZMD_{10} and wavelet moments are 54.21 percent, 51.07 percent and 48.79 percent respectively. Thus it is evident that retrieval rate of proposed ZM increased more than 3 percent than the traditional ZMD_{10} by using the same number of features. Average preci sion and recall for top 40 retrievals are depicted in Figure 15.6(a) and Figure 15.6(b) which also represents the similar trend of accuracy. Top 10 retrieval results corresponding to a

Table 15.2 Features with higher discrimination power for $p_{max} = 12$ (subject database)

p_{max}	Moments	Features	Features Set
2	$Z_{2,0}\ Z_{2,2}$	2	2
3	$Z_{3,1}\ Z_{3,3}$	2	4
4	$Z_{4,0}\ Z_{4,2}\ Z_{4,4}$	3	7
5	$Z_{5,1}\ Z_{5,3}\ Z_{5,5}$	3	10
6	$Z_{6,2}\ Z_{6,4}\ Z_{6,6}$	3	13
7	$Z_{7,3}\ Z_{7,5}\ Z_{7,7}$	3	16
8	$Z_{8,0}\ Z_{8,2}\ Z_{8,4}\ Z_{8,6}\ Z_{8,8}$	5	21
9	$Z_{9,7}\ Z_{9,9}$	2	23
10	$Z_{10,2}\ Z_{10,6}\ Z_{10,8}\ Z_{10,10}$	4	27
11	$Z_{11,7}\ Z_{11,9}\ Z_{11,11}$	3	30
12	$Z_{12,2}\ Z_{12,4}\ Z_{12,6}\ Z_{12,12}$	4	34

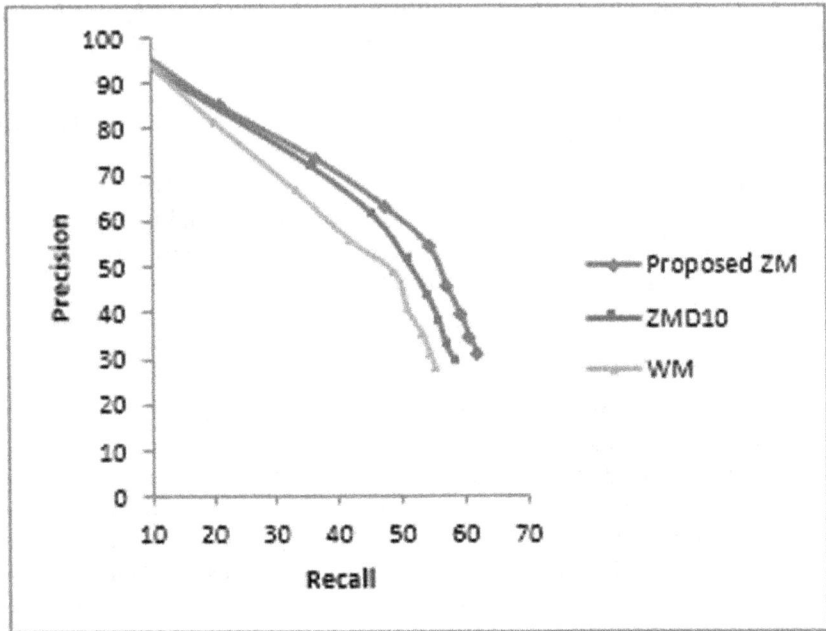

Figure 15.5 Precision and recall performance on MPEG subject database for Proposed ZM, ZMD_{10}, and WM.

query image for employed three methods are shown in Figure 15.7 along with the number of images retrieved by each method. Wavelet moments have the poor performance by retrieving merely two similar images. While comparing retrieval performance of proposed ZM and ZMD_{10}, we see that proposed ZM retrieves consecutive 3 similar images, whereas in case of ZMD_{10} and WM variation occurs at 3rd image.

Another set of experiments are performed on rotation database in which 140 images are passed as queries to the proposed method, traditional ZMD_{10} and WM. Selected coefficients with higher discrimination power for $p_{max} = 12$ are presented in Table 15.3. $P - R$ graph representing the behavior of applied methods is shown in Figure 15.8, which gives the evidence that proposed ZM, traditional ZMD_{10} and WM are giving best performance as being rotation invariant and their graph is overlapped. Average precision and average recall for rotation database are given in Figure 15.9(a) and Figure 15.9(b) respectively. The top 10 retrievals on rotation database are displayed in Figure 15.10, which demonstrates that the proposed ZM, ZMD_{10} and WM are retrieving 10 similar images from the database.

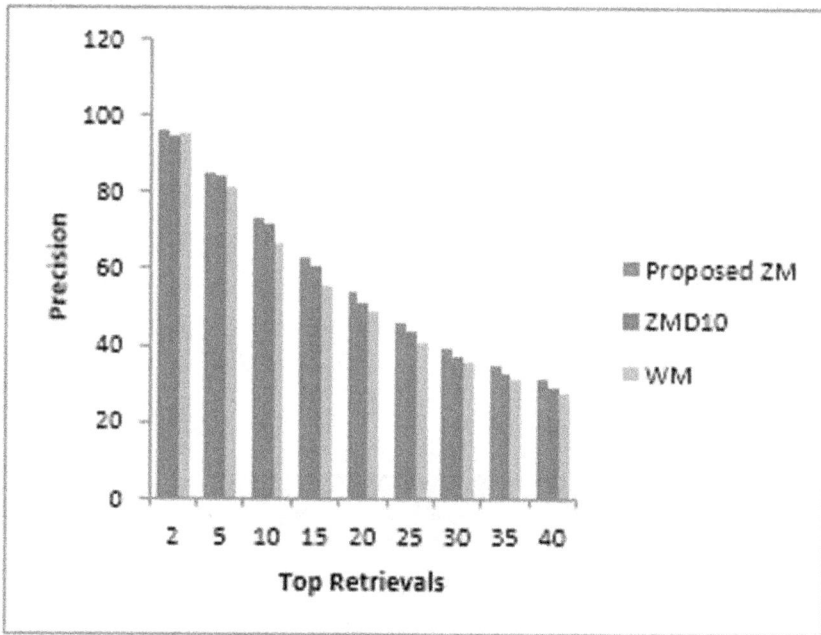

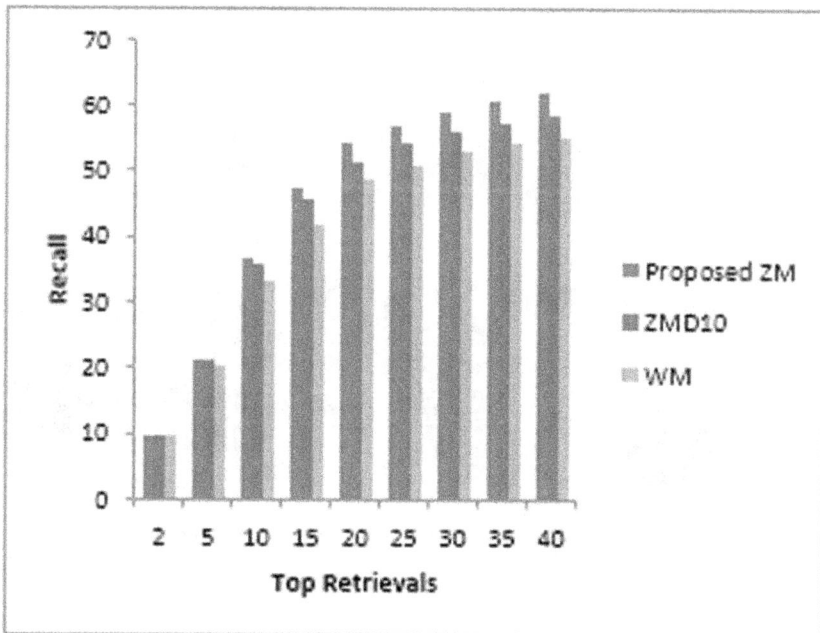

Figure 15.6 Top 40 retrievals from subject database (a) Average precision (b) Average recall.

Table 15.3 Features with higher discrimination power for $p$max = 12 (rotation database)

p_{max}	Moments	Features	Features Set
2	$Z_{2,0} \, Z_{2,2}$	2	2
3	$Z_{3,1} \, Z_{3,3}$	2	4
4	$Z_{4,0} \, Z_{4,2} \, Z_{4,4}$	3	7
5	$Z_{5,1} \, Z_{5,3}$	2	9
6	$Z_{6,0} \, Z_{6,2} \, Z_{6,4} \, Z_{6,6}$	4	13
7	$Z_{7,1} \, Z_{7,3} \, Z_{7,7}$	3	16
8	$Z_{8,0} \, Z_{8,2} \, Z_{8,4} \, Z_{8,6} \, Z_{8,8}$	5	21
9	$Z_{9,3} \, Z_{9,7}$	2	23
10	$Z_{10,0} \, Z_{10,2} \, Z_{10,4} \, Z_{10,6} \, Z_{10,8}$	5	28
11	$Z_{11,7}$	1	29
12	$Z_{12,0} \, Z_{12,2} \, Z_{12,4} \, Z_{12,6} \, Z_{12,8}$	5	34

Figure 15.7 (a) Query image (b) Top 10 retrieval results by proposed ZM, ZMD$_{10}$, and WM (subject database).

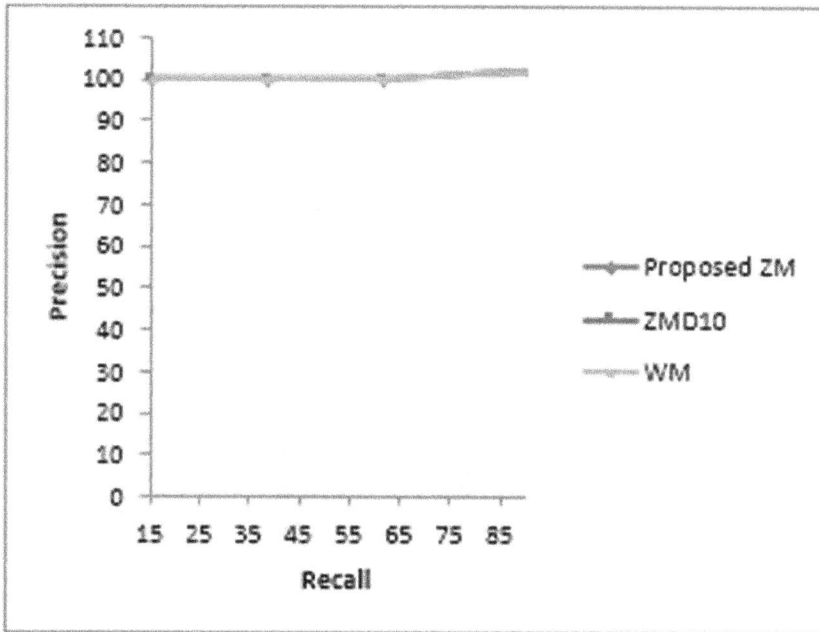

Figure 15.8 Precision and recall performance on MPEG subject database for proposed ZM, ZMD_{10} and WM.

15.6 DISCUSSIONS AND CONCLUSIONS

Through the performance observed in the experimental section, we see that proposed ZM on the subject database performs superior to conventional ZM. Moreover, ZMs are intrinsically rotation invariant and perform fairly well on every angle of rotation, and confer 100 percent accuracy. On the other hand wavelet moments are also rotation invariant and provide 100 percent accuracy on rotation database, nevertheless their performance reduces on subject database and gives only 48.79 percent accuracy. Retrieval accuracy of proposed ZM increases by 3 percent than traditional ZMD by using a similar number of features in both.

In future work we will strive to improve or propose a new classifier that can augment the recognition rate. More moment invariants will also be studied in order to improve the accuracy of the image retrieval system.

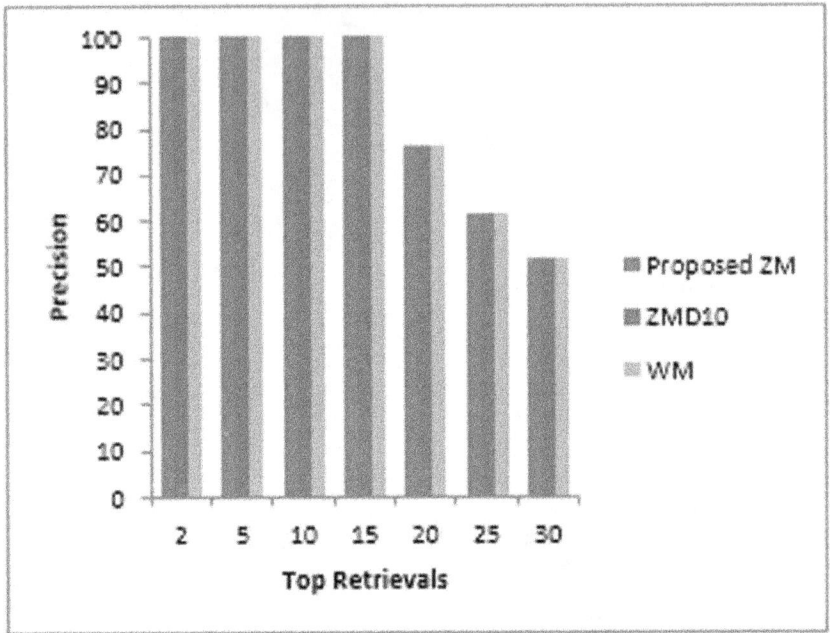

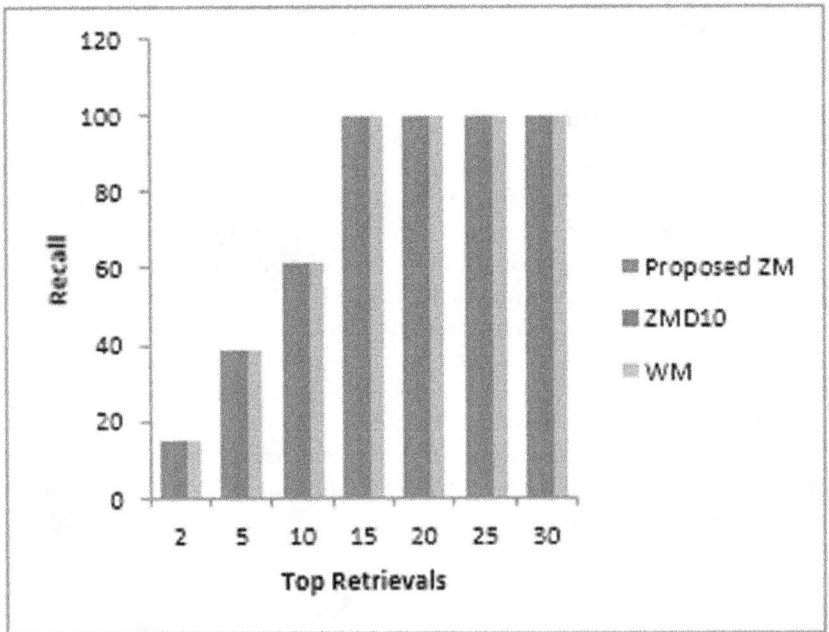

Figure 15.9 Top 30 retrievals from rotation database (a) Average precision (b) Average recall.

Figure 15.10 (a) Query Image (b) Top 10 retrieval results by proposed ZM, ZMD$_{10}$ and WM (rotation database).

REFERENCES

[1] Li, X., Yang, J., Ma, J. (2021). Recent developments of content-based image retrieval (CBIR). Neurocomputing 452, 675–689.

[2] Zhu, X., Wang, H., Liu P., Yang, Z., Qian, J. (2021). Graph-based reasoning attention pooling with curriculum design for content-based image retrieval. *Image and Vision Computing*, 115, 104289.

[3] Surendranadh, J., Srinivasa Rao, Ch. (2020). Exponential Fourier Moment-Based CBIR System: A Comparative Study. *Microelectronics, Electromagnetics and Telecommunications*, 757–767.

[4] Alaeia, F., Alaei, A., Pal, U., Blumensteind, M. (2019). A comparative study of different texture features for document image retrieval. *Expert Systems with Applications*, 121, 97–114.

[5] Datta, R., Joshi, D., Li, J., Wang, J. Z. (2008). Image retrieval: ideas, influences, and trends of the new age. *ACM Computing Surveys*, 40 (2), 1–60.

[6] Hu, N., An-An, H., Liu, et al. (2022). Collaborative Distribution Alignment for 2D image-based 3D shape retrieval. *Journal of Visual Communication and Image Representation*, 83, 103426.

[7] Li, H., Su, Z., Li, N., Liu, X. (2020). Non-rigid 3D shape retrieval based on multi-scale graphical image and joint Bayesian. *Computer Aided Geometric Design*, 81, 101910.

[8] Iqbal, K., Odetayo, M., O., James, A. (2002). Content-based image retrieval approach for biometric security using colour, texture and shape features controlled by fuzzy heuristics. *Journal of Computer and System Sciences*, 35 (1), 55–67.

[9] Wang, Y. (2003). Image indexing and similarity retrieval based on spatial relationship model. *Information Sciences*, 154 (1–2), 39–58.

[10] Zhou, X., Huang, T. (2002). Relevance feedback in content-based image retrieval: some recent advances. *Information Sciences*, 148 (1–4), 129–137.

[11] Teague M. R. (1980). Image analysis via the general theory of moments. *Journal of Optical Society of America*. 70, 920–930.

[12] Pawlak, M. (1992). On the reconstruction aspect of moment descriptors. *IEEE Transactions on Information Theory*, 38 (6), 1698–1708.

[13] Liao, S.X., Pawlak, M. (1996). On image analysis by moments. *IEEE Transactions on Pattern Analysis and Machine Intelligence*, 18 (3), 254–266.

[14] Belkasim, S.,O., Shridhar, M., Ahmadi, M. (1991). Pattern recognition with moment invariants: a comparative study and new results. *Pattern Recognition*, 24 (12), 1117–1138.

[15] Mukundan R., and Ramakrishnan, K.R. (1998). *Moment Functions in Image Analysis*. World Scientific Publishing, Singapore.

[16] Teh, C., H., Chin, R.,T. (1988). On image analysis by the methods of moments, *IEEE Transactions on Pattern Analysis and Machine Intelligence*. 10 (4), 496–512.

[17] Li, S., Lee, M., C., Pun, C., M. (2009). Complex Zernike moments features for shape-based image retrieval. *IEEE Transactions on Systems, Man, and Cybernetics*. 39 (1) 227–237.

[18] Kim, H., K., Kim, J., D., Sim, D., G,. Oh, D. (2000). A modified Zernike moment based shape descriptor invariant to translation, rotation and scale for similarity based image retrieval. IEEE Int. Conf. on Multimedia and Expo, 1307–1310.

[19] Kumar, Y., Aggarwal, A., Tiwari, S., Singh, K. (2018). An efficient and robust approach for biomedical image retrieval using Zernike moments. *Biomedical Signal Processing and Control*. 39, 459–473.

[20] An-Wen, D., Chih-Ying, G. (2018). Efficient computations for generalized Zernike moments and image recovery. *Applied Mathematics and Computation*. 339. 308–322.

[21] Vargas-Varga, H., JoséSáez-Landetea, et al. (2022). Validation of solid mechanics models using modern computation techniques of Zernike moments. *Mechanical Systems and Signal Processing*. 173, 109019.

[22] Kim, H., Kim, J . (2000). Region-based shape descriptor invariant to rotation, scale and translation. *Signal Processing: Image Communication*. 16, 87–93.

[23] Wei, C. –H., Li, Y., Chau W. –Y., Li, C. –T. (2009). Trademark image retrieval using synthetic features for describing global shape and interior structure. *Pattern Recognition*. 42, 386–394.

[24] Su, Z., Zhang H., Li, S., Ma, S. (2003). Relevance feedback in content based image retrieval: Bayesian framework, feature subspaces, and progressive learning. *IEEE Transactions on Image Processing.* 12(8), 924–937.

[25] Park, S. -S., Seo, K., -K., and Jang, D. -S. (2005). Expert system based on artificial neural networks for content-based image retrieval. *Missing Values Imputation Techniques.* 29(3), 589–597.

[26] Pakkanen, J., Iivarinen J., Oja, E. (2004). The evolving tree – A novel self-organizing network for data analysis. *Neural Processing Letters.* 20(3), 199–211.

[27] Koskela, M., Laaksonen, J., Oja, E. (2004). Use of image subset features in image retrieval with self-organizing maps. *LNCS.* 3115, 508–516.

[28] Fournier, J., Cord, M., and Philipp-Foliguet, S. (2001). Back-propagation algorithm for relevance feedback in image retrieval. In IEEE International conference in image processing (ICIP'01) 1, 686–689.

[29] Kumar M., A., Gopal, M. (2009). Least squares twin support vector machines for pattern classification. *Expert systems with applications.* 36, 7535–7543.

[30] Seo, K. –K. (2007). An application of one class support vector machines in content-based image retrieval. *Expert systems with applications.* 33, 491–498.

[31] Wee, C. Y., Paramseran, R. (2007). On the computational aspects of Zernike moments. *Image and Vision Computing.* 25, 967–980.

[32] Xin, Y., Pawlak, M., Liao, S. (2007). Accurate calculation of moments in polar co-ordinates. IEEE Transactions on Image Processing. 16, 581–587.

[33] Singh C., Walia, E. (2009). Computation of Zernike Moments in improved polar configuration, *IET Journal of Image Processing.* 3, 217–227.

[34] Khotanzad, A., Hong, Y. H. (1990). Invariant image recognition by Zernike moments. *IEEE Transactions on Pattern Analysis and Machine Intelligence.* 12 (5), 489–497.

[35] Shen D., Ip, H. H. S. (1999). Discriminative wavelet shape descriptors for recognition of 2-D patterns. Pattern Recognition. 32, 151–165.

Chapter 16

Corrected components of Zernike Moments for improved content based image retrieval

A comprehensive study

Pooja Sharma

16.1 INTRODUCTION

Digital images are a convenient medium for describing information contained in a variety of domains such as medical images in medical diagnosis, architectural designs, trademark logos, finger prints, military systems, geographical images, satellite/aerial images in remote sensing, and so forth. A typical database may consist of hundreds of thousands of images. Therefore, an efficient and automatic approach is required for indexing and retrieving images from large databases. Traditionally, image annotations and labeling with keywords heavily rely on manual labor. The keywords are inherently subjective and not unique. As the size of the image database grows, the use of keywords becomes cumbersome and inadequate to represent the image content [1,2]. Hence, content based image retrieval (CBIR) has drawn substantial attention during the last decade. CBIR usually indexes images with low level visual features such as color, texture and shape. The extraction of good visual features, which compactly represent the image, is one of the important tasks in CBIR. A color histogram is the most widely used color descriptor in CBIR; while colors are easy to compute, they represent large feature vectors that are difficult to index and have high search and retrieval costs [3]. Texture features do not provide semantic information [4]. Shape is considered a very important visual feature in object recognition and retrieval system, since shape features are associated with a particular object in an image [5,6]. A good shape representation should be compact and retain the essential characteristics of the image. Moreover, invariance to rotation, scale, and translation is required because such transforms are consistent with human perception. A good method should also deal with photometric transformations such as noise, blur, distortion, partial occlusion, JPEG compression, and so forth.

Various shape representations and description techniques have been proposed during the last decade [7]. In shape description, features are generally classified into two types: the region based descriptors and the contour based descriptors. In region based descriptors, features are extracted

DOI: 10.1201/9781003453406-16

from the interior of the shape and represent the global aspect of the image. The region based descriptors include geometric moments [15], moment invariants (MI) [16], a generic Fourier descriptor (GFD) [17], Zernike Moments descriptors (ZMD) [18]. In the contour based descriptors, features are extracted from the shape boundary points only. The contour based descriptors include Fourier descriptors (FD) [8], curvature scale space [9], contour flexibility [10], shape context [11], histograms of centroid distance [12], contour point distribution histograms (CPDH) [13], Histograms of Spatially Distributed Points, and Angular Radial Transform [33], Weber's local descriptors (WLD) [14], and so forth. Both the region and contour based methods are complimentary to each other as one method provides the global characteristics, while the other provides the local change in an image. Therefore, we exploit both local and global features of images to propose a novel and improved approach to an effective image retrieval system. Teague introduced [19] the notion of orthogonal moments to recover the image from moments based on the theory of orthogonal polynomials using Zernike Moments (ZMs), which are capable of reconstructing an image and exhibit minimum information redundancy. The magnitudes of ZMs have been used as global features in many applications [20–30], due to their rotation invariant property. Since ZMs are inherently complex, therefore, the real and imaginary coefficients possess significant image representation and description capability. The phase coefficients are considered to be very effective during signal reconstruction as demonstrated by [28–31]. However, the phase coefficients are not rotationally invariant, which is illustrated as follows:

Let Z_{pq} and Z_{pq}^r be the ZMs of original and rotated images, respectively, with order p and repetition q, then the two moments are related by:

$$Z_{pq}^r = Z_{pq}e^{-jq\theta}, \tag{16.1}$$

where θ is the angle by which the original image is rotated and $j = \sqrt{-1}$. The phase relationship is given by

$$\psi_{pq}^r = \psi_{pq} - q\theta, \tag{16.2}$$

or

$$q\theta = (\psi_{pq} - \psi_{pq}^r)\bmod 2\pi, \tag{16.3}$$

where ψ_{pq} and ψ_{pq}^r are the phase coefficients of original and rotated images, respectively. Therefore, in our approach we use the relationship given by Equation 16.3 to compute $q\theta$ from the phase coefficients of original and rotated images at each order p and repetition q. The computed $q\theta$ is then used to correct real and imaginary coefficients of ZMs of rotated query image, thereby making them rotation invariant features. The proposed approach eliminates the step of estimating rotation angle between query and database image and then correcting phase coefficients, which is followed by [28–30]. Thus, in our approach we use improved corrected real and imaginary coefficients of ZMs as global image features.

We propose histograms of local image features, which are derived from lines detected using Hough Transform (HT) [32]. The histograms of centroid distances from boundary points have been used widely as a local image feature [33]. However, those histograms do not provide relationship among adjacent boundary points. Therefore, lines are the best methods to link and associate adjacent boundary points, which we perform using Hough Transform. The rest of the chapter, organized as Section 2, proposes the improved ZMs and HT based region and contour based descriptors, respectively. Section 3 discusses the similarity metrics for image matching. In Section 4, the proposed system is analyzed and evaluated by performing widespread experiments and compared against existing major descriptors. Discussion and conclusion are given in Section 5.

16.2 PROPOSED DESCRIPTORS

In this section, we first introduce the invariant method for global features extraction using ZMs, and later we propose histograms of centroid distances to linear edges as local features based on HT.

16.2.1 Invariant region based descriptor using corrected ZMs features

In the proposed approach, we use both real and imaginary components of ZMs individually rather than computing magnitude. In case of a query image, which is the rotated version of the database image, then its real and imaginary coefficients need to be corrected based on the phase coefficients of ZMs. In some previous approaches [28–30], the rotation angle is estimated in order to correct the phase of query image. However, in our proposed solution we eliminate the step of rotation angle estimation and directly correct the real and imaginary coefficients described as follows:

Let Z_{pq} and Z_{pq}^r be the ZMs of original and rotated images, respectively, with order p and repetition q, then the two moments are related by:

$$Z_{pq}^r = Z_{pq}e^{-iq\theta}, \tag{16.4}$$

where θ is the angle of rotation. Since the ZMs magnitudes are rotation invariant therefore, we have

$$\left| Z_{pq}^{r} \right| = \left| Z_{pq} e^{-jq\theta} \right| = \left| Z_{pq} \right|, \tag{16.5}$$

where $\left| Z_{pq}^{r} \right|$ and $\left| Z_{pq} \right|$ are the magnitudes of rotated and original images, respectively. However, the ZMs phase coefficients are not rotation invariant and are related by:

$$\psi_{pq}^{r} = \psi_{pq} - q\theta \tag{16.6}$$

or

$$q\theta = \psi_{pq} - \psi_{pq}^{r}, \tag{16.7}$$

$$\psi_{pq}^{r} = \tan^{-1}\left(\frac{I(Z_{pq}^{r})}{R(Z_{pq}^{r})}\right), \ \psi_{pq} = \tan^{-1}\left(\frac{I(Z_{pq})}{R(Z_{pq})}\right), \tag{16.8}$$

where ψ_{pq}^{r} and ψ_{pq} are the phase coefficients of rotated and original image respectively, and $I(\cdot)$ and $R(\cdot)$ are the real and imaginary coefficients of ZMs. Using Equation 16.4 let Z_{pq}^{c} be the corrected ZMs derived from the rotated version of ZMs as follows:

$$Z_{pq}^{c} = Z_{pq}^{r} e^{jq\theta}, \tag{16.9}$$

$$R(Z_{pq}^{c}) + jI(Z_{pq}^{c}) = \left(R(Z_{pq}^{r}) + jI(Z_{pq}^{r})\right) \times \left(\cos(q\theta) + j\sin(q\theta)\right), \tag{16.10}$$

using Equation 16.7 we get,

$$R(Z_{pq}^{c}) + jI(Z_{pq}^{c}) = \left(R(Z_{pq}^{r}) + jI(Z_{pq}^{r})\right) \times (\cos(\psi_{pq} - \psi_{pq}^{r}) + j\sin(\psi_{pq} - \psi_{pq}^{r})), \tag{16.11}$$

Let $\alpha = \psi_{pq} - \psi_{pq}^{r}$ then Equation 16.11 becomes

$$R(Z_{pq}^{c}) + jI(Z_{pq}^{c}) = \left(R(Z_{pq}^{r}) + jI(Z_{pq}^{r})\right) \times (\cos(\alpha) + j\sin(\alpha)), \tag{16.12}$$

or

$$\begin{aligned} R(Z_{pq}^{c}) &= R(Z_{pq}^{r})\cos(\alpha) - I(Z_{pq}^{r})\sin(\alpha) \\ I(Z_{pq}^{c}) &= R(Z_{pq}^{r})\sin(\alpha) + I(Z_{pq}^{r})\cos(\alpha) \end{aligned} \tag{16.13}$$

If the two images are similar but rotated by an angle θ, then the ZMs of the rotated images are modified according to the Equation 16.23, then we have

$$R(Z_{pq}) = R(Z_{pq}^c) \text{ and } I(Z_{pq}) = I(Z_{pq}^c) \qquad (16.14)$$

From the Equation 16.14 we obtain two corrected invariant real and imaginary coefficients of ZMs, which we use as global features for images in the proposed system. The real and imaginary coefficients are corrected by the phase difference between original and rotated images. Thus, in the proposed system we use real and imaginary coefficients of ZMs individually rather than using the single features set obtained through ZMs magnitudes. We consider four similar images rotated at different angles and four dissimilar images given in Figure 16.1 from MPEG-7 database, and evaluate the discrimination power of real and imaginary components. It is observed from Figures 16.2(a) and 16.2(c) that, while considering similar images rotated at different angles, no variation perceived in real and imaginary coefficients. On the other hand, significant variation occurs among real and imaginary coefficients in case of dissimilar images as depicted in Figures 16.2(b) and 16.2(d).

16.2.2 Selection of appropriate features

The selection of appropriate and optimal number of features is an important task for an effective image retrieval system. A small number of features do not provide satisfactory results, while the high number of features prone to "overtraining" and reduce the computation efficiency. In addition, higher order ZMs are numerically instable. Therefore, three moment orders 10, 12 and 16 are common in image retrieval applications. The selection of order $p_{max} = 12$ is a tradeoff between the computational complexity and the descriptor performance [29]. In the proposed solution, we take $p_{max} = 12$,

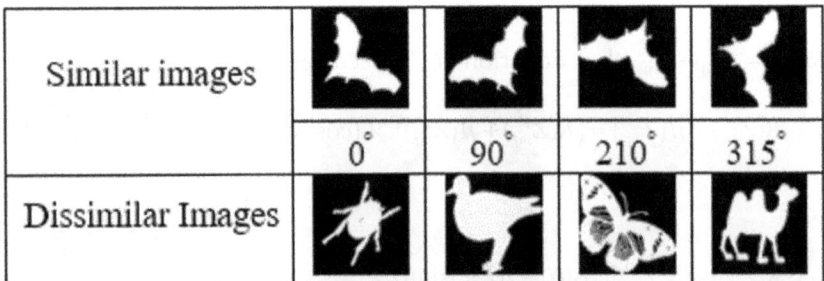

Similar images				
	$0°$	$90°$	$210°$	$315°$
Dissimilar Images				

Figure 16.1 (a) Similar images (b) dissimilar images.

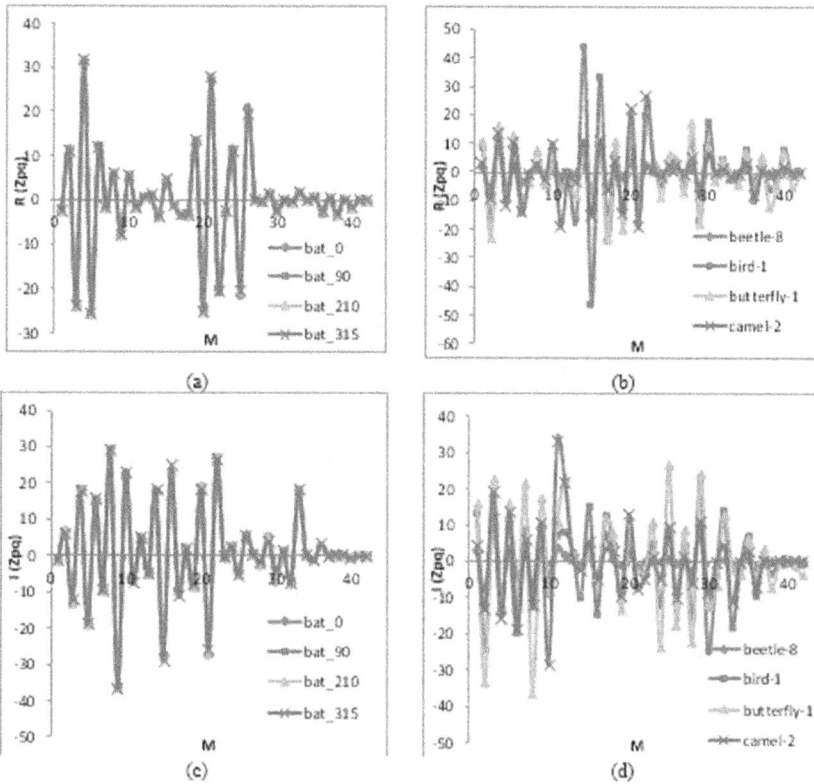

Figure 16.2 (a) ZMs real coefficients of similar images (b) ZMs real coefficients of dis-similar images (c) ZMs imaginary coefficients of similar images, and (d) ZMs imaginary coefficients of dissimilar images (*M* represents total number of moments used).

which generates 41 moments given by Table 16.1. The moments $Z_{0,0}$ and $Z_{1,1}$ are excluded from the features set as $Z_{0,0}$ signifies an average gray value of image and $Z_{1,1}$ is the first order moment, which is zero if the centroid of the image falls on the center of the disc.

Figure 16.2(a) ZMs real coefficients of similar images Figure 16.2(b) ZMs real coefficients of dissimilar images (c) ZMs imaginary coefficients of similar images, and (d) ZMs imaginary coefficients of dissimilar images (*M* represents total number of moments used).

16.2.3 Invariant contour based descriptor using HT

Edges characterize the gray scale discontinuities in an image. Detection of these discontinuities is an essential step in object recognition as edges

Table 16.1 Features set up to $p_{max} = 12$

p_{max}	ZMs	M (ZMs index)
2	$Z_{2,2}$	0
3	$Z_{3,1}\ Z_{3,3}$	1, 2
4	$Z_{4,2}\ Z_{4,4}$	3, 4
5	$Z_{5,1}\ Z_{5,3}\ Z_{5,5}$	5, 6, 7
6	$Z_{6,2}\ Z_{6,4}\ Z_{6,6}$	8, 9, 10
7	$Z_{7,1}\ Z_{7,3}\ Z_{7,5}\ Z_{7,7}$	11, 12, 13, 14
8	$Z_{8,2}\ Z_{8,4}\ Z_{8,6}\ Z_{8,8}$	15, 16, 17, 18
9	$Z_{9,1}\ Z_{9,3}\ Z_{9,5}\ Z_{9,7}\ Z_{9,9}$	19, 20, ..., 23
10	$Z_{10,2}\ Z_{10,4}\ Z_{10,6}\ Z_{10,8}\ Z_{10,10}$	24, 25, ..., 28
11	$Z_{11,1}\ Z_{11,3}\ Z_{11,5}\ Z_{11,7}\ Z_{11,9}\ Z_{11,11}$	29, 30, ..., 34
12	$Z_{12,2}\ Z_{12,4}\ Z_{12,6}\ Z_{12,8}\ Z_{12,10}\ Z_{12,12}$	35, 36, ..., 40

are one of the most effective features of an image. However, in general an image contains noise. Therefore, to eliminate spurious edges caused by noise, we use the Canny edge detector, which has also been adopted by [34, 35] for binary and gray scale images. The Canny edge detector is based on three basic objectives: (a) low error rate (b) localized edge points, and (c) single edge point response. Canny edge detector produces a binary edge map, that is, it only detects the edge points and does not link them in linear edges. The edge points are linked by many approaches, and HT is one of the best approaches to provide a linear edge map of the edge points. It links the edge points accurately while being robust against noise. If an image is rotated by an angle, the edges remain unaffected, although their orientation will change. The fundamentals of HT are already described in Section 3.2. Therefore, the complete procedure of local features extraction is described in [38]. Thus, we acquire 10 local features $\{p_i\}$, which are invariant to translation, rotation, and scale. The normalized histograms of four edge images of the "bell" class are presented in Figure 16.3(a), which depicts minor variations in features of instances of similar class, whereas Figure 16.3(b) displays a large variation among histograms of edge images of different classes ("apple", "device9", "dog" and "chicken").

16.3 SIMILARITY METRICS

In the above section, methods are proposed for the extraction of global and local features. An effective similarity measure is one that can preserve the discrimination powers of features and match them appropriately. In existing methods, the Euclidean Distance (ED) metric, also called L_2 norm, is used the most frequently. In ED, distances in each dimension are squared

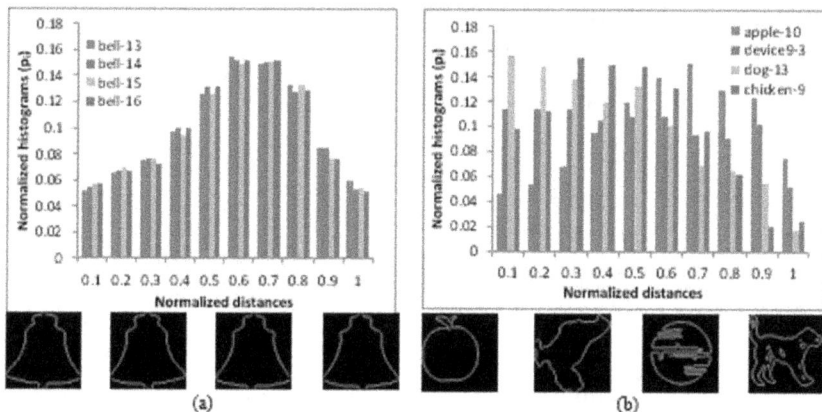

Figure 16.3 (a) Normalized histograms (p_i) for four images of similar class (b) Normalized histograms (p_i) for four images of dissimilar class.

before summation, which puts greater emphasis on those features for which the dissimilarity is large. Therefore, to overcome this issue we suggest city block (CB) distance, also called L_1 norm and Bray-Curtis (BC) also called Sorensen's distance metrics [37, 38]. The BC metric normalizes the feature values by dividing the summation of absolute differences of corresponding feature vectors by the summation of their absolute sums. We analyze the performance of the proposed system using the three distance metrics ED, CB, and BC, and analyze that the BC similarity metric outperforms rest of the metrics. The ED, CB and BC metrics for the proposed region based descriptor are given as:

$$d_r^{ED}(Q,D) = \sqrt{\sum_{j=0}^{M-1} \left[(R(Z_i^Q) - R(Z_i^D))^2 + (I(Z_i^Q) - I(Z_i^D))^2 \right]} \qquad (16.15)$$

$$d_r^{CB}(Q,D) = \sqrt{\sum_{j=0}^{M-1} \left[\left| R(Z_i^Q) - R(Z_i^D) \right| + \left| I(Z_i^Q) - I(Z_i^D) \right| \right]}, \qquad (16.16)$$

$$d_r^{BC}(Q, D) = \frac{\displaystyle\sum_{i=0}^{M-1} \left[\left| R(Z_i^Q) - R(Z_i^D) \right| + \left| I(Z_i^Q) - I(Z_i^D) \right| \right]}{\displaystyle\sum_{i=0}^{M-1} \left[\left| R(Z_i^Q) + R(Z_i^D) \right| + \left| I(Z_i^Q) + I(Z_i^D) \right| \right]}, \qquad (16.17)$$

where Z_i^Q and Z_i^P are the ZMs features of the query and database images, respectively and $M = 42$. The ED, CB and BC metrics for the proposed contour based descriptor are given as:

$$d_c^{ED}(Q,D) = \sqrt{\sum_{i=0}^{H-1}[f_i(Q) - f(D)]^2} \qquad (16.18)$$

$$d_c^{CB}(Q,D) = \sqrt{\sum_{i=0}^{H-1}\left[\left|f_i(Q) - f_i(D)\right|\right]}, \qquad (16.19)$$

$$d_c^{BC}(Q,D) = \frac{\displaystyle\sum_{i=0}^{H-1}\left|f_i(Q) - f_i(D)\right|}{\displaystyle\sum_{i=0}^{H-1}\left|f_i(Q) - f_i(D)\right|}, \qquad (16.20)$$

where $f_i(Q)$ and $f_i(D)$ represent the feature vectors of the query and database images, respectively, and H is the number of features, which is 10 for contour based features. Since we consider both global and local features to describe the shape, therefore, the above corresponding similarity metrics are combined to compute the overall similarity, given as:

$$d^{ED}(Q,D) = w_r d_r^{ED}(Q,D) + w_c d_c^{ED}(Q,D), \qquad (16.21)$$

$$d^{CB}(Q,D) = w_r d_r^{CB}(Q,D) + w_c d_c^{CB}(Q,D), \qquad (16.22)$$

$$d^{BC}(Q,D) = w_r d_r^{BC}(Q,D) + w_c d_c^{BC}(Q,D), \qquad (16.23)$$

where w_c and w_r represent the weight factors of the contour based and region based similarity measures, respectively. In our experiments, we assume that each feature component (contour and region) provides approximately equivalent contribution and hence we set, $w_c = w_r = 0.5$. For evaluating the classification performance of the above mentioned similarity metrics, we experiment on Kimia-99 and MPEG-7 shape databases to measure the retrieval accuracy of the proposed system, using $(P - R)$ curves and present the results in Figure 16.4. It is observed from Figure 16.4(a) and 16.4(b) that the BC similarity metric classifies more similar images because it normalizes the computed distance by the division of summation of absolute subtractions with summation of absolute additions of feature vectors. The performance of CB and ED are almost similar. However, CB

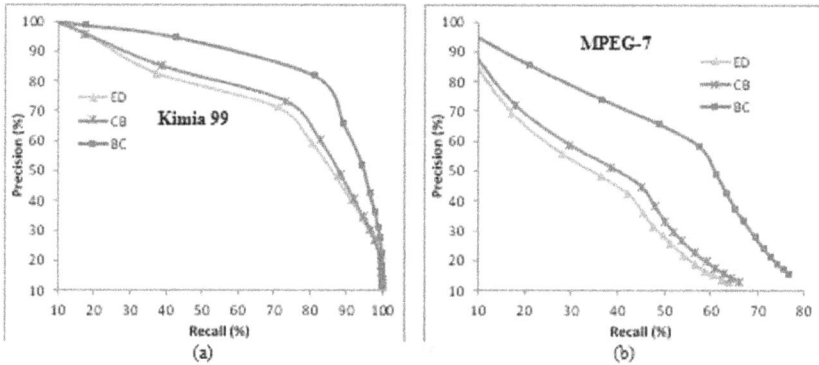

Figure 16.4 The performance of similarity metrics ED, CB, and BC for (a) Kimia-99 and (b) MPEG-7 shape databases.

is slightly superior to ED. Keeping this analysis in view, we use Bray-Curtis similarity metric in rest of the experiments.

16.4 EXPERIMENTAL STUDY AND PERFORMANCE EVALUATION

Our goal is to present the user a subset of most relevant images that are similar to query image. In our system, the global features are extracted using the corrected coefficients of real and imaginary coefficients of ZMs and local features are extracted using histograms of distances of centroid to linear edges, where linear edges are found using HT. The similarity value is obtained by combining the global and local features using Bray-Curtis similarity metric, which provides the normalized similarity values. The similarity values are sorted in increasing order and the images with closer similarity are presented to the user as most relevant images to query image. For performance evaluation of the proposed system, we compare it against three contour based descriptors such as FD [8], WLD [14], and CPDH [13], and three region based descriptors MI [16], ZMD [18] where ZMs magnitude is considered, and GFD [17]. Besides, we compare the proposed ZMs solution with complex Zernike Moments (CZM) [28], optimal similarity [30], and adjacent phase [29] approaches. We consider ten databases to review the system performance under various conditions, which include rotation, scale, translation, partial occlusion, noise, blur, JPEG compression, image type (binary and gray), 3D objects, and texture images, the detailed description of these databases can be found in [38] Kimia99, Brodatz, Columbia Object Image Library, MPEG-7 CE shape-1 Part B, Rotation, Scale, Translation, Noise, Blur, JPEG Compression.

16.4.1 Measurement of retrieval accuracy

The System retrieval accuracy is determined with respect to precision and recall, where these two components are computed as follows:

$$P = \frac{n_Q}{T_Q} \times 100, \ R = \frac{n_Q}{D_Q} \times 100, \tag{16.24}$$

where n_Q represents the number of similar images retrieved from the database, T_Q represents total number of images retrieved, D_Q represents number of images in database similar to query image Q. The system performance is also measured using Bull's Eye Performance (BEP).

16.4.2 Performance comparison and experiment results

In order to assess the system retrieval performance for each kind of image, all the images in the database are served as query. We analyze the system performance on all the ten databases, which represent images of various types. The retrieval accuracy is first presented by the $P - R$ curves and then by the average BEP. As mentioned earlier, we compare the proposed solution with ZMs phase based methods and other contour and region based methods. The results are given as follows:

(1) Kimia-99: The performance of the proposed system is examined for Kimia-99 database, which includes distorted and partial occluded shapes, and the results are given in Figure 16.5(a) for phase based methods, which show that CZM performs worse than the other methods. The proposed corrected ZMs overpower the other existing methods, followed by optimal similarity and adjacent phase methods. When both ZMs and HT based features are taken into account the performance of the proposed system is highly improved. The $P - R$ curves for comparison with contour and region based techniques are given in

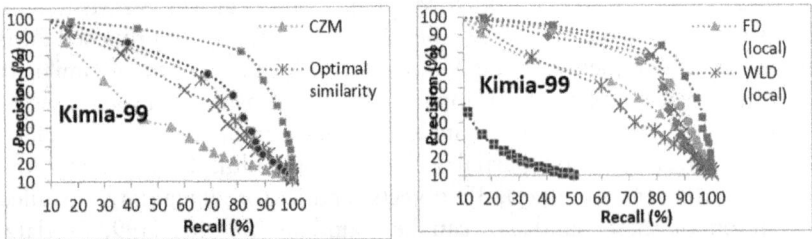

Figure 16.5 The performance comparison using $P - R$ curves for Kimia-99 database with (a) phase based methods and (b) contour and region based methods.

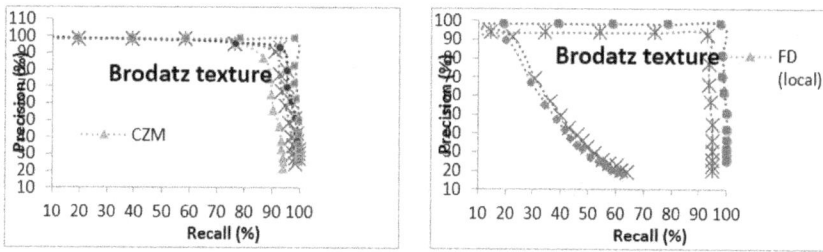

Figure 16.6 The performance comparison using P – R curves for Brodatz texture database with (a) phase based methods and (b) contour and region based methods.

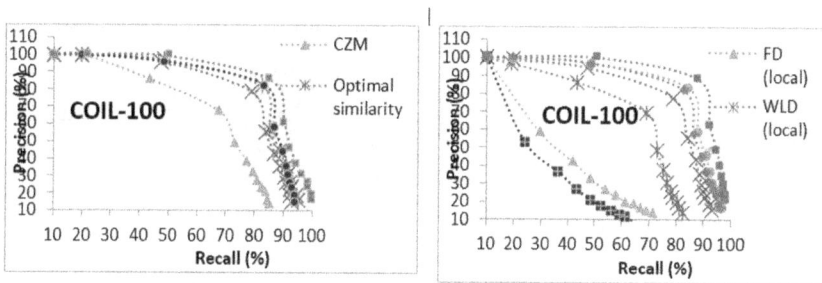

Figure 16.7 The performance comparison using P – R curves for COIL-100 database with (a) phase based methods and (b) contour and region based methods.

Figure 16.5(b), which depicts that the MI represents the overall worst performance. However, the proposed system outperforms all the region and contour based descriptors followed by CPDH and GFD with slight variation in their performance.

(2) Brodatz texture: The next set of experiments is performed over texture database and the results are given in Figure 16.6(a) for phase based methods. It is observed that proposed ZMs and optimal similarity methods have almost similar performance followed by adjacent phase and CZM methods. However, the proposed ZMs and HT based approach outperforms all the methods. The performance of MI and FD is lowest on the texture database as their P – R curves are not emerged in the graph as can be seen from Figure 16.6(b). The performances of GFD and CPDH are highly reduced. Nevertheless, the proposed system and ZMD represent very high performance and overlap with each other, followed by WLD.

(3) COIL-100: This database represents 3D images rotated at different angles and the performance for this database is given in Figure 16.7(a) for phase based approaches. The performance is similar as for

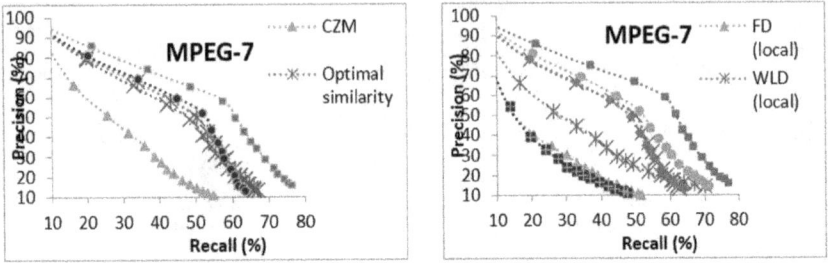

Figure 16.8 The performance comparison using P – R curves for MPEG-7 database with (a) phase based methods and (b) contour and region based methods.

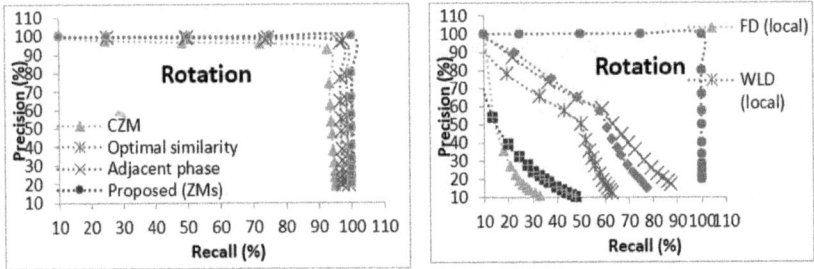

Figure 16.9 The performance comparison using P – R curves for rotation database with (a) phase based methods and (b) contour and region based methods.

Kimia-99 database. The proposed ZMs outperform the other phase based approaches and when both ZMs and HT based features are considered the system performance is improved further. While considering contour and region based approaches the proposed system overpowers the rest of the methods followed by ZMD, GFD, CPDH, WLD, FD, and MI as shown in Figure 16.7(b).

(4) MPEG-7: The performance for this database is given in Figure 16.8(a) for phase based methods. It is observed that the proposed ZMs method performs better than the optimal similarity method but its performance slightly reduced afterwards. Nevertheless, when both ZMs and HT based features are considered the performance of the proposed system is highly improved than other methods. The P – R curves for contour and region based descriptors are given in Figure 16.8(b). It is observed that region based methods ZMD and GFD perform better than WLD, FD, and MI. The performance of CPDH is comparable to GFD. But the proposed approach overcomes other methods and preserves its superiority.

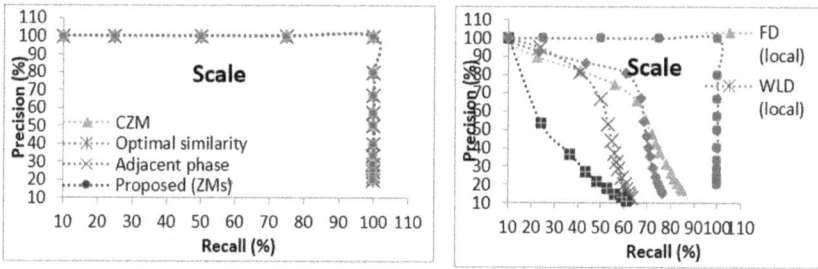

Figure 16.10 The performance comparison using *P − R* curves for scale database with (a) phase based methods and (b) contour and region based methods.

(5) Rotation: This database represents geometric transformation and the *P − R* curves for phase based approaches are given in Figure 16.9(a), which shows that all the methods give good results for rotation invariance. However, the retrieval accuracy of proposed approaches is almost 100 percent, followed by optimal similarity and adjacent phase methods. On the other hand, when region and contour based methods are considered, as shown is Figure 16.9(b), the performance of FD and MI is worse than all other methods and the ZMD and proposed methods represent high retrieval accuracy and overlapped with each other.

(6) Scale: This test is performed to examine the scale invariance of the proposed solution. The *P − R* curves are given in Figure 16.10(a), which shows that all the ZMs based methods represent a complete scale invariance including the proposed method and overlapped among each other. The scale invariance analysis for contour based methods represent that they are not scale invariant and WLD has the worst performance as the *P − R* curve is not emerge for this method as shown in Figure 16.10(b). In region based methods the ZMD overlaps with the proposed solution and gives 100 percent scale invariance followed by GFD and the MI represent poor performance.

(7) Translation: The translation invariance is presented in Figure 16.11(a) for phase based methods. We see that the proposed corrected ZMs based method is better than the other methods, whereas the retrieval accuracy is highly improved by considering both ZMs and HT features. The comparison with contour and region based methods is given in Figure 16.11(b), which demonstrates that the contour based methods represent high invariance to translation as compared to region based methods. The proposed solution performance is better than ZMD, GFD and MI but poorer than CPDH and WLD and comparable to FD.

(8) Noise: The next test is performed over a noise database to observe system robustness to noise, that is, photometric transformed images.

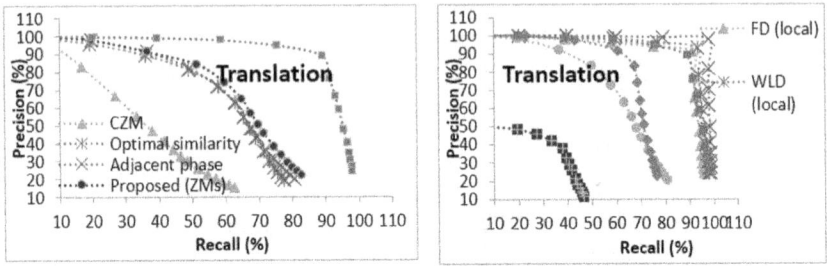

Figure 16.11 The performance comparison using $P - R$ curves for translation database with (a) phase based methods and (b) contour and region based methods.

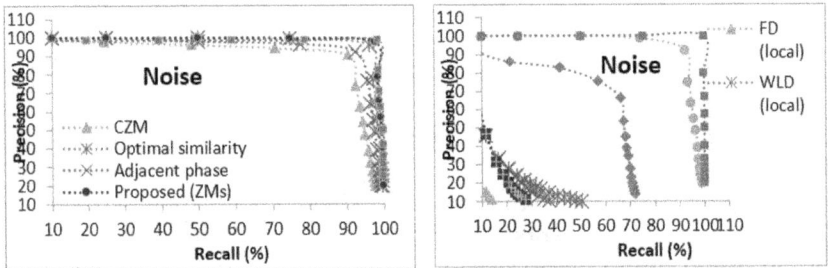

Figure 16.12 The performance comparison using $P - R$ curves for noise database with (a) phase based methods and (b) contour and region based methods.

The $P - R$ curves for phase based methods are given in Figure 16.12(a), which shows that the proposed solution is highly robust to noise followed by optimal similarity method. The CZM has the lowest performance among others. The Figure 16.12(b) represents $P - R$ curves for contour and region based descriptors, which shows that the region based descriptors are more robust to noise as compared to contour based descriptors. The proposed system represents 100 percent robustness to noise followed by ZMD and GFD. In contour based descriptors FD has the worst performance.

(9) Blur: Another photometric test is performed over blurred images. The results are given in Figure 16.13(a) for phase based methods, which represents that all the methods are highly robust to blur transformation, including the proposed approach. When contour and region based methods are considered, the proposed system still preserves its highest retrieval accuracy, followed by CPDH, FD, and WLD. The MI represent their worst performance over blur images, as can be seen from Figure 16.13(b).

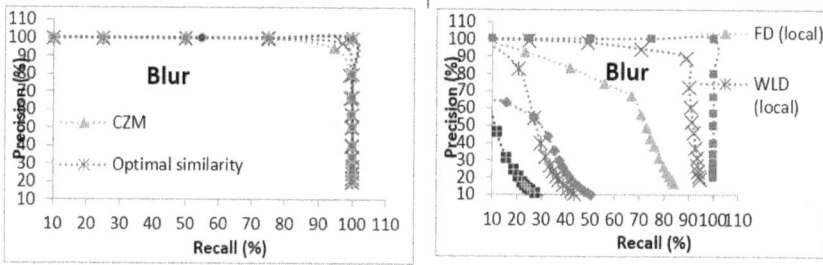

Figure 16.13 The performance comparison using P − R curves for blur database with (a) phase based methods and (b) contour and region based methods.

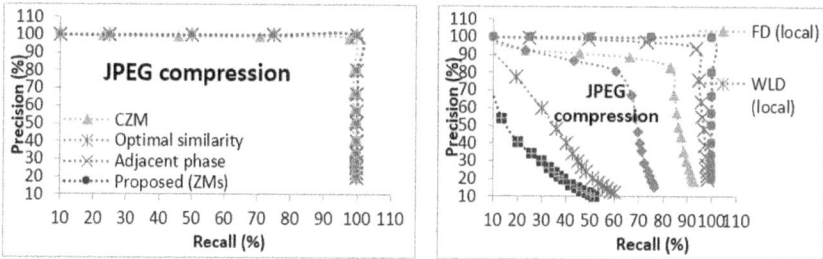

Figure 16.14 The performance comparison using P − R curves for JPEG-compressed database with (a) phase based methods and (b) contour and region based methods.

10) JPEG compression: The results of photometric transformed compressed images are given in Figure 16.14(a) for phase based methods, which also represent similar performance as that for blur database. The P − R curves for contour and region based methods are given in Figure 16.14(b), which shows that the proposed solution is highly robust to JPEG compressed images followed by CPDH. Whereas, MI has the worst performance.

The BEP in percentage for contour and region based methods is given in Table 16.2. It represents that the proposed method outperforms all other methods and gives an average retrieval accuracy of 95.49 percent. While comparing the average BEP with phase based approaches, the proposed ZMs based technique has higher retrieval rate than the CZM, adjacent phase, and optimal similarity methods as presented in Table 16.3.

Table 16.2 The comparison of average *BEP* of the proposed and other contour and region based methods

	FD	WLD	CPDH	MI	GFD	ZMD	Proposed (ZMs+HT)
Kimia-99	63.45	59.5	78.6	8.73	77.5	74.65	99.54
Brodatz texture	9.33	93.07	43.13	7.3	41.43	98.16	98.2
COIL-100	57.94	74.98	86.95	51.43	89.62	84.01	92.71
MPEG-7	36.54	31.89	55.97	34.24	55.59	58.24	77.27
Rotation	22.79	50.14	58.69	27.69	57.71	100	100
Scale	65.77	6.85	52.93	36.62	67.17	100	100
Translation	91.09	93.19	97.83	37.79	73.98	63.09	89.01
Noise	16.66	33.55	30.23	25.73	69.06	92.05	98.2
Blur	66.93	31.71	88.86	20.58	35.42	94.39	100
JPEG	82.93	40.19	93.59	29.98	67.17	100	100
Average	51.34	51.51	68.68	28.01	63.46	86.46	95.49

Table 16.3 The comparison of average *BEP* of the proposed and phase based methods

	CZM	Optimal similarity	Adjacent phase	Proposed (ZMs)
Kimia-99	68.28	87.17	83.74	90.4
Brodatz texture	86.99	92.99	90.98	96.46
COIL-100	67.55	82.24	78.9	83.05
MPEG-7	43.7	57.98	56.92	60.35
Rotation	92.65	97.37	96.41	100
Scale	100	100	100	100
Translation	41.39	62.92	62.71	64.81
Noise	90.31	95.96	92.28	97.5
Blur	94.39	100	96.98	99.5
JPEG	97.78	100	100	100
Average	78.3	87.66	85.89	89.2

16.5 DISCUSSION AND CONCLUSION

After performing a wide range of experiments on several kinds of databases, it is observed that among the global descriptors, MI have the overall worst performance. The MI were derived from lower order of moments, which are not sufficient enough to accurately describe the shape. Although, the performance of GFD is better than a few other methods, their performance degrades up to a large extent for texture, geometric and photometric transformed databases. In GFD, features are extracted from the spectral domain by applying 2D Fourier transform on polar raster sampled shape image. However, the unstable log polar mapping makes it sensitive to noise, blur, JPEG compressed and texture images. ZMD performs better than GFD and other methods due to the incorporation of sinusoid function in the radial kernel, and they have similar properties of spectral features [7]. On the other hand, among the local descriptors FD has the worst performance. CPDH performs better than WLD, and its performance is comparable to ZMD or to GFD in some of the databases. The performance of WLD is better for texture images. It is worth noticing that the contour based descriptors are translation invariant as compared to region based descriptors. This is due to the fact that centroid of the image is used while computing contour based features for making them translation invariant. While comparing the phase based methods with the proposed solution, the proposed corrected real and imaginary coefficients of ZMs perform better than them. The proposed ZMs approach eliminates the step of estimation of rotation angle in order to correct the phase coefficients of ZMs and making it rotation invariant. In fact, it directly uses the relationship among original and rotated phase coefficients of ZMs to compute $q\theta$. The value of $q\theta$ is used to correct real and imaginary coefficients of ZMs individually, rather than using them to compute magnitude. When both ZMs and HT based features are combined the performance of the proposed system extremely supersedes the existing approaches.

Hence, in this chapter we provide a novel solution to the image retrieval system in which ZMs based global features and HT based local features are utilized. The corrected real and imaginary coefficients of ZMs are used as feature vectors representing the global aspect of images. On the other hand, the histograms of distances between linear edges and centroid of image represent local feature vectors. Both global and local features are combined by the Bray-Curtis similarity measure to compute the overall similarity among images. The experimental results reveal that the proposed ZMs and ZMs+HT methods outperform existing recent region and contour based descriptors. The vast analyses also reveal that the proposed system is robust to geometric and photometric transformations. The average retrieval performance over all the databases represents that the proposed (ZMs+HT) attains 95.49 percent and proposed (ZMs) attains 89.2 percent accuracy rate.

REFERENCES

[1]   Alsmadi, M. K. (2020). Content-Based Image Retrieval Using Color, Shape and Texture Descriptors and Features, Arabian *Journal for Science and Engineering*. 45, 3317–3330.

[2]   Zhang. X., Shen, M., Li., X., Fang, F. (2022). A deformable CNN-based triplet model for fine-grained sketch-based image retrieval. *Pattern Recognition*. 125, 108508.

[3] Datta, R., Joshi, D., Li, J., Wang, J.Z. (2008). Image retrieval: ideas, influences, and trends of the new age. *ACM Computing Surveys.* 40 (2), 1–60.

[4] Chun, Y.D., Kim, N.C. , Jang, I.H. (2008). Content-based image retrieval using multiresolution color and texture features. *IEEE Transactions on Multimedia.* 10 (6), 1073–1084.

[5] Yang. C., Yu., Q. (2019). Multiscale Fourier descriptor based on triangular features for shape retrieval. *Signal Processing: Image Communication,* 71, 110–119.

[6] Qin, J., Yuan, S. et al. (2022). SHREC'22 track: Sketch-based 3D shape retrieval in the wild. *Computers & Graphics.* 107, 104–115.

[7] Zhang, D.S., Lu, G.J. (2004). Review of shape representation and description techniques. *Pattern Recognition.* 37, 1–19.

[8] Rui, Y., She A., Huang, T.S. (1998). A modified Fourier descriptor for shape matching in MARS. Image Databases and Multimedia Search. 8, 165–180.

[9] Mokhtarian, F., Mackworth, A.K. (1992). A theory of multiscale, curvature based shape representation for planar curves. *IEEE Transactions on Pattern Analysis and Machine Intelligence.* 14, 789–805.

[10] Xu, C.J., Liu, J.Z., Tang, X. (2009). 2D shape matching by contour flexibility. *IEEE Transactions on Pattern Analysis and Machine Intelligence.* 31 (1), 180–186.

[11] Belongie, S., Malik, J., Puzicha, J. (2002). Shape matching and object recognition using shape contexts. *IEEE Transactions on Pattern Analysis and Machine Intelligence.* 24 (4), 509–522.

[12] Zhang, D., Lu, G. (2002). A comparative study of Fourier descriptors for shape representation and retrieval, in: The 5th Asian Conference on Computer Vision (ACCV02).

[13] Shu, X., Wu, X.-J. (2011). A novel contour descriptor for 2D shape matching and its application to image retrieval. *Image and Vision Computing.* 29 (4) 286–294.

[14] Chen, J., Shan, S., He, C., Zhao, G., Pietikainen, M., Chen, X., Gao, W. (2010). WLD: A robust image local descriptor. *IEEE Transactions on Pattern Analysis and Machine Intelligence.* 32 (9), 1705–1720.

[15] Sonka, M., Hlavac, V., Boyle, R. (1993). *Image Processing, Analysis and Machine Vision,* Springer, NY. 193–242.

[16] Hu, M.-K. (1962). Visual pattern recognition by moment invariants. *IEEE Transactions on Information Theory.* 8 (2), 179–187.

[17] Zhang, D., Lu, G. (2002). Shape-based image retrieval using generic Fourier descriptor, *Signal Processing: Image Communication.* 17 (10), 825–848.

[18] Kim, W.-Y., Kim, Y.-S. (2000). A region based shape descriptor using Zernike moments, *Signal Processing: Image Communication.* 16, 95–102.

[19] Teague M. R. (1980). Image analysis via the general theory of moments. *Journal of Optical Society of America.* 70, 920–930.

[20] Singh, C., Pooja (2012). Local and global features based image retrieval system using orthogonal radial moments. *Optics and Lasers in Engineering.* 55, 655–667.

[21] Gope, C. , Kehtarnavaz, N., Hillman, G. (2004). Zernike moment invariants based photo-identification using Fisher discriminant model, in: Proceedings of IEEE International Conference Engineering in Medicine and Biology Society. 1, 1455–1458.

[22] Kim, H.-K., Kim, J.-D., Sim, D.-G., Oh, D.-I. (2000). A modified Zernike moment shape descriptor invariant to translation, rotation and scale for similarity-based image retrieval, in: Proceedings of IEEE International Conference on Multimedia and Expo, 1, 307–310.

[23] Maaoui, C., Laurent, H., Rosenberger, C. (2005). 2D color shape recognition using Zernike moments. in: Proceedings of IEEE International Conference on Image Processing. 3, 976–979.

[24] Teh, C.-H., Chin, R.T. (1988). On image analysis by the methods of moments. IEEE Transactions on Pattern Analysis & Machine Intelligence. 10 (4), 496–513.

[25] Terrillon, J. C., David, M., Akamatsu, S. (1998). Automatic detection of human faces in natural scene images by use of a skin color model and of invariant moments, in: Proceedings of 3rd IEEE International Conference on Automatic Face and Gesture Recognition, Apr. 14–16, 112–117.

[26] Wallin, A., Kubler, O. (1995). Complete sets of complex Zernike moment invariants and the role of the pseudo invariants. IEEE Transactions on Pattern Analysis Machine Intelligence. 17 (11) 1106–1110.

[27] Zhang, D. S., Lu, G. (2001). Content-based shape retrieval using different shape descriptors: A comparative study, in: Proceedings of IEEE ICME, Tokyo, 317–320.

[28] S. Li, M.-C. Lee, C.-M. Pun (2009). Complex Zernike moments features for shape based image retrieval, IEEE Transactions on System, Man and Cybernetics.-Part A: Systems and Humans 39 (1) 227–237.

[29] Chen, Z., Sun, S.-K. (2010). A Zernike moment phase-based descriptor for local image representation and matching. IEEE Transactions on Image Processing. 19 (1), 205–219.

[30] Revaud, J., Lavoue, G., Baskurt, A. (2009). Improving Zernike moments comparison for optimal similarity and rotation angle retrieval. IEEE Transactions Pattern Analysis Machine Intelligence 31 (4), 627–636.

[31] Oppenheim, A.V., Lim, J. S. (1981). *The importance of phase in signals*, in: Proceedings of IEEE 69 (5), 529–550.

[32] Singh, C., Bhatia, N., Kaur, A. (2008). Hough transform based fast skew detection and accurate skew correction methods. *Pattern Recognition*. 41, 3528–3546.

[33] Sharma, P. (2017). Improved shape matching and retrieval using robust histograms of spatially distributed points and angular radial transform. *Optik*. 140, 346–364.

[34] Wee, C.Y., Paramseran, R. (2007). On the computational aspects of Zernike moments. *Image and Vision Computing*. 25, 967–980.

[35] Wei, C.H., Li, Y., Chau, W.-Y., Li, C.-T. (2008). Trademark image retrieval using synthetic features for describing global shape and interior structure. *Pattern Recognition*. 42, 386–394.

[36] Qi, H., Li, K., Shen, Y., Qu, W. (2010). An effective solution for trademark image retrieval by combining shape description and feature matching. *Pattern Recognition.* 43, 2017–2027. https://doi.org/10.1016/j.patcog.2010.01.007

[37] Kokare, M., Chatterji, B.N., Biswas, P.K. (2003). Comparison of similarity metrics for texture image retrieval, TENCON Conference on Convergent Technologies for Asia-Pacific Regio, 2, 571–575.

[38] C Singh, Pooja (2011). Improving image retrieval using combined features of Hough transform and Zernike moments, *Optics and Lasers in Engineering*, 49 (12) 1384–1396.

Chapter 17

Translate and recreate text in an image

S. Suriya, K. C. Ridhi, Sanjo J. Adwin, S. Sasank,
A. Jayabharathi, and G. Gopisankar

17.1 INTRODUCTION

Nowadays, most people use language translation software, such as Google Translate and Microsoft Translator, to translate texts from one language into another. For example, if we were in some foreign country where we did not know the native language (foreign language) of the people. It is necessary for us to be able to communicate in their native (foreign) language in order to eat. In order to deal with this concern, a translator would have to assist them in their communication.

There have been significant advances brought about by Google's research into neural machine translation along with other competitors in recent years, providing positive prospective for the industry. As a result of recent advances in computer vision and speech recognition, machine translation can now do more than translate raw texts, since various kinds of data (pictures, videos, audio) are available across multiple languages. With the help of language translator applications, the user can translate the text to their own language, however, the picture cannot be connected with the text. If the user wants the picture with the translated text, then the user needs a language translator and image editing software to replace the translated text with the image. Our proposed work presents a method for translating text from one language to another, as well as leaving the background image unaltered. Essentially, this will allow users to connect with the image in their native language.

This chapter presents the design of a system that includes three basic modules: text extraction, machine translation, and inpainting. As well as these modules, adding the spelling correction network after the text extraction layer can solve the problem of spelling mistakes in OCR extracted text, since it highlights misspelled words and corrects them before translation. In this module, the input image is sequentially processed in order to translate the Tamil text into English text with the same background.

A method of improving image text translation for Tamil is presented in this work. For ancient languages like Tamil, OCR engines perform poorly

DOI: 10.1201/9781003453406-17

when compared to modern languages such as English, French, and so forth. The reason is that because of its complex nature, few reports about the language have been written in comparison to languages such as English and other Latin-based languages. The solution is to introduce a new module that checks the spelling and autocorrects the errors for OCR retrieved text. Currently there is no OCR spell-checking software compatible with Tamil, especially in an open source format. This project proposes to add contextual correction tools to an existing spell checker using a hybrid approach.

17.2 LITERATURE SURVEY

In using the proposed method, separating texts from a textured background with similar color to texts is performed [1]. Experimentation is carried out in their own data set containing 300 image blocks in which several challenges such as generating images manually by adding text on top of relatively complicated background. From experimentation with other methods, the proposed method achieves more accurate result, that is, precision of 95 percent, recall of 92.5 percent and F1 score of 93.7 percent are obtained, and the proposed algorithm is robust to the initialized value of variables.

Using LBP base feature, separation of text and non-text from handwritten document images is proposed [2]. Texture based features like Grey Level Co-occurrence Matrix (GLCM) are proposed for classifying the segmented regions. In this a detailed analysis of how accurately features are extracted by different variants of local binary pattern (LBP) operator is given, a database of 104 handwritten engineering lab copies and class notes collected from an engineering college are used for experimentation purposes. For classification of text and non-text, Nave Bayes (NB), Multi-layer perceptron (MLP), K-nearest neighbor (KNN), Random forest (RF), Support vector machine (SVM) classifiers are used. It is observed that Rotation Invariant Uniform Local Binary Pattern (RIULBP), performed better than the other feature extraction methods.

The authors proposed [4] a robust Uyghur text localization method in complex background images, which provide a CPU-GPU various parallelization system. A two stage component classification system is used to filter out non-text components, and a component connected graph algorithm is used for the constructed text lines. Experimentation is conducted on UICBI400 dataset; the proposed algorithm achieves the best performance which is 12.5 times faster.

In the chapter, Google's Multilingual Neural Machine Translation System [5] proposes a Neural Machine Translation (NMT) model to translate between multiple languages. This approach enables Multilingual NMT systems with a single model by using a shared word piece vocabulary. The models stated in this work can learn to perform implicit bridging between language pairs never seen explicitly during training, which shows that transfer learning and zero shot translation is possible in Neural Machine Translation. This work introduces a method to translate between multiple languages using a single model by taking the advantage of multilingual data to improve NMT for all languages involved without major changes in the existing NMT architecture. The multilingual model architecture is similar to the Google's Neural Machine Translation (GNMT) system. To enable multilingual data in a single model, it recommends modification in the input data by adding an artificial token at the start of all the input sequences which indicates the target language. The model is trained with all the multilingual data after adding the artificial tokens. To overcome the issue of translation of unknown words and to restrict the vocabulary of language efficiency in computing, a shared word piece model is used across all the source and target training data. This chapter suggests that training a model across several languages can improve the performance of individual languages and also increases the effectiveness of zero-shot translation.

In the work, Simple, Scalable Adaptation for Neural Machine Translation [6] proposes a simple and effective method for adaptation of Neural Machine Translation. This work recommends a method of injecting small task specific adapter layers into a pre-trained model. The task specific adapters are smaller compared to the original model, but they can adapt the model to multiple individual tasks at the same time. This approach consists of two phases: Training a generic base model and adapting it to new data, by adding small adapter modules added. The adapter modules are designed to be simple and flexible. When adapters are used for domain adaptation, a two-step approach is followed: (1) The NMT model is pretrained on a large open domain corpus and the trained parameters are fixed. (2) A set of domain specific adapter layers are injected for every target domain, these adapters are fine-tuned to maximize the performance on the corresponding domains. This proposed approach is evaluated in two tasks: Domain adaptation and massively multilingual NMT. Experiments revealed that this approach is on par with fully fine-tuned models on various datasets and domains. This adaptation approach bridges the gap between individual bilingual models and one substantial multilingual model.

In chapter [7], Tan et al. describe the development of a framework that clusters languages into different groups and trains the languages in one multilingual model for each cluster. This work studies two methods for language clustering, such as: (1) Using prior knowledge, where the languages are grouped according to the language family, and (2) Using language embedding where all the languages are represented by an embedding vector and the languages are clustered in the embedding space. The embedding vectors of all the languages are obtained by training a universal neural machine translation model. The experiments conducted on both approaches revealed that the first approach is simple and easy, but translation accuracy is lesser, whereas the second approach captures most of the relationships among the languages and provides better overall translation accuracy. The multilingual machine translation model translates multiple languages using a single model and so it simplifies the training process, reduces the online maintenance costs, and improves the low resource and zero-shot translation. The results for experiments conducted on 23 languages shows that language embedding clustering outperforms the prior knowledge based clustering in terms of BLEU scores, as the language embeddings can sufficiently characterize the similarity between the languages.

In the chapter [8] Liu et al. show that multilingual denoising pre-training produces remarkable performance gains across several Machine Translation (MT) tasks. This work presents a Seq2Seq (sequence-to-sequence) denoising auto-encoder that is pre-trained on a massive monolingual corpus using the BART objective called the mBART. Pre-training a complete model allows it to be directly optimized for supervised and unsupervised machine translation. Noises are added to the input texts by masking phrases and permuting the texts, and a single transformer model is learned to recover the input sentences, unlike other pre-training approaches for machine translation. Training mBART once for all languages produces a set of parameters that can be fine-tuned for any number of language pairs in supervised and unsupervised settings. This work demonstrates that by adding mBART initialization produces a performance increase of up to 12 BLEU points for low resource Machine Translation and over 5 BLEU points for several document-level and unsupervised models.

The work, Learning Deep Transformer Models for Machine Translation [9], claims that a true deep transformer model can surpass the transformer-big by using layer normalization properly and by using a new way of passing the combination of previous layers to the next. Recently, the systems based on multi-layer self-attention have shown promising results on several small scale tasks. Learning deeper networks for vanilla transformers is a difficult task as there is a deeper model already in use, and optimizing such deeper networks is a difficult task due to the gradient exploding problem. This work states that the appropriate use of layer normalization is the key to learning deep encoders, and so the deep network of the encoder can be optimized smoothly by repositioning the layer normalization unit. It proposes an approach based on Dynamic Linear Combination of Layers (DLCL) to store the features extracted from all the earlier layers, which eliminates the problems with the standard residual network where a residual connection depends only on the output of one ahead layer and may neglect the previous layers. The studies of this work have shown that the deep transformer models can be optimized by appropriate use of layer normalization. It also proposes an approach based on a dynamic linear combination of layers, and it trains a 30-layer Transformer system. The size of this model is 1.6 times smaller, it requires 3 times lesser training epochs, and it is 10 percent faster for inference.

In the work [10] Pan et al., aims to develop a many-to-many translation system that mainly focuses on the quality of non-English-language directions. This is based on the hypothesis that a universal cross language representation produces better multilingual translations. This work proposes mRASP2 training method to achieve a unified multilingual translation model; mRASP2 is enabled by two techniques: (1) contrastive learning schemes; (2) data augmentation on both parallel and monolingual data; mRASP2 combines monolingual and parallel corpora with contrast learning. The aim of mRASP2 is to ensure that the model represents analogous sentences in different languages in a shared space by using the encoder to minimize the representational distance of analogous sentences; mRASP2 outperforms the existing best unified model and attains competitive performance or even outperforms the pre-trained, fine-tuned mBART model for English-centric directions. In non-English directions, MRASP2 attains an improvement of average 10+ BLEU score compared to multilingual transformer baseline. This work reveals that contrast learning can considerably enhance the zero shot machine translation directions.

In the chapter [11], MRASP2 studies the issues related to the uncertainty of the translation task as there would be multiple valid translations for a single sentence and, also, due to the uncertainty caused by noisy training data. This work proposes tools and metrics to evaluate how the uncertainty is captured by the model distribution and its effect on the search strategies for generating translations. It also proposes tools for model calibration and to show how to easily fix the drawbacks of the current model. One of the fundamental challenges in translation is uncertainty, as there are multiple correct translations. In the experiment conducted in this work, the amount of uncertainty in the model's output is quantified and compared between two search strategies: sampling and beam search. This study concludes that the NMT model is well calibrated at the token and sentence level, but it tends to diffuse the probability mass excessively. The excessive probability spread causes low quality samples from the model. This issue is linked to a form of extrinsic uncertainty that causes reduced accuracy with larger beams.

"Design and Implementation of NLP-Based Spell Checker for the Tamil Language" [12]. In this work, advanced Natural language processing is used to detect a misspelled word in a text. In this model, the minimum edit distance (MED) algorithm is used to recommend correct spelling suggestions for misspelled words and the probability of their occurrence. It is customized for the Tamil vocabulary and offers correct ideas for missing words. A distance matrix is constructed between each misspelled word and its possible permutations. Finally, dynamic programming is used to to suggest the closest possible word of the misspelled word. This model not only detects the wrongly spelled words, but also predicts the possible suggestions of the correct words that the user might want to write. However, there is no benchmark dataset for the Tamil language as there is for other languages. This is one of the reasons for inefficient spell checkers in Tamil, as there is no proper dataset to test and validate the accuracy of the system.

"DDSpell, A Data Driven Spell Checker and Suggestion Generator for the Tamil Language" [13]. This is an application developed using a data-driven and language-independent approach. The proposed spell checker and suggestion generator can be used to check misspelled words. The model uses a dictionary of 4 million Tamil words, created from various sources, to check the spelling. Spelling correction and suggestion are done by a character level bi-gram similarity matching, minimum edit distance measures and word frequencies. In addition, the techniques like hash keys and hash table were used to improve the processing speed of spell checking and suggestion generation.

In the chapter, "Vartani Spellcheck Distance Automatic Spelling Error Detection and Context-Sensitive Error Correction" [14] can be used to improve accuracy by post-processing the text generated by these OCR systems. This model uses the context-sensitive approach for spelling correction of Hindi text using a state-of-the-art transformer – BERT – in conjunction with the Levenshtein distance algorithm. It uses a lookup dictionary and context-based named entity recognition (NER) for detection of possible spelling errors in the text.

This work, "Deep Learning Based Spell Checker for Malayalam Language," [15] is a novel attempt, and the first of its kind to focus on implementing a spell checker for Malayalam using deep learning. The spell checker comprises two processes: error detection and error correction. The error detection section employs a LSTM based neural network, which is trained to identify the misspelled words and the position where the error has occurred. The error detection accuracy is measured using the F1 score. Error correction is achieved by the selecting the most probable word from the candidate word suggestion.

This study, "Systematic Review of Spell-Checkers for Highly Inflectional Languages," [16] analyzes articles based on certain criteria to identify the factors that make spellchecking an effective tool. The literature analyzed regarding spellchecking is divided into key sub-areas according to the language in use. Each sub-area is described based on the type of spellchecking technique in use at the time. It also highlights the major challenges faced by researchers, along with the future areas for research in the field of spell-checkers using technologies from other domains such as morphology, parts-of-speech, chunking, stemming, hash table, and so forth.

In this work, "Improvement of Extract and Recognizes Text in Natural Scene Images" [17], spell checking is employed in the proposed device in order to correct any spelling issues that may also arise while optical personality recognition is taking place. For recognizing unknown characters, optical persona cognizance spell checking is applied. The optical persona cognizance spell checker application tries to shape phrases with unknown letters into phrases with comparable letters.

Three different spell checking systems are presented and tested in the paper [18] "Improvement of Extract and Recognizes Text in Natural Scene Images": bloom-filter, symspell, and LSTM based spell checkers. Symspell is extremely fast for validation and looking up, while LSTM based spell checking is very accurate. LSTM implementations are not accurate enough to be useful in day-to-day work, Bloom-filter is very fast, but cannot provide suggestions for misspelled words. Symspell can flag misspelled words and generate suggestions. In addition, the system is not dynamic, meaning that it cannot suggest new words when there is a greater edit distance than what it was generated with. The memory requirement increases of course with edit distance.

In the paper [19] "Inpainting Approaches, A Review," proposes two major techniques to inpaint an image. Here, inpainting is the remodeling of certain degenerated regions of an image or a video. The main algorithm was based on the idea that the direction of the slope gives the direction of the degenerated regions which need to be remodeled. The first technique 2D image implanting in turn has several approaches based on structure, texture or a blend of both wherein the first one used distance based inpainting techniques to form structures, and later the texture synthesis was performed. Alternative methods include Erosion and Multiscale Structure propagation. The second technique, "Depth Map Inpainting," generates a third visualization and then is followed by energy minimization using descriptors, generating sharp edges alongside noise elimination. There are also other alternative methods using Depth Map such as Fusion based inpainting and selective median filter and background estimator.

In a journal [20], "Image Inpainting: A Review," gives an overview of the existing three categories; image inpainting techniques with their own subcategories based on distortion(image restoration, object removal, blocks recovery) in images. Inpainting depends on the reconstruction of an image, which is based on type of distortion found and techniques needed to rectify. The article further elaborates on the sequential based, CNN and GAN based approaches. IN sequential, the distortion is reconstructed using neural networks without the usage of deep learning, either by patch method or diffusion-based method. Deep learning is utilized in a CNN approach, where notable methods include features like LDI and MRF. GAN based technique uses generative adversarial networks to train the inpainting model, and it is the much-preferred technique over the others.

A journal article by [21] Jireh et al., gives an in-depth analysis of traditional and current image inpainting techniques in "A comprehensive review of past and present image inpainting methods." Notable traditional techniques include texture synthesis techniques which derive image statistics in turn used in generating texture like the input image and other methods like diffusion-based techniques which are mainly used on images with scratches or straight-line curves as they bound good pixels of the surrounding area with the inpainting image gap. In deep learning, GAN and CNN techniques in depth analysis working were discussed and how they provide the best inpainting output when compared to the previous methods were elaborated. Choosing the correct dataset is critical as far as inpainting is concerned.

In a paper [22] Raymond et al. use DCGAN technique for inpainting where initially a loss function is defined where it preserves the contextual loss between the corrupted image with that of the output image and a perceptual loss that ensures a practical output image. In "Semantic image inpainting with deep generative models," the authors refer to the concept of predicting the contents of the corrupted areas based on the neighbouring pixel regions. The inpainting is done with pretrained GAN networks. The contextual loss and perceptual loss need to be defined properly. The proposed methods' framework is trained on CelebA and SVHN dataset with high percentage of corruption to get the best practical output.

In a paper titled "Large inpainting of face images with trainlets," [23] Jeremias et al. elaborate that the problem with inpainting becomes quite challenging when the corrupted region is large. Here a framework called trainlets is proposed which allows the model to design large portions of image from various face image datasets with the help of OSDL ("Online Sparse Dictionary Learning") algorithm. Primarily, the global model is obtained by sparse representation of the images. However, this is limited since it can be only used for small patches of inpainting. Hence, OSDL algorithm is applied to the sparse vectors to obtain the desired global model. In Inpainting formulation, the corrupted image is masked initially followed by back propagation continued with PCA basis and SEDIL algorithm is applied. This is passed on to the trainlet model to obtain the desired accurate image output.

In a paper [24], "Image inpainting for irregular holes using partial convolutions," Guilin et al., propose a partial convolution for image inpainting. Normal Inpainting uses CNN based on both valid pixels and mean value pixel of masked regions. This method is efficient as it creates color discrepancy or gives a blurred image. After-effects can be done on this, but they are costly and have a high chance of failure. The proposed model's Partial CNN is masked and renormalized on valid pixels of the input image. PyTouch Network is used along with UNet(similar) architecture which converts full CNN layers with partial CNN layers. The loss functions like perceptual loss functions are applied here. The model is trained on datasets like Irregular mask dataset.

In an article, "Non-local patch-based image inpainting," [25] Newson et al., proposes that the best solution for inpainting video is where the patches are similar to its nearest pixel region. Initially Nearest neighbor search propagation is applied followed by random search where each pixel is randomly assigned to a pixel in the patch region. Here, the Patch Match algorithm is applied, which accelerated the search for ANN. Secondarily, the patch is further refined with initialization using an onion-peel approach. In tertiary, the texture features are reconstructed in the similar manner. A pyramid scheme is implemented with multiple levels consisting of different texture choices from which the best one is chosen.

17.3 EXISTING SYSTEM

In traditional machine translation tools, a combination of a first level detector network, a second level recognizer network, and a third level NMT network is employed.

NETWORK	PURPOSE
Text detector network	Region proposal network (RPN) which uses variants of Faster-RCNN and SSD to find character, word, and line bounding boxes.
Text extraction network	Convolution neural network (CNN) with an additional quantized long shorter memory (LSTM) network to identify text inside the bounding box.
Neural machine translation	Sequence-to-sequence network, which uses variants of LSTM and Transformers to translate the identified text to target language.

Modules existing system and their purposes.

It is rather difficult to evaluate the effectiveness of existing systems, particularly for languages like Tamil. The benchmark accuracy for OCR is considerably lower for Tamil than it is for Western languages. As a result, quality of translation can be a real problem.

17.4 PROPOSED SYSTEM

The proposed system consists of three basic modules, including text extraction, machine translation, and inpainting. In these modules, the input image

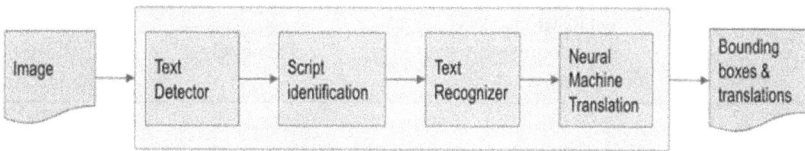

Figure 17.1 Flowchart of existing system.

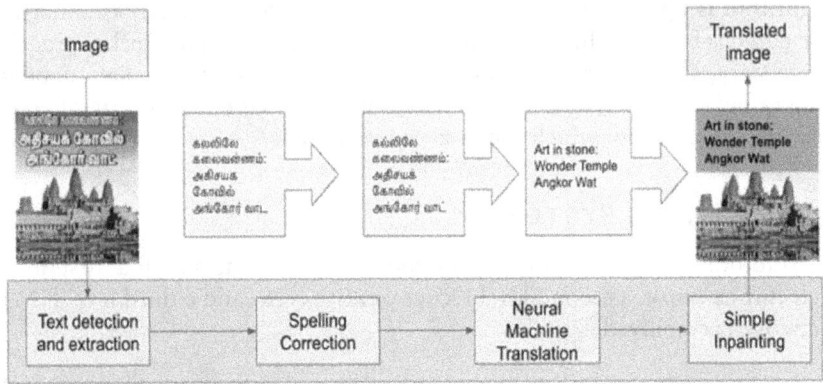

Figure 17.2 Flowchart of proposed system.

is sequentially processed to translate the Tamil text to English text with the same background color. It is possible to address spelling mistakes in OCR extracted text by adding a spelling correction network after the text extraction layer, as it highlights misspelled words and corrects them before translation. In this section we provide the specifications of the design, such as the architecture and use cases in both the system and the model trained. Figure 17.2 depicts the architecture diagram of the system used for recreating text in an image.

17.4.1 Flow chart

MODULE	PURPOSE	EXISTING AVAILABILITY
Text detection and extraction	It can be used to determine the bounding box for a line, word, or character and identify the text inside the box.	Easy OCRTesseract
Auto Spelling correction	Correct the incorrect word by identifying the them and making the corrections	Seq2Seq Encoder Decoder Model
Neural machine translation	Translate the identified text to target language. (Tamil to English)	Python deep translator
In painting	Superimpose the text in an image	OpenCV

Modules in proposed system and their purposes.

17.4.2 Experimental setup

In terms of hardware, the following requirements are required:

- Operating Systems: Windows or Mac currently available
- CPU cores: 6 (Intel i5 or AMD Ryzen 7)
- GPU: Minimum of 2 GB of RAM (Nvidia or any other GPU of similar capability)
- RAM System Requirements: 16 GB (Google Collaboratory)

In terms of software, there are the following requirements:
Google Collaboratory

- PyCharm community version
- Python deep learning, image processing, NLTK, deep translator, OpenCV libraries

17.4.3 Dataset

Pm-India corpus dataset

- The corpus contains the releases of the PM-India project. It is a parallel corpus derived from the news updates of the prime minister of India in 13 Indian languages as well as English.
- In addition, it pairs 13 major languages of India with English, resulting in parallel sentences. Each language pair contains approximately 56 000 sentences.

For this work, the Seq2Seq encoder decoder model is trained using a corpus of Tamil sentences. It is possible to obtain corrupted error data for this training by creating a picture that includes the sentences that were included in the PM-corpus dataset. The images are then fed into OCR to produce the OCR extracted text, which contains errors due to the low accuracy of OCR for Tamil in comparison to other Western languages. The error corpus and error free data corpus are then used to train the Seq2SeqAuto spell correction model.

17.4.4 Text detection and extraction

Since text detection and extraction module assumes that the input text image is fairly clean, we need to make sure the image is appropriately pre-processed in order to improve the accuracy. The pre-processing steps include rescaling, blurring, noise removal and so forth. After pre-processing, the image is fed into the OCR model, which extract the text.

Error sentance : கடசயன துணைத்ததலைல ௭ா கவாஸ ஹூசேன அகமத சென்ற
Correct Sentence : கட்சியின் துணைத்தலைவர் க்வாஸி ஹூசேன் அகமத் சென்ற மாதம்
Error sentance : சமபகாலததல சல தகவலகள யூலயஸ ரோசனபேௗ ொ ௧க ஒ
Correct Sentence : சம்பகாலத்தில் சில தகவல்கள் யூலியஸ் ரோசன்பேர்க் ஒரு வித உளவுச்செய்தியை
Error sentance : ஆசோ ௭ா சாதோகைகப பெற்றான சாதோககு ஆகைமப பெற்றான ஆகம
Correct Sentence : ஆசார் சாதோக்கைப் பெற்றான் சாதோக்கு ஆக்கைமப் பெற்றான் ஆகீம் எலியூதைப்
Error sentance : எனை நடககறது எனபது தமககு தெ ௨ாயும எனறும ஆனால
Correct Sentence : என்ன நடக்கிறது என்பது தமக்கு தெரியும் என்றும் ஆனால் தம்மால்
Error sentance : நாை தபபநடநதது மெயயானாலும என தபபதம எனனோடேதான இ ௨க்கறது
Correct Sentence : நான் தப்பிநடந்தது மெய்யானாலும் என் தப்பிதம் என்னோடேதான் இருக்கிறது
Error sentance : டால ௨ாமயானூடைய அறவஜஷவத தொடுவானததறகு அபபால எவவளவோ தொலைலல இ
Correct Sentence : டால்ரிம்பிளினுடைய அறிவுஜீவித தொடுவானத்திற்கு அப்பால் எவ்வளவோ தொலைவில் இருந்தன
Error sentance : இநத அமைபபுககளை British Tamil Forum La Maison du
Correct Sentence : இந்த அமைப்புகளில் British Tamil Forum La Maison du
Error sentance : தைது மனைவ மகளுடன அஞசாதே படைசாதே ௨சததா ௭ா
Correct Sentence : தனது மனைவி மகளுடன் அஞ்சாதே படத்தை ரசித்தார் விஜய்
Error sentance : முவா ௨ாப மறறும வாஜபாய இ ௨ாவ ௭ா ௩மே அவ
Correct Sentence : முவ்ராப் மற்றும் வாஜபாயி இருவருமே அவர்கள்து நாடுகளில் அவர்களின் ஆட்சிக்கு
Error sentance : ஆனால எனை ஆைது
Correct Sentence : ஆனால் என்ன ஆனது

Figure 17.3 Image of sample text in dataset.

17.4.5 Auto spelling correction

Sequence to sequence models map a variable length input sequence to a variable length output sequence. They consist of an encoder and a decoder block as shown in Figure 17.4. In the figure, the model converts misspelled input letter sequence into letter sequence which forms a valid.

As a pre-processing stage, sentences are split into words and the context words are coded into three letter sequences with their first consonants. Then, the letter sequence is fed into the seq2seq spelling corrector. The output is post-processed to produce the final output.

17.4.6 Machine translation and inpainting

The extracted auto spell corrected text from seq2seq model is translated using python deep translator. In in-painting module the mask image is created from input image by masking the detected text in an image. In next step both original and masked images are used to recreate the image portion at masked area. Finally, the translated text is superimposing into the image at specific position by bounding box coordinates of detected text.

17.5 IMPLEMENTATION

17.5.1 Text detection and extraction

Text extraction from an image can be done using OCR models. There exist two open source OCRs that extract the Tamil text from image, namely, Easy OCR and Tesseract OCR. This work uses the Easy OCR based on the

IMAGE

Image resizing, blurring, varying threshold

Preprocessing of the Image

Contours detection to locate text location in image

Text Localization

Text recognition and extraction by tesseract model

Text extraction

TXT

THE LAST CAR
THAT PARKED HERE
IS STILL MISSING

THE LAST CAR
THAT PARKED HERE
IS STILL MISSING

THE LAST CAR
THAT PARKED HERE
IS STILL MISSING

THE LAST CAR
THAT PARKED HERE
IS STILL MISSING

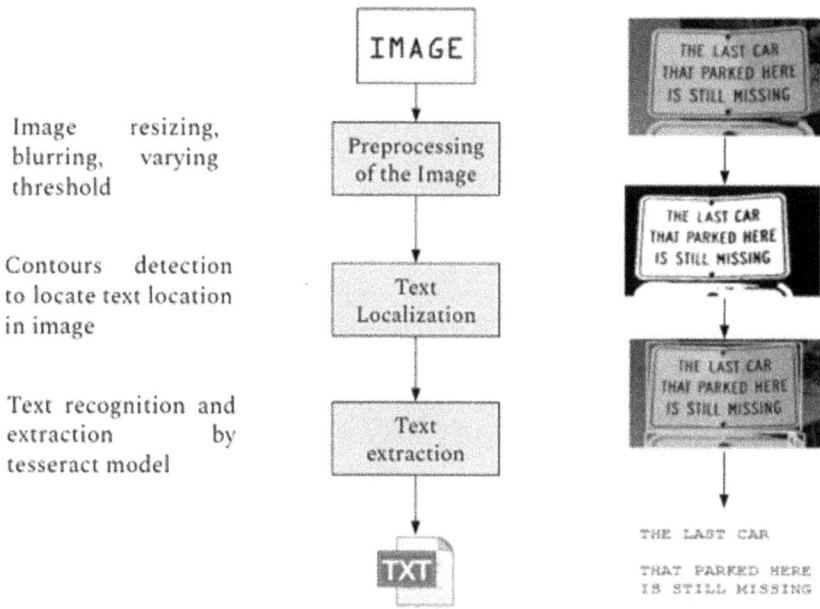

Figure 17.4 Flowchart of text dection and extraction.

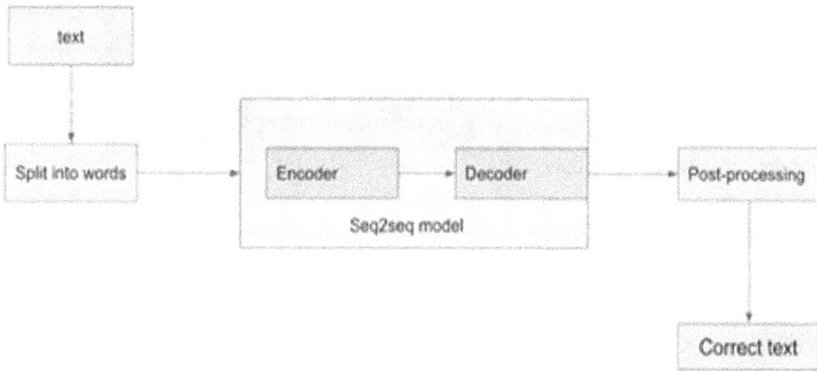

text

Split into words

Encoder → Decoder

Seq2seq model

Post-processing

Correct text

Figure 17.5 Flowchart of Autospell correction.

test result. We performed tests on the two models with Pm-India corpus dataset to decide which one performs well for Tamil language. For this test we converted the dataset text into images (Figure 17.7) contain a random 1,000 words from the dataset and experimented on both models. Error

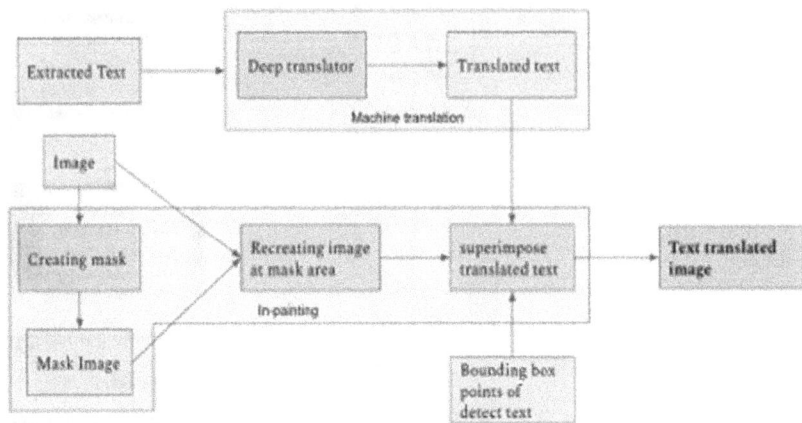

Figure 17.6 Flowchart of Machine translation and inpainting.

Figure 17.7 Created image of word in dataset.

rate and time consumption of each model with/without GPU is tabulated below:

Model	Error rate	Time with GPU	Time with CPU
Easy OCR	17.6%	2.18 min	38.5 min
Tesseract OCR	17.8%	2.6 min	2.75 min

17.5.2 Auto spelling correction

The field of natural language processing has seen significant advances in the last few decades, but it is still early in its development for languages like Tamil. For the purpose of retrieving the text from an image, optical character recognition (OCR) is developed since Tamil characters are more complex than characters in other languages (with a few variations in character of one character becoming completely different characters with a unique sound). In order to correct these misconceptions, we developed the Auto spelling corrector for Tamil, which converts OCR-derived misspellings into the correct spellings.

This model is created by combining multiple sequences of the keras sequence layer, and tested for accuracy.

- Simple RNN
- Embed RNN
- Bidirectional LSTM
- Encoder decoder with LSTM
- Encoder decoder with Bidirectional LSTM + Levenshtein Distance

17.5.2.1 Simple RNN

Using a simple RNN model with two layers (as shown in Figure 17.8), we started training the model using a try-and-error method. First, the Gated Recurrent Unit (GRU) receives the sequence of training input, and the weights have been changed during training (as illustrated in Figure 17.9).

In short, GRUs work the same as RNNs but have different operations and gates. GRU incorporates two gate operating mechanisms, called the Update Gate and the Reset Gate, to address the problem posed by standard RNN.

```
Model: "sequential_2"

 Layer (type)                Output Shape           Param #
=================================================================
 gru_2 (GRU)                 (None, 76, 128)        50304

 time_distributed_2 (TimeDis (None, 76, 9332)       1203828
 tributed)

=================================================================
Total params: 1,254,132
Trainable params: 1,254,132
Non-trainable params: 0
```

Figure 17.8 Simple RNN structure and layers.

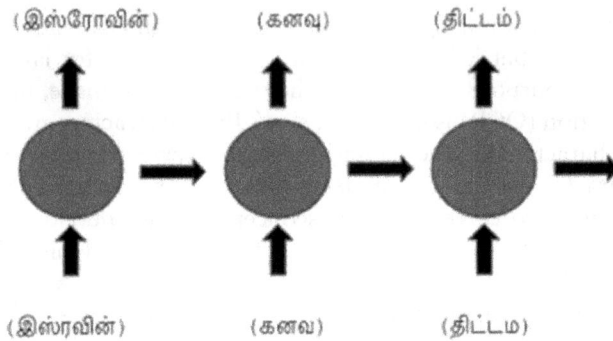

(இஸ்ரோவின்)   (கனவு)   (திட்டம்)

(இஸ்ரவின்)   (கனவ)   (திட்டம)

Figure 17.9 Simple RNN diagram.

```
Model: "sequential_3"
_____
Layer (type)                 Output Shape              Param #
=======================================================================
embedding (Embedding)        (None, 76, 128)           1194368

lstm (LSTM)                  (None, 76, 128)           131584

dense_3 (Dense)              (None, 76, 9331)          1203699

=======================================================================
Total params: 2,529,651
Trainable params: 2,529,651
Non-trainable params: 0
```

Figure 17.10 Embed RNN structure and layers.

17.5.2.2 Embed RNN

In the embedded RNN model (as shown in Figure 17.10), the first layer makes the process more efficient and faster. A LSTM (Long Short Term Memory) layer is a kind of RNN that can learn long-term sequences. This layer was specifically designed to protect against long-term dependency concerns. As a result of working with long sequences for very long periods of time, it is able to remember them.

Figure 17.11 illustrates how the embedding of RNNs introduces a new layer which accepts input sequences and converts them to vectors (as illustrated).

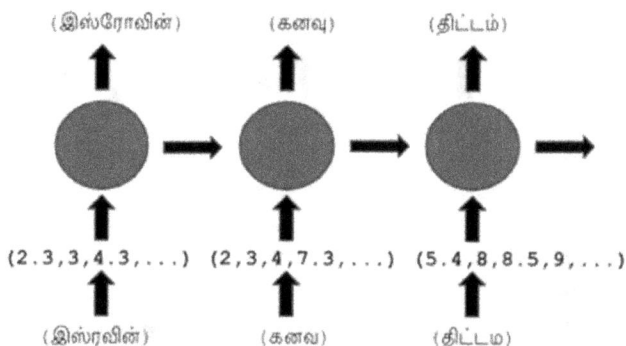

(இஸ்ரோவின்)   (கனவு)   (திட்டம்)

(2.3,3,4.3,...) (2,3,4,7.3,...) (5.4,8,8.5,9,...)

(இஸ்ரவின்)   (கனவு)   (திட்டம)

Figure 17.11 Embed RNN diagram.

```
Model: "sequential_5"

Layer (type)                Output Shape            Param #
=================================================================
bidirectional_1 (Bidirectio  (None, 76, 256)         133120
nal)

time_distributed_4 (TimeDis  (None, 76, 9331)        2398067
tributed)

=================================================================
Total params: 2,531,187
Trainable params: 2,531,187
Non-trainable params: 0
```

Figure 17.12 Bidirectional LSTM structure and layers.

17.5.2.3 Bidirectional LSTM

The implementation of BI-LSTMs (the architecture layer of model shown in Figure 17.12) in neural networks can allow them to process sequence information both backward (from the future to the past) and forward (from the past to the present) as mentioned in Figure 17.13.

17.5.2.4 Encoder decoder with LSTM

Encoder and decoder models can be viewed as two separate blocks connected by a vector (referred to as a context vector) representing the state of the encoder and the decoder, as shown in Figure 17.14. Input-sequences are encoded one by one by the encoder. Using this procedure, all the information about the input sequence is compiled into a single vector of fixed

(இஸ்ரோவின்)   (கனவு)   (திட்டம்)

(இஸ்ரவின்)   (கனவ)   (திட்டம)

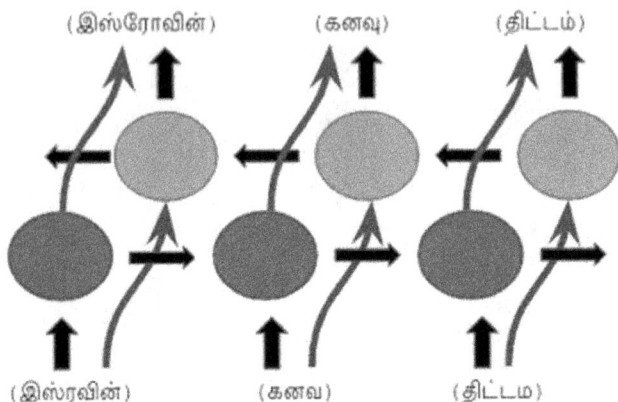

Figure 17.13 Bidirectional LSTM diagram.

```
Model: "sequential_6"

Layer (type)                  Output Shape            Param #
==================================================================

embedding_1 (Embedding)       (None, 76, 128)         1194368

lstm_3 (LSTM)                 (None, 128)             131584

repeat_vector (RepeatVector   (None, 76, 128)         0
)

lstm_4 (LSTM)                 (None, 76, 128)         131584

time_distributed_5 (TimeDis   (None, 76, 9331)        1203699
tributed)

==================================================================
Total params: 2,661,235
Trainable params: 2,661,235
Non-trainable params: 0
```

Figure 17.14 Encoder Decodee LSTM structure and layers.

length, called the "context vector." This vector is passed from the encoder to the decoder once it has processed all the tokens. Target sequences are predicted token by token using context vectors read from the decoder.

Figure 17.15 shows how vectors are built so that they will help the decoder make accurate predictions by containing the full meaning of the input sequence.

Output

| Encoder LSTM | → | Decoder LSTM |

Internal state

While training

(இஸ்ரவின்), (கனவ), (திட்டம)   (இஸ்ரோவின்), (கனவு), (திட்டம்)

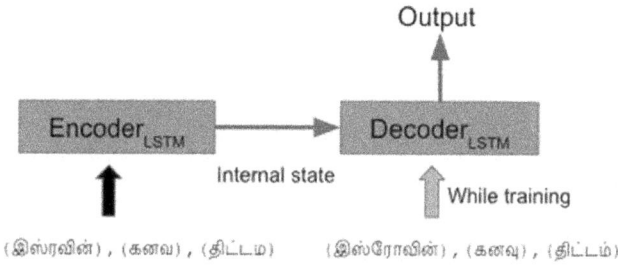

Figure 17.15 Encoder Decoder LSTM diagram.

```
Layer (type)                    Output Shape             Param #
=================================================================
embedding_3 (Embedding)         (None, 15, 128)          44032

bidirectional_3 (Bidirection    (None, 256)              197376

repeat_vector_3 (RepeatVecto    (None, 21, 256)          0

bidirectional_4 (Bidirection    (None, 21, 256)          295680

time_distributed_7 (TimeDist    (None, 21, 344)          88408
=================================================================
Total params: 625,496
Trainable params: 625,496
Non-trainable params: 0
```

Figure 17.16 Encoder Decoder Bidirectional LSTM structure and layers.

17.5.2.5 Encoder decoder with bidirectional LSTM + Levenshtein distance

We introduced bi-LSTMs to increase the amount of input information that can be provided to the network. The advantage of bi-LSTMs is that they can access future inputs from the current state without any time delays. Figures 17.16 and 17.17 illustrate the architecture of the bidirectional encoder we employ in our spellcheck system.

A distance matrix can be used to assess the accuracy and precision of a predictive model or to envision different kinds of predictive analytics. Based on the misspelled word (source) and the possible word suggestions (target word), a distance matrix is formed. It is used to calculate the cost of operations that must be performed to obtain the target word from the source word in order to achieve the target. Here is a diagram showing the distance matrix of the error phrase (தமிழ்நடு) relative to one of the suggestions (தமிழ்நாடு).

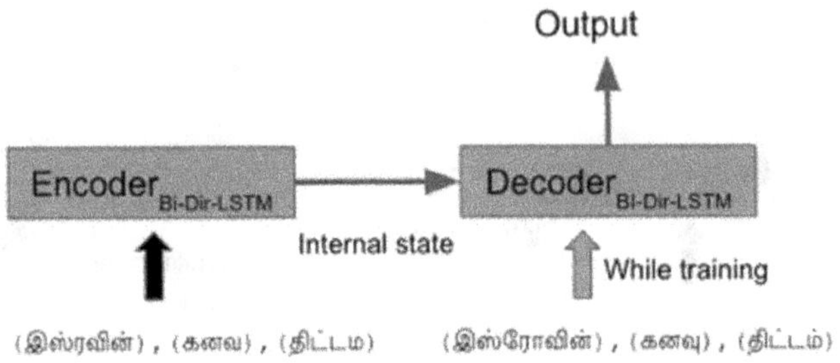
Output
Encoder Bi-Dir-LSTM
Internal state
Decoder Bi-Dir-LSTM
While training
(இஸ்ரவின்), (கனவ), (திட்டம)
(இஸ்ரோவின்), (கனவு), (திட்டம்)

Figure 17.17 Encoder Decoder Bidirectional LSTM diagram.

	#	த	ம	ி	ழ	ந	ட	ு
#	0	1	2	3	4	5	6	7
த	1	0	1	2	3	4	5	6
ம	2	1	0	1	2	3	4	5
ி	3	2	1	0	1	2	3	4
ழ	4	3	2	1	0	1	2	3
்	5	4	3	2	1	2	3	4
ந	6	5	4	3	2	1	2	3
ா	7	6	5	4	3	2	3	4
ட	8	7	6	5	4	3	2	3
ு	9	8	7	6	5	4	3	2

Figure 17.18 Edit distance matrix.

In order to construct the distance matrix, we apply the Levenshtein distance formula (show in equation 1). Levenshtein distance is a string metric used to measure the differences between two sequences.

$$lev(a,b) = \begin{cases} \max(a,b) & \text{if } \min(a,b) = 0 \\ \min \begin{cases} lev(a-1,b)+1 \\ lev(a,b-1)+1 \\ lev(a-1,b-1)+ \begin{cases} 1 & \text{if } a \neq b \\ 0 & \text{else} \end{cases} \end{cases} & \text{else} \end{cases}$$

Equation 1: Levenshtein distance formula

17.5.3 Machine translation

We would like to propose using Python Deep Translator for Tamil to English translation. The Deep Translator is a python package that allows users to translate between a variety of languages. Basically, the package aims to integrate multiple translators into a single, comprehensive system, such as Google Translator, DeepL, Pons, Linguee and others.

17.5.4 Inpainting

In the in-painting module, the mask image is created from the input image by masking the detected text in an image. In the next step both original and masked images are used to recreate the image portion at the masked area. Finally, the translated text is superimposed onto the image at specific position by bounding box coordinates of detected text.

The process of inserting new text into the image in place of the old text involves four steps:

1. Determine the bounding box coordinates for each text by analyzing the image with Easy-OCR;
2. To provide the algorithm with information about what areas of the image should be painted, each bounding box should be specified with a mask;
3. Create a text-free image using CV2 by applying an inpainting algorithm to the masked areas; and
4. Using OpenCV, the translated English text is replaced inside the image.

17.6 RESULT ANALYSIS

In this chapter, we explore the various variants of RNN network models in order to find the most suitable one for OCR spell checker and correction for Tamil. The result and analysis are mentioned below.

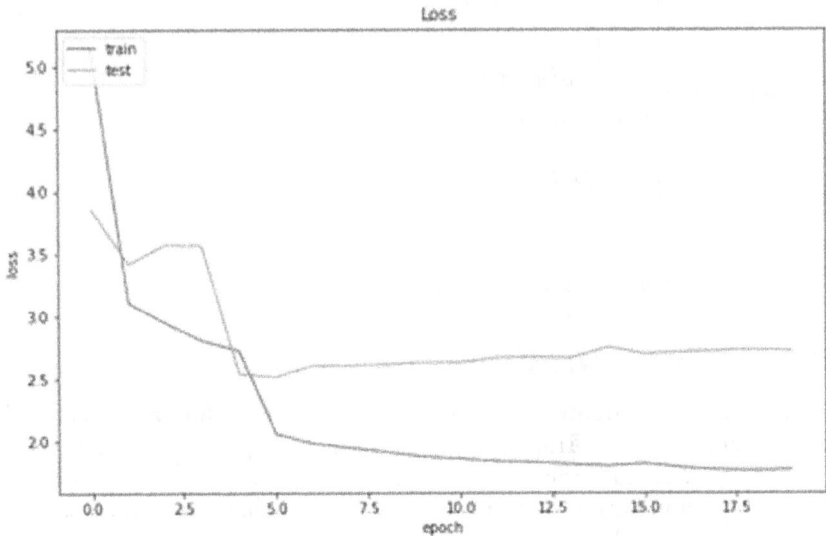

Figure 17.19 loss graph of simple RNN.

17.6.1 Simple RNN

Using the GRU network as the input for the simple RNN model, we can correct spelling by using GRU for spell check. Since GRU does not have any internal memory, it is only able to handle a limited number of datasets, which progressively decreases its training accuracy. The training loss results are mentioned in the graph (Figure 17.19).

17.6.2 Embed RNN

As GRU does not come equipped with an internal memory, Embed RNN replaces GRU with LSTM and adds a new input layer that converts input sequences into a dictionary of vectors. The modifications to the simple RNN make it faster, but the loss during training does not reduce. This is due to the conversion of the text to identical alphabets, a figure of the loss graph during training is shown in Figure 17.20.

17.6.3 Bidirectional LSTM

A bidirectional LSTM model can provide a greater level of detail about future data in order to analyze patterns. LSTMs can receive input in two directions, which makes them unique. LSTMs only allow input to flow

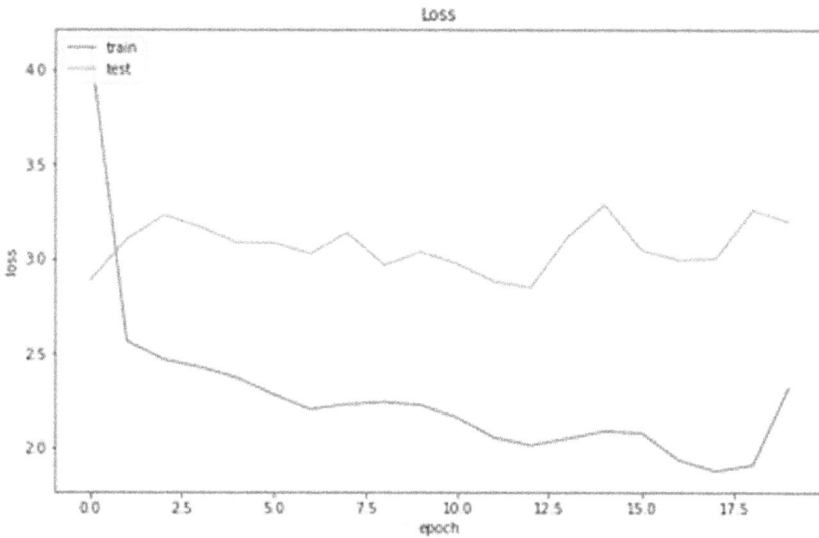

Figure 17.20 loss graph of Embed RNN.

in one direction, either backwards or forwards. An information network that is bi-directional preserves both past and future data by allowing input information to flow both ways. Although the network uses this information in both directions, the information is in an error state, making the model unsuitable for Tamil spelling check corrections. As shown in Figure 17.21, this model cannot train.

17.6.4 Encoder decoder with LSTM

The encoder encodes input sequences, while the decoder decodes encoded input sequences into target sequences, in this approach. But this model is not perfect to this problem because overfitting and validation accuracy are not changing over training as shown in the loss graph in Figure 17.22.

17.6.5 Encoder decoder with bidirectional LSTM + Levenshtein distance

Finally, the encoder decoder with bidirectional LSTM perform well over training data and loss starts to decrease during training as shown in Figure 17.23.

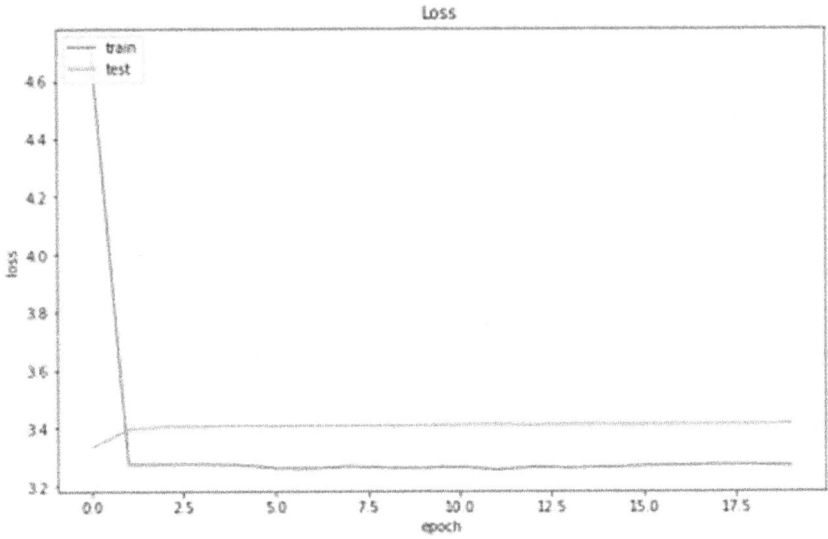

Figure 17.21 loss graph of Bidirectional LSTM.

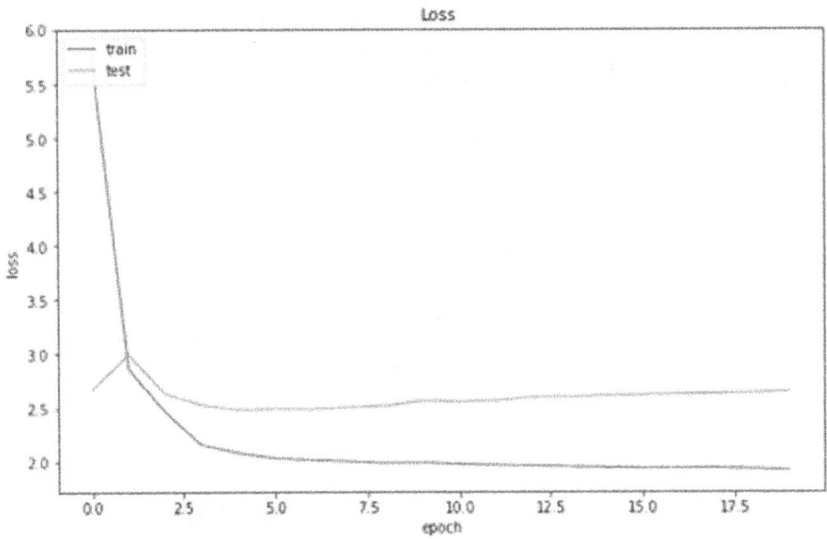

Figure 17.22 loss graph of encoder decoder with LSTM.

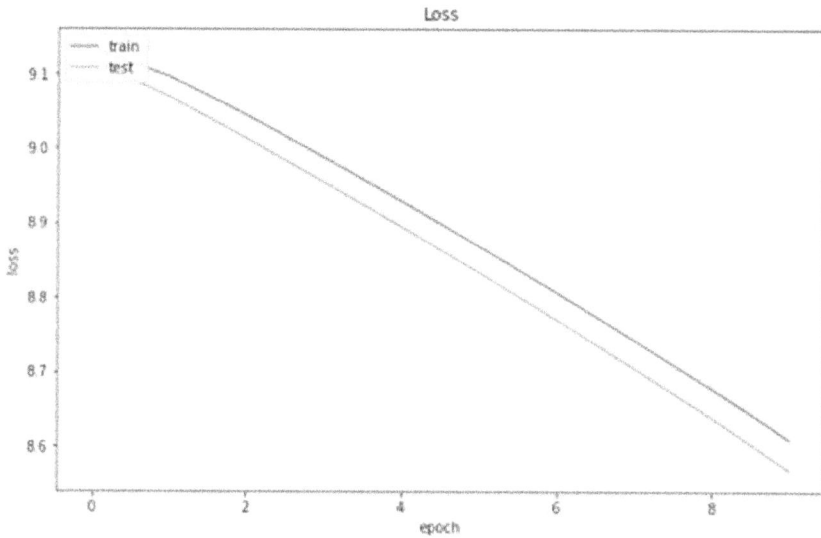

Figure 17.23 loss graph of encoder decoder with Bidirectional LSTM.

17.6.6 BLEU (bilingual evaluation understudy)

The BLEU is a metric used to determine how closely human translations resemble machine translations. This measure is used to evaluate the accuracy of automatic translation (error words into the correct spelling words). The formula for this measure is mentioned below.

Steps involved in calculating BLEU score

1. First step is to calculate clip count value based on predicted and correct word.

   Count clip = min(count, max_ref_count).

2. The modified precision score is calculated by dividing the clipped counts by the unclipped counts of candidate words.
3. Brevity penalty (BP) is an exponential decay that is calculated as follows. In this, r refers to the number of words in the reference document, and c refers to the number of words in the predicted document.

$$Bp = \begin{cases} 1 & if\ c > r \\ e^{\left(1-\frac{r}{c}\right)} & if\ c \leq r \end{cases}$$

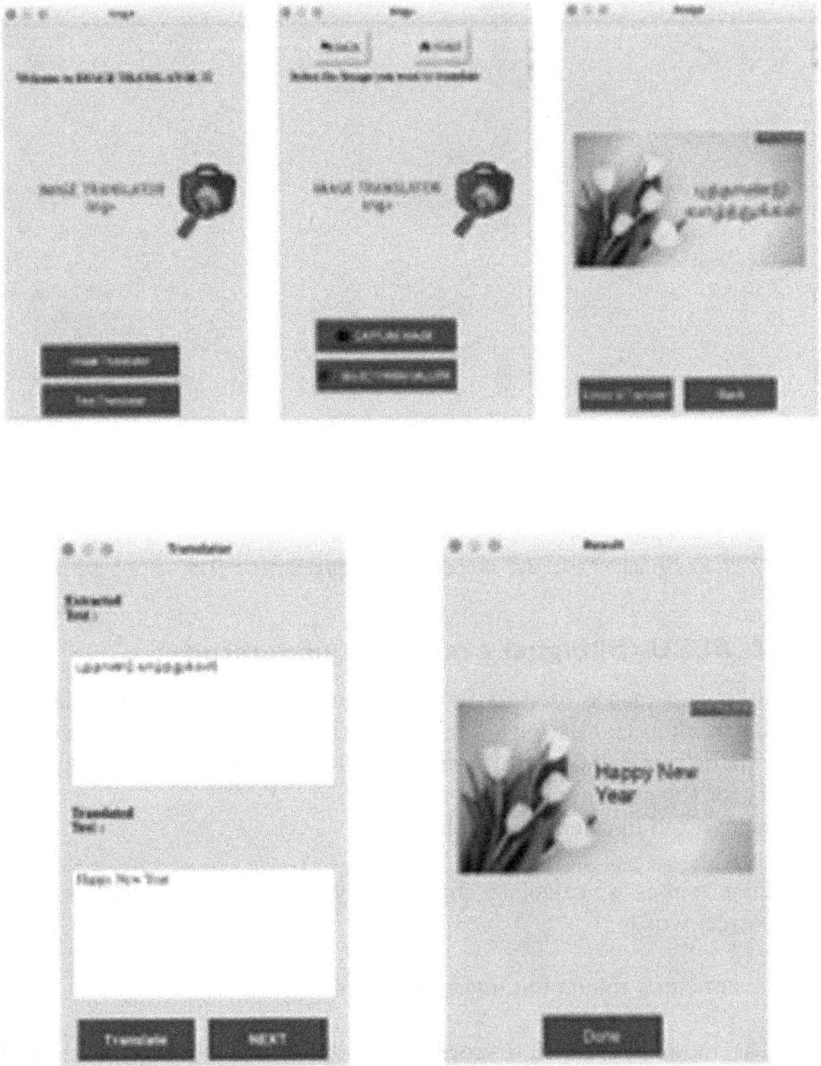

Figure 17.24 sample input and output of proposed work.

4. To calculate the BLUE score, the following formula is used: wn weight associated with n-grams

$$BLEU = BP.exp\left(\sum_{n=1}^{N} wn.\log pn\right)$$

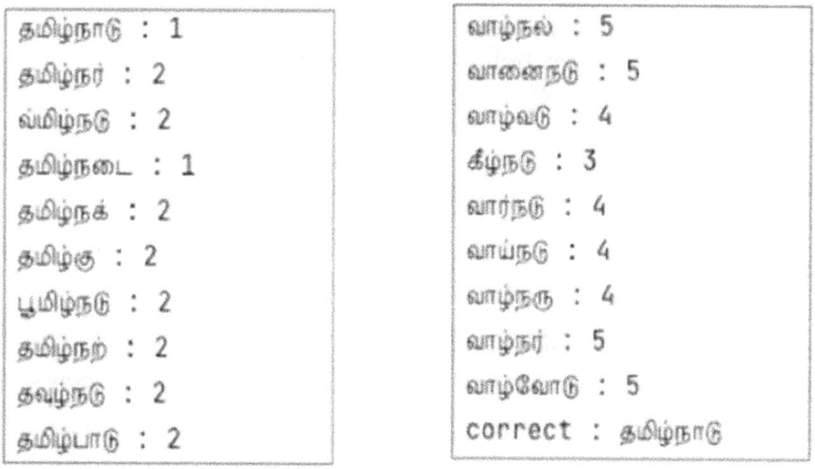

தமிழ்நாடு : 1	வாழ்நல் : 5
தமிழ்நர் : 2	வாஊைநடு : 5
ஸ்மிழ்நடு : 2	வாழ்வடு : 4
தமிழ்நடை : 1	கீழ்நடு : 3
தமிழ்நக் : 2	வார்நடு : 4
தமிழ்கு : 2	வாய்நடு : 4
பூமிழ்நடு : 2	வாழ்நரு : 4
தமிழ்நற் : 2	வாழ்நர் : 5
தஇழ்நடு : 2	வாழ்வோடு : 5
தமிழ்பாடு : 2	correct : தமிழ்நாடு

Figure 17.25 Sample spelling suggestions for misspelled word.

As a result of the procedure mentioned above, our mode has a BLEU score of 0.67. The sample input and output are shown below.

17.7 CONCLUSION

The chapter presents a computer vision and seq2seq encoder decoder model used to translate the information contained in an image (Tamil) to English and put it on the same image background. There are several modules incorporated into the system for extracting, spelling checking, translating, and inpainting text. Machine Translation can be improved through modules such as text extraction with easy OCR and spelling correction with the Seq2Seq model.

It requires improvement in two areas, the first one being that the system currently only handles Tamil to English translation. By adding several Indian languages as well as Western languages, the scope of the project can be expanded. The PM-India dataset is currently available for all Indian languages. Furthermore, if new words not used in training are encountered in any phase of the process, it will lead to mistranslation and overcorrection that ultimately leads to incorrect results. To improve the process, it is recommended that the dataset be expanded and new words made part of it.

ACKNOWLEDGMENTS

We would like to express our sincere gratitude to our respected principal-in-charge, PSG College of Technology, Coimbatore Dr. K. Prakasan for providing us the opportunity and facilities to carry out our work.

We also like to express our sincere thanks to Dr. G. Sudha Sadasivam, head of Computer Science and Engineering department for her guidance and support given to complete our work.

Our sincere thanks to our program coordinator, Dr. G. R. Karpagam, professor, Department of Computer Science and Engineering for guiding and encouraging us in completing our project.

Our sincere gratitude to our mentor, Dr. S. Suriya, associate professor, Department of Computer Science and Engineering, for her valuable guidance, additional knowledge, and mentorship throughout the course of our work.

We would also like to extend our gratitude to our tutor Dr. C. Kavitha, assistant professor, Department of Computer Science and Engineering, for her continuous support and evaluation throughout our work.

REFERENCES

[1] Shervin Minaee and Yao Wang, Fellow. Text Extraction From Texture Images Using Masked Signal Decomposition, IEEE(Global SIP), pp. 01–05 (2017).

[2] Sourav Ghosh, Dibyadwati Lahiri, Showmik Bhowmik, Ergina Kavallieratou, and Ram Sarkar. Text/Non-Text Separation from Handwritten Document Images Using LBP Based Features: An Empirical Study, mdpi, pp. 01–15 (2018).

[3] Frank D. Julca-Aguilar, Ana L. L. M. Maia and Nina S. T. Hirata. Text/non-text classification of connected components in document images, SIBGRAPI, pp. 01–06 (2017).

[4] Yun Song, Jianjun Chen, Hongtao Xie, Zhineng Chen, Xingyu Gao Xi Chen. Robust and parallel Uyghur text localization in complex background images, *Machine Vision and Applications*, vol. 28, pp. 755769 (2017).

[5] Johnson, Melvin, Mike Schuster, Quoc V. Le, Maxim Krikun, Yonghui Wu, Zhifeng Chen, Nikhil Thorat et al. "Google's multilingual neural machine translation system: Enabling zero-shot translation." Transactions of the Association for Computational Linguistics 5 (2017).

[6] Bapna, Ankur, Naveen Arivazhagan, and Orhan Firat. "Simple, scalable adaptation for neural machine translation." *arXiv preprint arXiv*:1909.08478 (2019).

[7] Tan, Xu, Jiale Chen, Di He, Yingce Xia, Tao Qin, and Tie-Yan Liu. "Multilingual neural machine translation with language clustering." *arXiv preprint arXiv*:1908.09324 (2019).

[8] Liu, Yinhan, Jiatao Gu, Naman Goyal, Xian Li, Sergey Edunov, Marjan Ghazvininejad, Mike Lewis, and Luke Zettlemoyer. "Multilingual

denoising pre-training for neural machine translation." Transactions of the Association for Computational Linguistics 8 (2020).

[9] Wang, Qiang, Bei Li, Tong Xiao, Jingbo Zhu, Changliang Li, Derek F. Wong, and Lidia S. Chao. "Learning deep transformer models for machine translation." arXiv preprint arXiv:1906.01787 (2019).

[10] Pan, Xiao, Mingxuan Wang, Liwei Wu, and Lei Li. "Contrastive learning for many-to-many multilingual neural machine translation." *arXiv preprint arXiv:2105.09501* (2021).

[11] Ott, Myle, Michael Auli, David Grangier, and Marc'Aurelio Ranzato. "Analyzing uncertainty in neural machine translation." In International Conference on Machine Learning, pp. 3956–3965. PMLR (2018).

[12] Kumar, Pawan, Abishek Kannan, and Nikita Goel. "Design and Implementation of NLP-Based Spell Checker for the Tamil Language." In Presented at 1st International Electronic Conference on Applied Sciences, vol. 10, p. 30 (2020).

[13] Uthayamoorthy, Keerthana, Kirshika Kanthasamy, Thavarasa Senthaalan, Kengatharaiyer Sarveswaran, and Gihan Dias. "DDSpell – A Data Driven Spell Checker and Suggestion Generator for the Tamil Language." In 2019 19th International Conference on Advances in ICT for Emerging Regions (ICTer), vol. 250, pp. 1–6. IEEE (2019).

[14] Pal, Aditya, and Abhijit Mustafi. "Vartani Spellcheck--Automatic Context-Sensitive Spelling Correction of OCR-generated Hindi Text Using BERT and Levenshtein Distance." *arXiv preprint arXiv:2012.07652* (2020).

[15] Sooraj, S., K. Manjusha, M. Anand Kumar, and K. P. Soman. "Deep learning based spell checker for Malayalam language." *Journal of Intelligent & Fuzzy Systems* 34, no. 3 (2018).

[16] Singh, Shashank, and Shailendra Singh. "Systematic Review of Spell-Checkers for Highly Inflectional Languages." *Artificial Intelligence Review* 53, no. 6 (2020)

[17] Hameed, M.H., Shawkat, S.A., Niu, Y. and Al_Barazanchi, I., "Improvement of Extract and Recognizes Text in Natural Scene Images." *Journal of Telecommunication Control and Intelligent System* (JTCIS) E-ISSN, 1(2) (2021).

[18] Murugan, Selvakumar, Tamil Arasan Bakthavatchalam, and Malaikannan Sankarasubbu. "SymSpell and LSTM based Spell-Checkers for Tamil." (2020).

[19] Pushpalwar, Rohit T., and Smriti H. Bhandari. "Image inpainting approaches-a review." In 2016 IEEE 6th International Conference on Advanced Computing (IACC), pp. 340–345. IEEE (2016).

[20] Elharrouss, Omar, Noor Almaadeed, Somaya Al-Maadeed, and Younes Akbari. "Image inpainting: A review." *Neural Processing Letters* 51, no. 2 (2020).

[21] Jam, Jireh, Connah Kendrick, Kevin Walker, Vincent Drouard, Jison Gee-Sern Hsu, and Moi Hoon Yap. "A comprehensive review of past and present image inpainting methods." *Computer Vision and Image Understanding* 203 (2021).

[22] Yeh, Raymond A., Chen Chen, Teck Yian Lim, Alexander G. Schwing, Mark Hasegawa-Johnson, and Minh N. Do. "Semantic image inpainting

with deep generative models." In Proceedings of the IEEE conference on computer vision and pattern recognition, pp. 5485–5493 (2017).

[23] Sulam, Jeremias, and Michael Elad. "Large inpainting of face images with trainlets." IEEE signal processing letters 23, no. 12 (2016).

[24] Liu, Guilin, Fitsum A. Reda, Kevin J. Shih, Ting-Chun Wang, Andrew Tao, and Bryan Catanzaro. "Image inpainting for irregular holes using partial convolutions." In Proceedings of the European conference on computer vision (ECCV), pp. 85–100 (2018).

[25] Newson, Alasdair, Andrés Almansa, Yann Gousseau, and Patrick Pérez. "Non-local patch-based image inpainting." *Image Processing On Line* 7 (2017).

Chapter 18

Multi-label Indian scene text language identification

Benchmark dataset and deep ensemble baseline

Veronica Naosekpam * *and Nilkanta Sahu*

18.1 INTRODUCTION

Language identification [1] deals with predicting the script of the text in a scene image. It is a sub-module of a scene text understanding system [2], as depicted in Figure 18.1. It is also taken as the successor of the text detection system [3,4] as well as a predecessor module of the scene text recognition system [5,6]. As text recognition algorithms are language-dependent, selecting a correct language model is essential. This is where our application will be essential. It is a prominent research area in the computer vision community [7] owing to its wide range of potential applications, such as language translation, image-to-text conversion, assistance for tourists, scene understanding [8], intelligent license reading systems, and product reading assistance for specially abled people in indoor environments, and so forth. Although two different languages can have the same script, we have used language and script interchangeably in this chapter.

While language identification from document analysis [9–11] is a well-explored problem, scene text language identification still remains an unexplored problem. Scene text comprises very few words, contrary to the presence of longer text passages in document images. Due to huge stroke structural differences, it is easy to classify using a simple classifier in cases such as identifying scripts between English and Chinese. However, it is quite cumbersome if the scripts have strong inter-class similarities, like Russian and English. The existing works on script identification in the wild are mainly dedicated to English, Chinese, Arabic, and a few East Asian languages [12–14] and have so far been limited to video overlaid text.

The challenges associated with the scene image text language identification task are: (1) enormous difference in the aspect ratios of the text images; (2) close similarity among scripts in appearance such as Kannada and Malayalam; (3) character sharing such as in English, Russian, and Greek; (4) variability in text fonts, scene complexity, and distortion; and (5) presence of two different languages per cropped text. Research in script identification on scene text images is scarce and, to the best of our knowledge, it is mainly

DOI: 10.1201/9781003453406-18

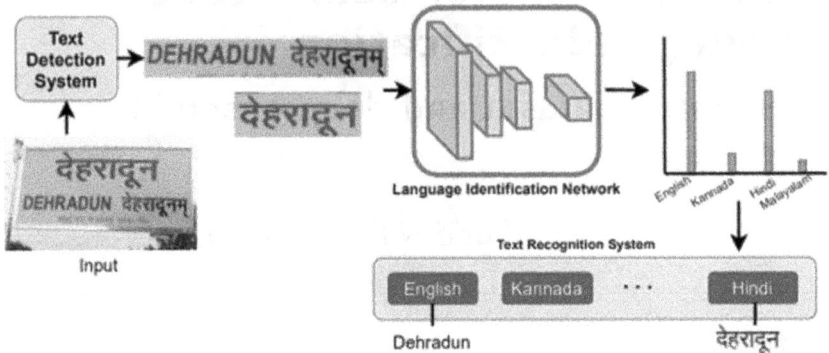

Figure 18.1 Scene text language identification module in the scene text understanding pipeline.

dedicated to English with some Chinese [15] and Korean texts [12]. It also mainly works on video overlaid text images [15,16] which are horizontal and clear.

India is a diverse country with 22 officially recognized scripts that are disproportionately distributed concerning the country's demographic. According to the Wikipedia source [17], Hindi is spoken by around 57 percent of the total population, whereas Bengali, Kannada and Malayalam are used by 8 percent, 4.8 percent and 2.9 percent, respectively, of the overall country's population. Due to this, there is lack of research in case of low resource Indian languages owing to its data scarcity. Though English is part of the official Indian languages, it is not prevalent in many parts of the Indian sub-continent. Furthermore, the existing datasets for script identification consists of a single language per image (SIW-13 [1], MLe2e [12], CVSI [15]) but, in reality, more than one language can be present in the scene text images. Hence, it is currently an open challenge for the research community in academia and industry. This motivates us to focus our area of research in this domain of multi-label regional Indian language identification.

To the best of our knowledge, there is not yet a work that deals with regional Indian scripts that involve many compound characters and structural similarities (for instance, Kannada and Malayalam). Although datasets are available for scene text script identification [12,15,18], they consist of one word per cropped image. In contrast, two or more world languages can occur in a scene image in a real environment.

To bridge the gap, we strive to solve the problems associated with script identification in the wild via multi-label learning, where a text image can be associated with multiple class labels simultaneously. We create a dataset called IIITG-MLRIT2022 consisting of two languages of text per image in different combinations. As the dataset is the first of its kind, we proposed a novel majority voting deep learning-based ensemble model as the baseline for the proposed dataset. We chose the ensemble learning technique as classification via ensembling [19,20] approaches are proven to be more accurate than the individual classifier through theoretical and empirical studies, and the same is also observed in the later sections.

The main contributions of this chapter are as follows:

- We put together a multi-lingual text image benchmark dataset called IIITG-MLRIT2022 for scene text language identification. Each image in the mentioned dataset consists of texts of two languages. The text scripts in the images are from five regional Indian languages.
- We propose a strong baseline model based on majority voting ensemble deep learning for multi-label language identification.
- We propose a weighted binary cross-entropy as the objective function to deal with the class imbalance problem.
- We present an experimental study to demonstrate the effectiveness of the ensemble-based learning and proposed objective function in terms of F1 and mAP scores.

The remainder of the chapter is organized as follows: Section 2 reviews the existing scene text script identification methods; Section 3 describes the proposed benchmark dataset, IIITG-MLRIT2022; Section 4 presents the proposed baseline scheme. Training and experimental settings of the proposed method are discussed in Section 5. In Section 6 we analyze and discuss the results obtained. Finally, we provide the conclusion and future research prospects in Section 7.

18.2 RELATED WORKS

The work of script or language identification initially started in the printed document text analysis domain. In the document image analysis domain, there are few established works for some Indian script identification. Samita et al. presented a script identification scheme [9] that could be used to identify 11 official Indian scripts from document text images. In their approach they first identify the text's skew before counter-rotating the text to make it oriented correctly. To locate the script, a multi-stage tree classifier is employed. By contrasting the appearance of the topmost and bottommost curvature of lines, edge detection is employed in Phan et al. [21] to identify scripts comprising English, Chinese, and Tamil languages. Singh et al. [22] extracted gray level co-occurrence matrix from handwritten document images and classify the scripts into Devanagari, Bangla, Telugu, and Roman using various classifiers such as SVM (Support Vector Machine), MLP (Multi-Layer Perceptron), Naive Bayes, Random Forest, and so forth, and concluded that the MLP classifier performed best among the classifiers on the dataset used. By redefining the issue as a sequence-to-label problem, Fuji et al. [23] proposed script identification at line level for document texts. A conventional method for cross-lingual text identification was introduced by Achint and Urmila in their paper [24]. When given images with texts in Hindi and English as input, the algorithm converts those texts into a single language (English). These conventional component analysis and binarization methods appear to be pretty unfit for images of natural scenes. Moreover, document text encompasses a series of words, whereas scene text images contain mostly less than two or three words.

The concept of script identification on scene text images was presumably originally introduced by Gllavata et al. [18]. In their research, a number of hand-crafted features were used to train an unsupervised classifier to distinguish between Latin and Chinese scripts. The subset of characters used in languages such as English, Greek, Latin, and Russian is the same. As a result, it is difficult to recognize the indicated scripts when using the hand-crafted features representation directly. Incorporating deep convolutional features [26–30] and spatial dependencies can solve differentiating such kinds of issues.

CNN and Naive-Bayes Nearest Neighbour classifier are combined in a multi-stage manner in Gomez et al. [12]. The images are first divided into patches and use a sliding window to extract stroke parts features. The features are fed into the CNN to obtain feature vectors that are further classified by using the traditional classifier. For features representation, deep features, and mid-level representation are merged in Shi et al. [14]. Discriminative clustering is carried out to learn the discriminative pattern called the codebook. They are optimized in a deep neural network called discriminative convolutional neural network. Mei et al. [13] combine CNN with Recurrent Neural Network (RNN) to identify scripts. The CNN structure comprising of convolutional and max-pooling layers without the fully connected layer are stacked up to extract the feature representations of the image. Then, these image representations are fed to the Long Short Term Memory (LSTM) layers. The outputs of the LSTM are amalgamated by average pooling. Finally, a softmax layer is built on top of LSTM layers to give out the normalized probabilities of each class. A novel semi-hybrid approach model integrates a BoVW (bag of visual words) with convolutional features in [14]. Local convolutional features in the form of more discriminative triplet descriptors are used for generating the code word dictionary. The merging of strong and weak descriptors increases the strength of the weak descriptors. Ankan et al. [32] developed script identification in natural scene image and video frames using Convolutional-LSTM network. The input to the network is image patches. It entails extracting local and global features using the CNN-LSTM framework and dynamically weights them for script identification. As far as we know, Keserwani et al. [33] were probably the first and the only researchers to propose a scene text script identification technique via few-shot learning. However, their method is based on multi-model approaches where word corpora of the text scripts are used along with text images for training to generate the global feature and semantic embedding vector.

So far, the works mentioned in the related work section for scene text script identification consider a single class label per cropped image. In contrast, in a real-time environment, more than one language can be present in scene images (for example, roads/railways' signboards). Therefore, it would be helpful to develop an application that can identify multiple languages simultaneously. The output of this application can be succeeded by passing it to its corresponding language recognizing system for further processing such as language translation, image-to-speech processing, and so forth.

18.3 IIITG-MLRIT2022

The objective of creating the dataset is to provide realistic and rigorous benchmark datasets to evaluate the multi-label language identification system, particularly for regional Indian scripts. As mentioned in the previous sections, the public benchmarks available [1,13,15] have concentrated primarily on English, European, and East Asian languages/scripts with little mention of diverse Indian languages. Moreover, the existing datasets have only one language per cropped image, unlike in reality, where there can exist more than one language per image. Hence, we present a novel dataset for multi-label language identification in the wild for regional Indian languages called the IIITG-MLRIT2022. It contains five languages (Hindi, Bengali, Malayalam, Kannada, and English) with the presence of two scripts per image (implying the multi-linguality).

The newly created dataset is diverse in nature with the existence of curved, perspective distorted, and multi-oriented text in addition to the horizontal text. This diversity is achieved by applying various image transformation techniques such as affine, arcs, and perspective distortion with different angular degrees. The dataset is harvested from multiple sources: captured from mobile cameras, existing datasets, and web sources. We illustrate sample images both with regular (Row 1 and Row 2) and irregular (Row 3 and Row 4) texts in Figure 18.2. The language combinations of the proposed dataset are: (Bengali, Hindi); (English, Kannada); and (English, Malayalam). The collected images are first concatenation into pairs and then resized to a fixed dimension while passing as an input to the proposed baseline. We show the statistical distribution of the language text pairs in Table 18.1.

The IIITG-MLRIT2022 dataset has a total of 1,385 text images that are cropped from scene images. The dataset classes are slightly imbalanced owing to the existence of more combinations of English with other regional

Figure 18.2 Sample images from the proposed benchmark IIITG-MLRIT2022 for regional Indian scene text script identification.

Table 18.1 Statistics of the IIITG-MLRIT2022 dataset

Class label combination	Train	Validation	Test
Bengali, Hindi	360	40	101
English, Kannada	347	39	99
English, Malayalam	300	33	67

Indian languages. A pie chart is shown in Figure 18.3 (a), representing the class distribution.

We also illustrate the co-occurrence matrix of the data labels in Figure 18.3 (b). It represents the number of times entities in the row appear in the same contexts as each entity in the columns. It also defines the number of times the combination occurs via color-coding. We hope that the proposed IIIT-MLRITS2021 will serve as a benchmark for multi-label language identification in the scene images.

18.4 PROPOSED METHODOLOGY

CNN can be used to implement a model $g(i, \phi)$ for multi-label classification with an input I and a C-dimensional score vector s as the output. In deep neural network, feature extraction and classification are integrated in a single framework, thus, enabling end-to-end learning. Multi-label language identification is defined as the task of generating sequential labels and forecasting the possible labels in a given scene image containing text.

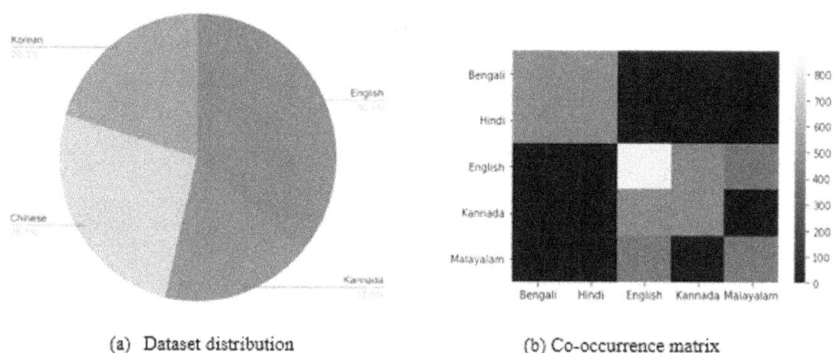

(a) Dataset distribution   (b) Co-occurrence matrix

Figure 18.3 Summarization of the proposed IIITG-MLRIT2022 via distribution plot and co-occurrence matrix.

As a set of training images, we are given $I=(i_1,i_2,...i_M)$ and corresponding labels $Y=(y_1,y_2,...,y_M)$ with M denoting the number of training images. The corresponding labels of the ith image i_i is $y_i=(y_{i1},y_{i2},...,y_{iC})$ where C is the number of labels. If $y_{ik}=1$, it means that the image i_i contains the label k otherwise $y_{ik}=0$.

Deep ensemble learning is the technique of integrating various deep learning models to achieve better results. While in a simple machine learning approach, a single hypothesis is learned from the training data, whereas ensemble methods try to construct a set of hypotheses and combine to use them. One of the fundamental design questions is what combination rule to use. Majority voting is a widespread choice when the individual model gives label output [34]. The merit of applying the majority voting is that it does not require further parameter tuning once the individual classifiers are trained. We build an end-to-end majority voting deep ensemble model and learn a mapping from image to label *g: I-> Y*. During the inference time, an image will be given, and the labels are predicted by the mapping *g* from the learned model. to label *g: I-> Y*. During the inference time, an image will be given, and the labels are predicted by the mapping *g* from the learned model.

In total, we explore five CNN's as base learners. They are divided into two categories according to their training strategy: (1) Fine-tune standard CNNs pre-trained on ImageNet dataset via transfer learning; and (2) three layers vanilla CNN. As the task is a multi-label classification, the number of neurons in the output layer of the individual network is matched with the number of class labels present in the dataset. Figure 18.4 shows the schematic of the ensemble deep learning approach implemented in this study.

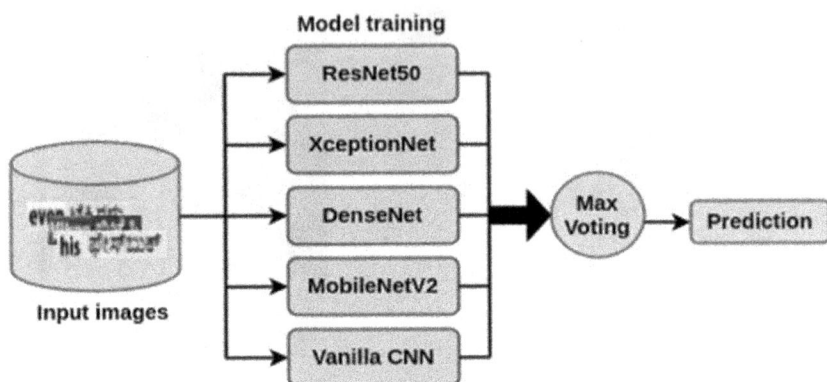

Figure 18.4 Proposed max-voting deep ensemble architecture.

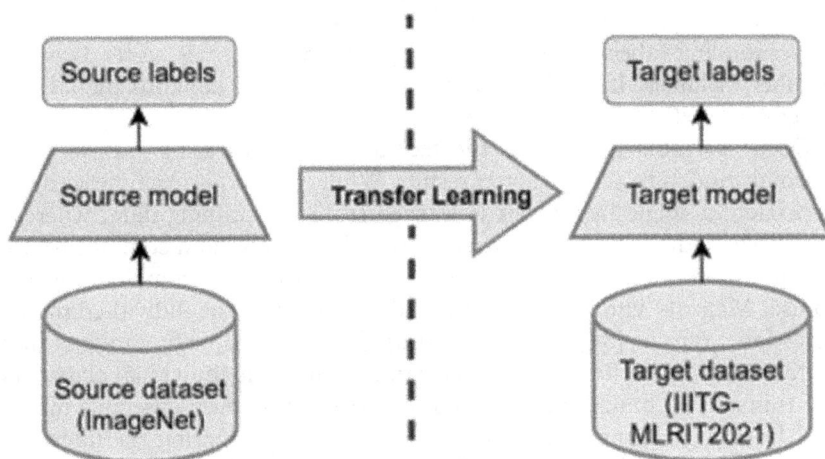

Figure 18.5 Transfer learning.

18.4.1 Transfer learning

The principal philosophy behind transfer learning is using knowledge learned from performing a particular task to solve problems in a different field (refer to Figure 18.5).

If a learning task Γ_t is given based on D_t, we can get help from D_s for the learning task Γ_t. $D_{(.)}$ denotes the domain of the respective, tasks which is made up of two parts: the feature space and the edge probability distribution. A task is represented by a pair: a label space and target prediction function $f_{(.)}$. Transfer learning tries to improve the predictive function $f_{t(.)}$ for a task Γ_t by discovering and transferring the latent knowledge from D_s and Γ_s with $D_t \neq D_s$ and $\Gamma_t \neq \Gamma_s$. In majority of the cases, the size of D_s is much larger than D_t.

Transfer learning is used for different tasks in deep learning, such as feature extraction, pre-training, and fine-tuning [35]. The initial layers of the CNN extract feature like corners, edges, and colors. As the layers get deeper, data-dependent features like textures emerge. In fine-tuning a pre-trained model, some of the prior convolutional layers' weights are frozen during the training process and are not updated.

In this research, transfer learning is achieved by leveraging the pre-trained weights of four standard deep CNN architectures: MobileNetV2 [36], ResNet50 [37], DenseNet [38], and XceptionNet [39]. A brief description of each of the component model is given below.

18.4.1.1 ResNet50 [37]

ResNet50 consists of five stages, each with a convolution and identity block. Each convolution block is made up of three convolution layers, and each identity block also has three convolution layers. ResNet 50 has over 20 million trainable parameters. ResNet fixes the problem of exploding/vanishing gradient, which is a common issue that occurs during the backpropagation step as the network gets deeper via skip connection. This short connection (also known as identity mapping) skips the training from a few layers and connects directly to the output. The intuition behind the ResNet is that instead of allowing the layer to learn the actual layer by layer mapping, say $L(x)$ where L is the output of the layer, and the network will fit, $H(x) = L(x) + x$ (refer Figure 18.6). Therefore, during the

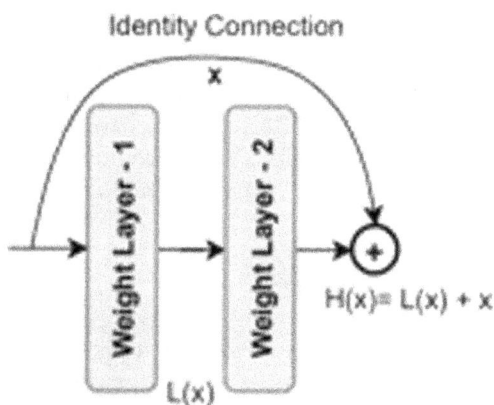

Figure 18.6 Residual Block of ResNet.

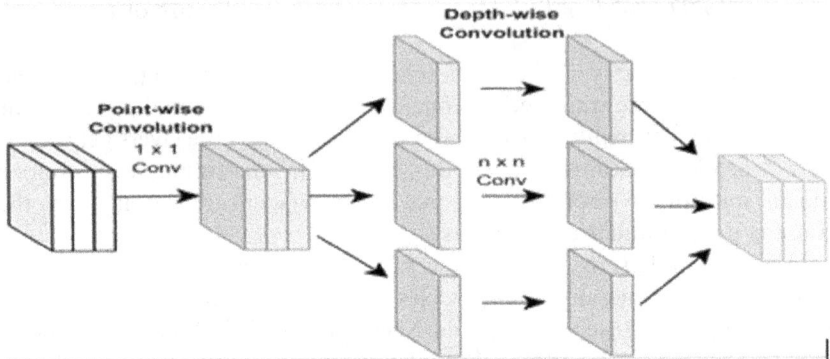

Figure 18.7 The depthwise separable convolution used as an Inception module in Xception with n=3.

backpropagation, while traversing through the residual block, the gradients can transit to the input layer via two pathways.

18.4.1.2 XceptionNet [39]

It is an extreme version of GoogleNet that stretches the inception concept. It introduces advanced inception layers called the modified depth-wise separable convolution that first performs a point-wise convolution layer followed by the depth-wise convolution (as illustrated in Figure 18.7). Point-wise convolution is a 1 x 1 convolution used for dimension change, and depth-wise convolution performs channel-wise n x n spatial convolution. For instance, for depthwise convolution, instead of applying convolution of size n x n x C, where C is the number of channels, convolution of size n x n x 1 is applied. As convolution is not required to be performed across all the channels, the computational cost is reduced as well as the memory requirements. It consists of 36 layers disguised in 14 modules. The architecture of XceptionNet is created by depth-wise separable convolution blocks and Maxpooling that are connected via shortcuts as in ResNet. Except for the first and the last module, the other modules have linear residual connections.

18.4.1.3 DenseNet [38]

DenseNet introduced an architecture with a straightforward connectivity pattern to warrant a maximum flow of information between the layers of the CNN. The network layers are connected so that a layer acquires auxiliary inputs from each and every previous layer (known as "collective knowledge") and sends its own feature vectors to all the successive layers via concatenation. It increases the efficiency of the network. It lowers the vanishing gradient problem with improved feature propagation in the forward and backward directions.

DenseNet is composed of N layers. Every layer implements a non-linear transformation $T_n[x_0,x_1,...,x_{n-1}]$ where n refers to the index of the layer; $T_n(.)$ is a composite function such as a combination of Batch normalization, ReLU, pooling, and convolutional layers; $[x_0, x_1, ..., x_{n-1}]$ is the feature vectors concatenation produced in layers 0 to $n-1$. To ease the process of down-sampling in the network, the entire architecture is divided into multiple compactly connected dense blocks, and the layers between these blocks are the transition layer that do convolution and pooling. DenseNet121 is used in our case.

18.4.1.4 MobileNetV2 [36]

MobileNetV2 belongs to the family of lightweight, low-latency and low-power CNN designed for deployment in mobile devices. The model was created to efficiently maximize accuracy while caring for the limited application resources. MobileNetV2 consists of two types of blocks: (1) a residual block of stride 1, and (2) a block of stride 2 for downsizing. For both varieties of blocks, there are three levels. Eleven convolutions with ReLU6 activation are present in the first layer. The depth-wise convolution is the second layer. The third layer is another 11 convolutions without a non-linear activation function owing to the claim that reusing non-linear activation will make the network become a linear classifier on the non-zero volume part of the output.

18.4.2 Convolutional Neural Network

The basic vanilla CNN consists of convolution, max-pooling, and dropout layers with ReLU non-linear activation functions. Suppose if N is the input, Θ is the parameter of the composite non-linear function $Z(.|\Theta)$. The purpose of this function is to map N to output prediction, say Y:

$$Y = Z\left(N/_\theta\right) = zL\left(...z_3\left(z_2\left(N/_\theta\right)/\theta_3/\theta L\right)\right) \tag{18.1}$$

where $z_l(.|\Theta)$ is the layer l of the vanilla CNN. If the parameter of layer-l is $\Theta_l=[X,b]$ where X denotes the corresponding filter and b denotes the vector bias term. The convolution operation is represented as:

$$Y_l = z_l\left(N_l/_{\theta_l}\right) = h\left(X * N_l + b\right) \tag{18.2}$$

where * denotes the convolution operation and $h(.)$ represents the point-wise activation. Pooling layers aid in the multi-scale analysis and input image size reduction. Max pooling applies a max filter to (usually) non-overlapping sub-regions of the initial representation. The activation function of ReLU returns the value provided as input or 0 if the value provided is less than 0. We chose this function as it accelerates the gradient descent towards a global minimum of the loss function.

Our CNN architecture is constructed by using seven convolutional layers. Every two convolutional layers are followed by a max-pooling layer and a dropout rate ranging between 0.2 to 0.6. The filter size for each convolutional layer is set to 3 x 3 with a stride of 1. The pooling dimension for the max-pooling layer is set as 2 x 2. In addition to these layers, the network contains a linear layer of 1,024 neurons and a final layer of five classes. We used Adam optimizer for training the network.

18.4.3 Multi-label deep ensemble via majority voting

Majority voting trains various individual base models/learners, and their outputs are aggregated through several rules. From empirical and theoretical studies [19], the training performances of the different deep learning component varies, meaning that some networks achieved good results while some achieved average results. Therefore, in order to improve the model's accuracy, we chose deep ensemble learning via a majority voting benchmark architecture for the newly created IIITG-MLRIT2022 dataset. It is found [41] that this technique is rarely biased towards the output of a specific component model owing to the mitigation by the majority vote count. The prediction, by utilizing only the shallow networks, is more diverse [25] than using only the deep neural networks. Hence, we select a combination of deeper and shallow networks for our task. Thus ideally, it is achieving better performance than the individual component model used in generating the ensemble.

In majority voting, every individual component classifier votes for a specific class label, and the final predicted class is the one that obtains more than half the votes. This is the reason for choosing an odd number of component classifiers. Additionally, the individual classifiers contribute the same weightage of the votes (weight is set as 1 for each of the base learners). As the IIITG-MLRIT2022 has more than one class label per image, the approach is called majority voting deep ensemble multi-label learning. Let N represent the number of individual classifiers $k_1, k_2, ..., k_N$ and the k_i's are to be combined to predict the label from the set of c class labels $L = l_1, l_2, ..., l_c$. For an input instance x, each member classifier k_i will output a probability score (also called the decision) $k_i(x, l_j)$ for each label l_j in L. That is, the final outputs of the component classifier k_i is a c-dimensional label vector $(k\_i^{c(x)}, ..., k\_i^{c(x)})^N$, where $k_i^j(x)$ is the output of k_i for the label l_j. Then, $k_i^j \in \{0, 1\}$

meaning it takes the value 1 if a component model k_i predicts lj as the label of the class. If more than half the classifiers agree to a particular label, then it will be selected.

The majority voting approach can be outlined as follows:

$$\max_{1 \le j \le C} \sum_{i=1}^{N} k_{i,j} \tag{18.3}$$

18.4.4 Weighted binary cross entropy

As the task is a multi-label classification problem, an image will be associated with more than one label. There is a need to check which labels out of all the N possible labels should be selected. Hence, the sigmoid activation function $\sigma(z) = \dfrac{1}{1 + e^{-z}}$ is used for the prediction of the labels. Each class prediction will be normalized between the (0,1) range, and the sum of all the predicted outputs can be more than 1. The threshold value is set for the prediction task. An appropriate class label is predicted for an image if the model exceeds a probability score greater than the set threshold (it is fixed at 0.5 after rigorous trials with different values) for that particular class.

As we have seen in Figure 18.3 (a), the IIITG-MLRIT2022 dataset's class distribution is skewed with the presence of 850+ English labels as opposed to the rest of the languages (that is, Malayalam, Hindi, Bengali, and Kannada), whose contents are 400 and below. However, this class imbalance biases the learning algorithm's performance toward the class label with majority samples. To mitigate this problem, we introduce a weighted binary cross-entropy called the Modified Weighted Binary Cross-Entropy (MWBCE).

The weights calculation is performed by the given formula:

$$weight_j = \frac{s}{L \sum_{k}^{s} \{yk = j\}} \tag{18.4}$$

S refers to the number of samples, and L is the number of classes. The MWBCE is given by:

$$MWBCE = \frac{1}{L \sum_{k=1}^{s} \{y_{k=j}\}} \sum_{j} weight_j \times (y_j * \log \bar{y} + (1 - y_j) * \log(1 - \bar{y})) \tag{18.5}$$

where y and $\hat{y}$ are respectively the target label, and the predicted label's probability. *weight*$_i$ is the class weight. The frequently occurring label samples will have less incentive and vice versa. This objective function is used for training the component models separately. The individual model's parameters are updated iteratively in a direction that reduces this loss function.

18.5 TRAINING AND EXPERIMENT

In this section, we perform training of the preliminary models and the proposed majority voting baseline model for the IIITG-MLRIT2022 dataset. To mitigate over-fitting that may prevent good performance of the model due to the skewedness of the data and also improve the model's generalization capability, we apply various data modification techniques called augmentation. The training images are randomly augmented using random flipping up and down, shifting height and width by 0.2 range, and a random amount of rotations. We also resize the image to 150 x 40 pixels. For rotated images, the gaps are filled in through reflection. Adding such random distortions at each epoch prevents the model from coming across the same samples many times.

As part of the fine-tuning process, we begin with an initial classification threshold value of 0.1 across the classes and perform an exhaustive search to get the best uniform baseline threshold value of 0.5, which is then applied to all the classes concurrently. The deep ensemble learning baseline method's experiments are performed on the NVIDIA Tesla V100 work station with 4 GPU cards, each having 32GB RAM with CentOS 7 operating system. For the implementation, Python programming language and Keras with Tensorflow backend framework are used. The collected images are concatenated into pairs, and then they are resized to a fixed dimension of 150 x 40 pixels. So, the image dimension is 150 x 40 x 3. The first 100, 20, 60, and 149 layers of the ResNet50, MobileNetV2, XceptionNet and DenseNet (specifically DenseNet121) respectively are frozen. The remaining layers are then trained. The dimensions of the input image are changed per the pre-trained network's requirements except for the CNN component, where the image input size remains 150 x 40. Adam optimizer is used with a learning rate of 0.001 and decay of 0.0001. The batch size is set to between 16 and 64 based on the network's size. The component models are trained for 100 epochs. In addition to this, dropouts are used for regularization of the fully connected layers with a value chosen between 0.2 and 0.5. Besides the above settings, the L1 regularization technique is utilized in the fully connected (FC) layer by a value of 0.001 to further take care of the overfitting problem. For each image, the system chooses the class label with the maximum number of votes as the correct prediction.

Table 18.2 Experimental analysis on IIITG-MLRIT2022 using F1-score

Method	F1-score (MWBCE)	F1-score (BCE)
CNN	87.14%	77.14%
ResNet50	87.60%	81.70%
XceptionNet	84.07%	73.13%
DenseNet (DN)	85.75%	74.77%
MobileNetV2	80.04%	64.87%
Ensemble(CNN, ResNet50, DN)	88.04%	77.16%
Deep Ensemble Baseline (all models)	**88.40%**	**78.03%**

We chose F1-score and mAP as the metric for evaluation. F1-score is a particular case of the general F beta function. It merges the Precision and Recall into a single metric of evaluation that assesses the harmonic mean between the two. This score is helpful when there is an imbalance in the data.

Where *Prec* and *Rec* denote the Precision and Recall of the prediction. Table 18.2 summarizes the overall performance obtained for each component model along with the majority voting ensemble baseline obtained by applying the proposed objective function and the binary cross-entropy using F1-score. The majority voting ensemble baseline model is evaluated on the IIITG-MLRIT2022 dataset with 1,118 training sets and 267 testing sets.

18.6 RESULTS AND DISCUSSION

In this section, we discuss the results of the experiments performed in the previous section. We also compare the individual component models with the max voting ensemble to prove the ensemble's superiority. To show the effectiveness of the weighted objective function, we also exhibit the performance of the basic binary cross-entropy function. As the dataset is the first of its kind, we could not correlate with the state-of-the-art techniques that work with single-label data. As can be seen in Table 18.2 in terms of F1-score, the majority voting baseline ensemble model performs the highest. We also observed that when the best three highest performing deep learning models are combined via a majority voting ensemble, the prediction accuracy is slightly decreased. The proposed baseline is effective even when the text script is slanted, curved, or slightly disoriented (refer to Figure 18.11 (b, d and e). The F1-score performances of the model using the binary cross-entropy is lower, implying the model has a tendency to be biased towards the majority classes.

Following works like [31,40] we also evaluate the classification performance for multi-label data using the mean average precision (mAP) metric. Similarly, as in Table 18.2, we listed the mAP scores with MWBCE and BCE in Table 18.3. Although the proposed work outperforms the combination

Table 18.3 Experimental analysis on IIITG-MLRIT2022 using mAP-score

Method	mAP (MWBCE)	mAP (BCE)
CNN	72.07%	61.22%
ResNet50	81.85%	57.40%
XceptionNet	83.00%	58.71%
DenseNet (DN)	84.41%	62.57%
MobileNetV2	75.43%	48.63%
Ensemble(CNN, ResNet50, DN)	84.00%	**62.76%**
Deep Ensemble Baseline (all models)	**85.62%**	61.08%

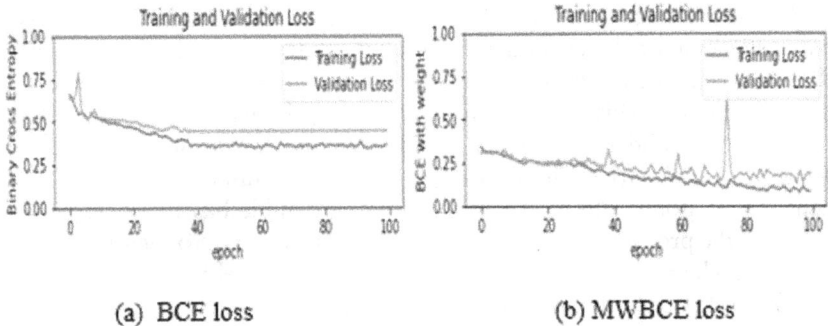

(a) BCE loss   (b) MWBCE loss

Figure 18.8 Loss plots for CNN component model.

of the three best models while utilizing the proposed objective function, it fails to overtake the five-model combination while utilizing the normal BCE. We also noticed that, overall, the mAP scores are slightly lower than the F1-score.

A comparison of the proposed loss function plot (MWBCE) with the ordinary BCE for the CNN and MobileNetV2 components with respect to the training and validation loss is depicted in Figure 18.8 and 18.9, respectively. The plot implies that the proposed objective function is more stable, the training convergence uniformly within the set epoch, and a much lower value of the loss is also obtained (Figure 18.8 (b) and Figure 18.9 (b)).

To have better insight into the F1-score distribution of the different class label combinations present in the IIITG-MLRIT2022, the performance of the individual models and the ensemble are checked for each combination, as viewed in Figure 18.10. As depicted in the plot, the English-Kannada text script combination label has the highest overall accuracy scores, while the Bengali-Hindi combination label has the least prediction score. This may be due to the similar appearance of Bengali and Hindi, owing to the fact that the two languages share the same origin. Of all the component models, the MobileNetV2 output the lowest prediction score. It is observed from the

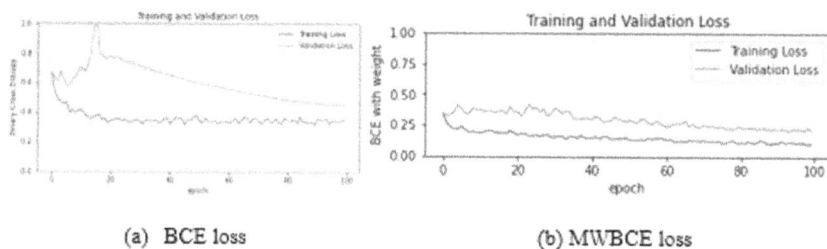

(a) BCE loss   (b) MWBCE loss

Figure 18.9 Loss plots for MobileNetV2 component model.

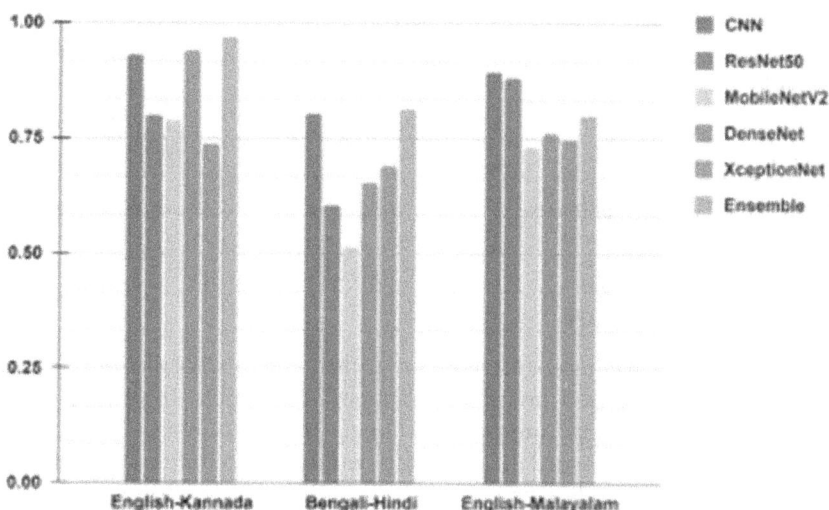

Figure 18.10 Comparison of language identification accuracies (pair-wise) among the component model and the ensemble baseline on the proposed IIITG-MLRIT2022 dataset.

experimental analysis that the English-Malayalam pair is often misclassified as the English-Kannada pair. The most likely reason can be that these scripts are structurally similar.

The intuition behind merging different models via the ensembling technique is that if a model fails to identify a label correctly, any other different model may be able to predict the expected label. It reflects an increase in the performance of the model. With the help of the majority voting deep ensemble scheme, the classification accuracy has been improved. In Figure 18.11 (b, d and f) are predicted correctly owing to the ensemble method.

	(a)	(b)	(c)
MobileNetV2	Bengali Hindi	Bengali Hindi	English Kannada
ResNet50	Bengali Hindi	Bengali Hindi	English Kannada
CNN	Bengali Hindi	Bengali Hindi	English Kannada
DenseNet	Bengali Hindi	English	English Kannada
XceptionNet	Bengali Hindi	English Malayalam	English Kannada

	(d)	(e)	(f)
MobileNetV2	English Malayalam	English Malayalam	English
ResNet50	English Kannada	English Malayalam	English Malayalam
CNN	English Kannada	English Malayalam	English Malayalam
DenseNet	English Kannada	English Kannada	Kannada
XceptionNet	English	English Kannada	English Malayalam

Figure 18.11 Correctly predicted samples as a result of majority voting deep ensemble learning scheme.

18.7 CONCLUSION

In this chapter, we introduced in an end-to-end multi-label scene text language's identification preliminary framework, which we believe is the first research to incorporate multiple languages in an image. We created multi-label scene text word images using two images of different languages. The word images are classified using a majority voting deep ensemble architecture that achieved better prediction accuracy than the individual component models. The ensemble model includes MobileNetV2, ResNet50, DenseNet, Xception Network and 7-layers of vanilla CNN. We further investigated the impacts of varying the number of base learners and its effect on the voting strategy. We found that the F1-score of combination of all the base learners is superior than the performance of combination of best three highest performing deep learning models on application of the proposed weighted objective function. We have also created a multi-label scene text language dataset called the IIITG-MLRIT2022, the first of its kind based on regional Indian languages. In future, The IIITG-MLRIT2022 dataset can be extended to more Indian language scene text images. Also, exploring an end-to-end multi-lingual natural scene text understanding system by emphasizing the regional Indian languages will be a good research direction.

REFERENCES

[1] Shi, B., Yao, C., Zhang, C., Guo, X., Huang, F., & Bai, X. (2015, August). Automatic script identification in the wild. In 2015 13th International Conference on Document Analysis and Recognition (ICDAR) (pp. 531–535). IEEE.

[2] Naosekpam, V., & Sahu, N. (2022). Text detection, recognition, and script identification in natural scene images: a Review. *International Journal of Multimedia Information Retrieval*, 1–24.

[3] Naosekpam, V., Kumar, N., & Sahu, N. (2020, December). Multi-lingual Indian text detector for mobile devices. In International Conference on Computer Vision and Image Processing (pp. 243–254). Springer, Singapore.

[4] Naosekpam, V., Aggarwal, S., & Sahu, N. (2022). UTextNet: A UNet Based Arbitrary Shaped Scene Text Detector. In International Conference on Intelligent Systems Design and Applications (pp. 368–378). Springer, Cham.

[5] Naosekpam, V., Shishir, A. S., & Sahu, N. (2021, December). Scene Text Recognition with Orientation Rectification via IC-STN. In TENCON 2021–2021 IEEE Region 10 Conference (TENCON) (pp. 664–669). IEEE.

[6] Sen, P., Das, A., & Sahu, N. (2021, December). End-to-End Scene Text Recognition System for Devanagari and Bengali Text. In International Conference on Intelligent Computing & Optimization (pp. 352–359). Springer, Cham.

[7] Naosekpam, V., Bhowmick, A., & Hazarika, S. M. (2019, December). Superpixel Correspondence for Non-parametric Scene Parsing of Natural Images. In International Conference on Pattern Recognition and Machine Intelligence (pp. 614–622). Springer, Cham.

[8] Naosekpam, V., Paul, N., & Bhowmick, A. (2019, September). Dense and Partial Correspondence in Non-parametric Scene Parsing. In International Conference on Machine Intelligence and Signal Processing (pp. 339–350). Springer, Singapore.

[9] Ghosh, S., & Chaudhuri, B. B. (2011, September). Composite script identification and orientation detection for indian text images. In 2011 International Conference on Document Analysis and Recognition (pp. 294–298). IEEE.

[10] Phan, T. Q., Shivakumara, P., Ding, Z., Lu, S., & Tan, C. L. (2011, September). Video script identification based on text lines. In 2011 International Conference on Document Analysis and Recognition (pp. 1240–1244). IEEE.

[11] Lui, M., Lau, J. H., & Baldwin, T. (2014). Automatic detection and language identification of multilingual documents. *Transactions of the Association for Computational Linguistics*, 2, 27–40.

[12] Gomez, L., & Karatzas, D. (2016, April). A fine-grained approach to scene text script identification. In 2016 12th IAPR workshop on document analysis systems (DAS) (pp. 192–197). IEEE.

[13] Mei, J., Dai, L., Shi, B., & Bai, X. (2016, December). Scene text script identification with convolutional recurrent neural networks. In 2016 23rd international conference on pattern recognition (ICPR) (pp. 4053–4058). IEEE.

[14]   Shi, B., Bai, X., & Yao, C. (2016). Script identification in the wild via discriminative convolutional neural network. *Pattern Recognition*, 52, 448–458.

[15]   Sharma, N., Mandal, R., Sharma, R., Pal, U., & Blumenstein, M. (2015, August). ICDAR2015 competition on video script identification (CVSI 2015). In 2015 13th international conference on document analysis and recognition (ICDAR) (pp. 1196–1200). IEEE.

[16]   Naosekpam, V., & Sahu, N. (2022, April). IFVSNet: Intermediate Features Fusion based CNN for Video Subtitles Identification. In 2022 IEEE 7th International conference for Convergence in Technology (I2CT) (pp. 1–6). IEEE.

[17]   Wikipedia contributors. (2022, September 8). List of languages by number of native speakers in India. In Wikipedia, The Free Encyclopedia. Retrieved 06:04, September 12, 2022, from https://en.wikipedia.org/w/index.php?title=List_of_languages_by_number_of_native_speakers_in_India&oldid=1109262324

[18]   Gllavata, J., & Freisleben, B. (2005, December). Script recognition in images with complex backgrounds. In *Proceedings of the Fifth IEEE International Symposium on Signal Processing and Information Technology*, 2005. (pp. 589–594). IEEE.

[19]   Sammut, C., & Webb, G. I. (Eds.). (2011). Encyclopedia of machine learning. Springer Science & Business Media.

[20]   Matan, O. (1996, April). On voting ensembles of classifiers. In *Proceedings of AAAI-96 workshop on integrating multiple learned models* (pp. 84–88).

[21]   Phan, T. Q., Shivakumara, P., Ding, Z., Lu, S., & Tan, C. L. (2011, September). Video script identification based on text lines. In *2011 International Conference on Document Analysis and Recognition* (pp. 1240–1244). IEEE.

[22]   Jetley, S., Mehrotra, K., Vaze, A., & Belhe, S. (2014, October). Multi-script identification from printed words. In *International Conference Image Analysis and Recognition* (pp. 359–368). Springer, Cham.

[23]   Fujii, Y., Driesen, K., Baccash, J., Hurst, A., & Popat, A. C. (2017, November). Sequence-to-label script identification for multilingual ocr. In 2017 14th IAPR international conference on document analysis and recognition (ICDAR) (Vol. 1, pp. 161–168). IEEE.

[24]   Kaur, A., & Shrawankar, U. (2017, February). Adverse conditions and techniques for cross-lingual text recognition. In 2017 International Conference on Innovative Mechanisms for Industry Applications (ICIMIA) (pp. 70–74). IEEE.

[25]   Choromanska, A., Henaff, M., Mathieu, M., Arous, G. B., & LeCun, Y. (2015, February). The loss surfaces of multilayer networks. In *Artificial intelligence and Statistics* (pp. 192–204). PMLR.

[26]   Mahajan, S., Abualigah, L., & Pandit, A. K. (2022). Hybrid arithmetic optimization algorithm with hunger games search for global optimization. *Multimedia Tools and Applications*, 1–24.

[27]   Mahajan, S., & Pandit, A. K. (2022). Image segmentation and optimization techniques: a short overview. *Medicon Eng Themes*, 2(2), 47–49.

[28] Mahajan, S., Abualigah, L., Pandit, A. K., & Altalhi, M. (2022). Hybrid Aquila optimizer with arithmetic optimization algorithm for global optimization tasks. *Soft Computing*, 26(10), 4863–4881.

[29] Mahajan, S., & Pandit, A. K. (2021). Hybrid method to supervise feature selection using signal processing and complex algebra techniques. *Multimedia Tools and Applications*, 1–22.

[30] Mahajan, S., Abualigah, L., Pandit, A. K., Nasar, A., Rustom, M., Alkhazaleh, H. A., & Altalhi, M. (2022). Fusion of modern meta-heuristic optimization methods using arithmetic optimization algorithm for global optimization tasks. *Soft Computing*, 1–15.

[31] Chen, Z. M., Wei, X. S., Wang, P., & Guo, Y. (2019). Multi-label image recognition with graph convolutional networks. In Proceedings of the IEEE/CVF conference on computer vision and pattern recognition (pp. 5177–5186).

[32] Bhunia, A. K., Konwer, A., Bhunia, A. K., Bhowmick, A., Roy, P. P., & Pal, U. (2019). Script identification in natural scene image and video frames using an attention based Convolutional-LSTM network. *Pattern Recognition*, 85, 172–184.

[33] Keserwani, P., De, K., Roy, P. P., & Pal, U. (2019, September). Zero shot learning based script identification in the wild. In 2019 International Conference on Document Analysis and Recognition (ICDAR) (pp. 987–992). IEEE.

[34] Kuncheva, L. I. (2014). *Combining Pattern Classifiers: Methods and Algorithms*. John Wiley & Sons.

[35] Yosinski, J., Clune, J., Bengio, Y., & Lipson, H. (2014). How transferable are features in deep neural networks?. *Advances in neural information processing systems*, 27.

[36] Sandler, M., Howard, A., Zhu, M., Zhmoginov, A., & Chen, L. C. (2018). Mobilenetv2: Inverted residuals and linear bottlenecks. In Proceedings of the IEEE conference on computer vision and pattern recognition (pp. 4510–4520).

[37] He, K., Zhang, X., Ren, S., & Sun, J. (2016). Deep residual learning for image recognition. In: Proceedings of the IEEE conference on computer vision and pattern recognition (pp. 770–778).

[38] Iandola, F., Moskewicz, M., Karayev, S., Girshick, R., Darrell, T., & Keutzer, K. (2014). Densenet: Implementing efficient convnet descriptor pyramids. *arXiv preprint* arXiv:1404.1869.

[39] Chollet, F. (2017). Xception: Deep learning with depthwise separable convolutions. In Proceedings of the IEEE conference on computer vision and pattern recognition (pp. 1251–1258).

[40] Wang, J., Yang, Y., Mao, J., Huang, Z., Huang, C., & Xu, W. (2016). Cnn-rnn: A unified framework for multi-label image classification. In Proceedings of the IEEE conference on computer vision and pattern recognition (pp. 2285–2294).

[41] Ganaie, M. A., & Hu, M. (2021). Ensemble deep learning: A review. *arXiv preprint* arXiv:2104.02395.

Chapter 19

AI based wearables for healthcare applications

A survey of smart watches

*Divneet Singh Kapoor, Anshul Sharma, Kiran Jot Singh,
Khushal Thakur, and Amit Kumar Kohli*

19.1 INTRODUCTION

The wrist watches have seen tremendous development towards being called
"smart," and mainly for the utilization in distant wellbeing checking and
versatile wellbeing [1]. Smart watches can be considered as important
an innovation as the smart phone (mobile phone), which highlights the
persistent information checking for advance wellbeing, for example,
step's count, pulse observing, energy use, and actual work levels [2].
They can give input to the users, who can screen their wellbeing, perform
mediations in the nick of time – for example, drug utilization dependent
on discussions, and direct correspondence with guardians and doctors
[3]. The extensive technology use in medical services and telemedicine is
restricted by the boundaries that are specific to the smart watches, such
as expense, wearability, and battery life [4]. Hence, the further mentioned
analysis studies the applications of the smart watch in the medical field
and its related characteristics leading to possible smart watch applications
to monitor health remotely. The medical services and telemedicine depend
on the utilization of cell (mobile) phones to empower distant wellbeing
observation of patients [5]–[12]. The instances of practical use of a smartphone in the medical sector are management of a prolonged disease at
home [5], discontinuation of the smoking habit [6], planning one's family
[7], mental health treatment [8], and various other applications of clinical
studies [9]–[12].

Smartphones contemplate nonstop intuitive correspondence from any
area, figuring the ability to help media programming applications, and
consistent observation of patients through remote detecting innovations
[12]. In any case, smartphones cannot be worn on the body (owing to their
large size) to give certain real-time pulse data, and are not generally carried
during the practices of interest, for example, during constrained physical
activities [13]. Smart watches, on the other hand, can be worn on the body
(especially the wrist) to gather real-time pulse and action data of the human
user, which makes it ideal for medical care and associated applications [14].

DOI: 10.1201/9781003453406-19

Smart watches can consolidate sensors' data – accelerometers, whirligigs, compasses, and pulse, along with Geographical Position System (GPS) information. A smart watch is especially favorable where a regular check on physical activity is needed, for recognizing sudden fluctuations and, henceforth, alert the users/ beneficiaries and suggest aid as per geographical region. Compared to a smartphone, a smart watch helps in effortlessly accessing messages and alarms because from this tiny device one gets sounds, texts, and vibrations as well. A smart watch has infinite scope in the medical sector. On top of this, the flexibility of the software applications also helps in customizing a smart watch according to the user's medical requirements.

Past surveys on smart watches [1, 3] have led to the discovery that despite the many studies involving smart watch utilization, not many have been tried past feasibility. In this survey, we evaluate whether any investigations have tried beyond feasibility and usability, and which medical care applications have used smart watches in their intercessions. Moreover, contrary to Lu et al. [1], the present survey thoroughly studies the technology used in every investigation, by inspecting the characteristics essential for categorization, the kind of smart watch utilized and various in-built sensors. Current smart watches on the market were also studied for medically useful characteristics of the watch, other sensing procedures, and their adaptability in present lab trial investigation design. All this was done because the adaptability of the present lab trials and the layout for lab interventions had been unexpanded in Reeder et al. [3].

The survey intends to explore the answers to the questions below:

- Determine if any research on smart watches for use in medical sector had been tried past achievability and ease of use.
- Recognize the kinds of medical service applications where smart watches have been used.
- Find the populaces and test states that used smart watches for past medical cases.
- Compare the sensing innovations, classification, and current features of the smart watch based medical care frameworks, and distinguish the different smart watch innovations and highlight the accessible smart watches for further exploration purposes.

These questions helped determine whether a smart watch fits the present lab trial blueprint for varied medical sector applications. These questions helped in finding the areas of the medical sector that would be benefitted by using a smart watch in their system layout, along with determining the sensing technologies and the best-fit smart watches to be utilized. The above-mentioned questions were answered via an organized study, which is explained below. The analysis studied the types of applications in the medical sector, explanations about the tests carried out, and the characteristics of this tiny device's technologies, which were the most significant of all. For analyzing the resultant innovative characteristics amongst the selected articles, a short search was carried out via search engines (available on the worldwide Web), so that the smart watches present in the market could be located along with their technical characteristics. Section 2 represents the systematic review, which is followed by discussion in Section 3 and finally a conclusion is given in Section 4.

19.2 SYSTEMATIC REVIEW

19.2.1 Criterion to select research

Papers published from 1998 to 2020 were looked for because the premier computer that was worn on a wrist, was made and owned in 1998. This premier computer was a wrist-watch having a Linux OS, which was developed by Steve Mann [15]. Research papers written in languages besides English, were selected on the condition of fulfilling the selection criterion along with translating the language via Google translate. Results of the study comprised of the type of review (e.g. randomly controlled test, achievability review, beneficial review), number of participants, test setup (lab or community), type of the population on whom the trial was conducted and kinds of uses in the medical sector. Also, each research with details about the smart watch used, like the OS, the kind of smart watch, essential sensors, characteristic types, and lastly the kind of connectivity (e.g. Bluetooth smartphone connectivity) was considered, and researches using both smart watch and a smartphone, were selected.
Selection of a research depended on

- The usage of smart watches;
- The reach of the research being related to the uses in the medical sector;
- The research being a randomly controlled trial, used to study the delivery of an intercession by utilizing smart watches; and
- Humans being tested.

Rejected researches were the ones that

- Depended completely on a cell phone or a smartphone only;
- Provided no health improvement application;
- Had been selected as an analytical study on past reported researches, editorials, abstract, poster or a case report;
- Excluded tests on humans; and
- Had no relation to medical applications.

19.2.2 Source of information

To dig out appropriate smart watch research for the medical sector, material for smart watch application in the mentioned field was looked for in the ACM Digital Library bank from 1998–2020, IEEE Xplore, Elsevier, Springer, EBSCO, and PubMed databases. The findings in every information bank was restricted to English and translated into to English papers. "Smart watch health care," "smart watch apps," "smart watch app" and "smart watch application," were the applied keywords to identify the required research papers. The short-listed papers from the above-mentioned information banks were then filtered for relevant research papers, as per the stated criteria. The papers falling under the eligibility criteria, were dispatched to two independent persons who evaluated and then verified the appropriate material. Also, reference lists from short-listed research were added in the findings, and analysis was done as per the eligibility criteria.

19.2.2.1 Search plan

The terms *smart watch* and *savvy* were searched for throughout EBSCO and PubMed. Similar words were searched for through ACM Digital Libraries and IEEE Xplore, in addition to the inquiry term *wellbeing*. The rule was to incorporate papers written in English or the ones that announced smart watch innovation utilization, focusing on medical purposes or wellbeing. Avoidance rules were considered that showed: suggested plans, structures, systems, and dummies; banner, meeting digests; work environment, driving wellbeing applications. The principal writer applied incorporation/prohibition measures to screen digests of the underlying outcomes corpus, leaving 33 papers for complete-text survey. Subsequent writers applied consideration/rejection models to audit the complete-text of these papers, and the first author checked article prohibition, which left 25 papers to be incorporated in the survey.

19.2.2.2 Data abstraction

The first and second writers delved into the accompanying information components from each included paper – distribution scene, evaluation center, gadget dummy, gadget type (customer grade, engineer gadget, or test model), member data, and specialized difficulties – to utilize smart watch functions for wellbeing purposes. Summarization of outcomes took place to delineate the present status of smart watch research for wellbeing related purposes.

19.2.3 Outcomes

A total of 1391 papers were considered in the pursuit (Figure 19.1 illustrates PRISMA interaction stream chart of the list items), and 610 exceptional

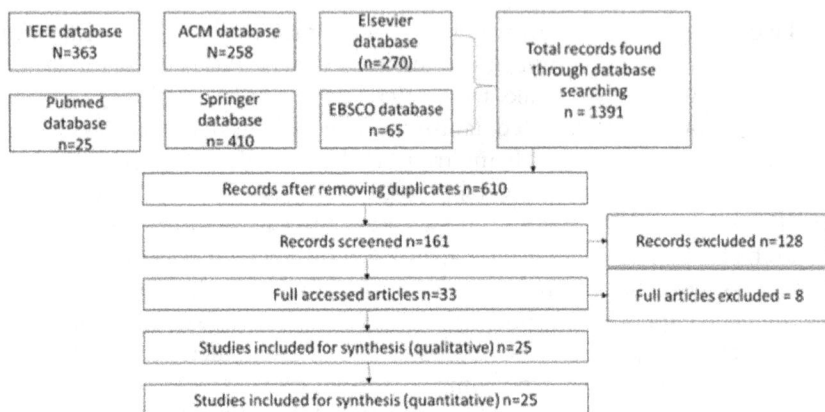

Figure 19.1 Flow diagram of the literature search result's PRISMA process.

articles stayed after duplicates were taken out. In the wake of assessing the articles, as per the incorporation standards mentioned, 161 articles were chosen for additional consideration, dependent of the above qualification measures. The screening cycle showed that 128 papers were rejected out of the 161 papers barred from additional investigation, on the grounds that a smart watch was not utilized. In the wake of playing out a theoretical survey, selection of 33 papers was made to audit the complete text. From these 33 papers, 25 were considered for the last investigation. Out of these, none of the papers had sufficient plan homogeneity for meta-examination.

Eight from the final survey were prohibited on the grounds that they utilized a half breed wellness band and smart watch: for example, the Microsoft watch was hazy whether they tried utilizing a smart watch or another wearable sensor or just examined the coordination of the framework and did not unambiguously clarify whether pilot information from human members was received. Besides, a few frameworks did not indicate which sort of innovation was used in the intercession, or permitted members to utilize either wellness band like a Fit Bit or a smart watch. Wellness groups were rejected from the survey, as they do not give comparable input and user interface highlights such as extraordinary accessible programming applications as smart watches.

19.2.4 Healthcare applications

Smart watches have been tried out in very few applications of clinical survey; 25 papers were scrutinized, and a majority of the assessments focused on utilizing smart watches to check different activity types and consistent self-organization applications. Various uses of smart watches comprised nursing or privately arranged wellbeing checking and clinical surveillance. Note the activity surveillance comprised of advance gathering along with routine tasks categorization. Additionally, research on managed chronic contamination were those that arranged an intervention based on smart watches, to address a particular group of sick people. Few of these papers had subresearches for physically fit individuals, who endorsed their task identification methods depending on the significance of the variable in their intercession.

A smart watch research paper summary and their applications in clinical consideration can be found in Table 19.1 [16]–[35]. In general, the table shows couple of papers on chronic ailment managed by self uses that focus on epilepsy or Parkinson's disease. The researches explicitly gave their structures a shot on geriatrics for private-care use or the people who experience the evil impacts of dementia. Although the majority of the task categorization researches ran their models on physically fit persons, there was one research attempted and facilitated toward uses for blind persons. As many examinations were possibility research, it is not astounding that the majority of the assessments across applications attempted their computation and structure plans on sound individuals before carrying out future testing on selected applications for clinical benefits.

19.2.4.1 Activity and human motion

Activity based surveillance examines how smart watches intended for possible trials in future medical service uses ordinarily tried their frameworks on sound people in the lab and local area. In any case, one investigation zeroed in on movement characterization for people having visual disabilities. One study proposed an intelligent route framework that supported blind people by perceiving present the assessment of highlights in a scene identifying the client's area and posture. Subsequently, it gave sound and haptic inputs to the user regarding present unknown surroundings, hence helping them perform everyday routines like shopping, and so forth. To test this framework plan, the smart watch navigation framework was tried for an investigation of one blind individual when that person was both inside and outside in the immediate environment.

Different investigations that zeroed in on movement acknowledgment tried the effectiveness of these frameworks on physically fit people. In these investigations, effectiveness testing was done on more than one member, ranging from 5–20 members, and these examinations tested the framework under both research center and local environment conditions. At long last, the principal objective of these investigations was to test characterization calculations and procedures for different sorts of exercises – for example,

Table 19.1 Studies conducted in lab, community, or medical centers with prospective cohort, feasibility, or usability assessments in context to wellbeing

Article	Objective of Pilot Study	Population Size	Environment Setting	Test Population Health Condition	Implementation Area
Ali 2016 [16]	Usability	1 (Nursing Home Staff)	Community	Healthy	Nursing or Home-Based Care
Årsand 2015 [17]	Usability	6	Community	Diabetic	Self Care (During Chronic Health Conditions)
Banos 2016 [18]	Feasibility	10 healthy, 6 medical experts	Community	Healthy	Self Care (During Chronic Health Conditions)
Boletsis 2015 [19]	Usability	1	Community	Dementia Affected	Nursing or Home-Based Care
Chippendale 2014 [20]	Feasibility	1	Community	Visually Impaired	Activity Monitoring
Dubey 2015 [21]	Feasibility	3	Laboratory	Parkinson's Disease Affected	Self Care (During Chronic Health Conditions)
Dubey 2015 [22]	Feasibility	3 healthy, 3 affected by Parkinson's Disease	Laboratory	Parkinson's Disease Affected; Healthy;	Self Care (During Chronic Health Conditions)
Duclos 2016 [23]	Feasibility	16	Laboratory and Community	Healthy	Activity Monitoring
Faye 2015 [24]	Feasibility	13	Community	Healthy	Activity Monitoring
Haescher 2015 [25]	Feasibility	14	Community	Healthy	Activity Monitoring
Jeong 2015 [26]	Feasibility	1	Laboratory	Healthy	Healthcare Education
Kalantarian 2015 [27]	Feasibility	10	Laboratory	Healthy	Self Care (During Chronic Health Conditions)
Lockman 2011 [28]	Usability	40	Laboratory and Community	Affected by Epilepsy	Self Care (During Chronic Health Conditions)
Lopez 2014 [29]	Feasibility	10	Laboratory	Parkinson's Disease Affected	Self Care (During Chronic Health Conditions)
Mortazavi 2015 [30]	Feasibility	20	Laboratory	Healthy	Activity Monitoring

Neto 2015 [31]	Feasibility	15 healthy, 11 low vision	Laboratory	Visually Impaired; Healthy	Self Care (During Chronic Health Conditions)
Panagopoulos 2015 [32]	Usability	26	Laboratory	Geriatric	Nursing or Home-Based Care
Sharma 2014 [33]	Feasibility	5	Laboratory	Parkinson's Disease Affected	Self Care (During Chronic Health Conditions)
Thomaz 2015 [34]	Feasibility	20	Laboratory and Community	Healthy	Self Care (During Chronic Health Conditions)
Vilarinho 2015 [35]	Feasibility	3	Laboratory	Healthy	Nursing or Home-Based Care

sitting, standing, resting, running, hopping, cycling, and other daily exercises, including working.

19.2.4.2 Healthcare education

The use of smart watches stressed their utilization for medical care schooling. Specifically, researchers used smart watches to assess, and show medical care experts, the technique of performing cardio-pneumonic revival (CPR). By estimating chest pressure or depth via accelerometers on the smart watch, the framework had the option to give input to the client about the degree of chest pressure being applied, along with guiding them through the CPR cycle according to American Heart Association rules. A few studies assessed the framework's possibilities for individuals and realized the smart watch's ability to give exact estimates of chest compressions. In contrast to cell phone application, the smart watch was discovered to be more user-friendly, because an individual did not have to grasp the gadget. Also, visual impedance was absent when the screen was covered by hands. Therefore, a smart watch has a few of the best qualities over mobile phones for medical care applications, which are evaluated below.

19.2.5 Ideal smart watch characteristics

As per the particular medical care applications discussed, smart watches clearly possess many significant highlights, making them perfect for utilization in medical care over cell phone applications. Smart watches constantly gather work information, along with other biosensor information, pulse for example, therefore making them perfect gadgets for intercessions, where attention is on movement and tracking walks/jogging/running. Moreover, where cell phones need to be in the person's hand or pockets, smart watches are worn during actual work and treatment that might need physical workouts or a significant degree of effort by the individual.

Along with nonstop bio-signal surveillance, smart watches are likewise ready to be worn and can be brilliant dependable watches locally and at home. Research has revealed the restricted life of batteries in field settings. Moreover, smart watches can give messages and warnings that afre accessed with ease by individuals, while exercising or having intercessions, because they utilize sound, text, and vibration to monitor the client. This gives more prompt correspondence with medical services experts. Lastly, since the new 5.1.1 Android Wear update, smart watches currently are improving calculation force and battery life, hence making it possible for the smart watch being utilized for the entire day, to constantly track data locally. These highlights gave way to successful possibility assessment, in the previously mentioned research, and if battery life is improved this will result in the introduction of valuable applications to be tried in preliminary clinical examinations.

19.2.5.1 Operating system

After analyzing the research papers, it was found that most medical service smart watch applications preferred Android-based smart watches over Tizen-based and iOS ones (see Table 19.2 [16]–[35]), as the Android OS is an open-source and smart watch. And running on Android is less expensive. Also, because the Apple smart watch running on iOS was launched in 2015, the majority of the research had no access to the iOS smart watch at the time their reports were published. Additionally, the type of watch used for research purposes was the Samsung Gear Live, due to its availability that time. The Samsung Gear Live dummy was utilized in 30 percent of the researches. The Samsung watch needed a matched cell phone or tablet for complete functioning, prior to the 5.1.1 Android Wear update. But currently, after the 5.1.1 update, having WiFi backing, the smart watches are utilized as the sole gadget for widely carrying out medical experiments.

19.2.5.2 Sensors

Table 19.3 [36]–[51] shows that financially accessible smart watches have a plethora of sensors. Accessible sensors include microphones, GPS, compasses, altimeters, gauges, pulse sensors, pedometers, magnetometers, proximity sensors, gyroscopes, and accelerometers. The modular smart watch portrayed in Table 19.3 is a custom smart watch based on the Android OS and permits scientists to pick detecting instruments they might want on the watch, which allows additional types of sensors, significant for wellness surveillance. These sensors incorporate sweat sensors, skin temperature sensor, ECG (electrocardiogram) and heartbeat oximeters. The smart watch may turn out to be especially helpful in future medical care applications, as it takes into consideration physiological sensors not commonly found in smart watches, allowing to persistently screen people locally and along these lines take into account more potential ailments to be checked and intercessions to be investigated.

19.3 DISCUSSION

Even though the majority of the features highlighted above are accessible on cell phones, there were a few strong reasons to include smart watches in these investigations. In the first place, as recently referenced by the researchers in multiple studies [13], [17], [33], [34], [52–58], [18], [19], [21], [22], [27–29], [31], with the use of inertial sensors, the smart watches

Table 19.2 Comparative analysis of smart watches used for different application areas in the literature

Article	Implementation Area	Smart watch Used	Classification Features	Sensors Available	Operating System	Connectivity
Ali 2016 [16]	Nursing or Home-Based Care	Samsung Gear Live	None – used as feedback only	Micro-phone	Wear OS	Not Specified
Årsand 2015 [17]	Self Care (During Chronic Health Conditions)	Pebble	Built-in step count	accelerometer ($\pm$4G)	Wear OS	Bluetooth to Mobile
Banos 2016 [18]	Self Care (During Chronic Health Conditions)	Not Specified	Not Specified	Accelerometer, gyroscope	Not Specified	Bluetooth to Cloud
Boletsis 2015 [19]	Nursing or Home-Based Care	Basis B1	Heart Rate Variations, Intensity of Movements, Sweat Intensity, Temperature of Body & Environment	Accelerometer; Heart rate, Temperature	Custom	Bluetooth to PC
Chippendale 2014 [20]	Activity Monitoring	Sony smart-watch	Vertical angular acceleration matched to stepping	Microphone, gyroscope ($\pm$8G)	Wear OS	WiFi to Mobile
Dubey 2015 [21]	Self Care (During Chronic Health Conditions)	ASUS ZenWatch	Kurtosis, negentropy	Microphone	Wear OS	Bluetooth to PC
Dubey 2015 [22]	Self Care (During Chronic Health Conditions)	ASUS ZenWatch	Fundamental frequency, Loudness	Microphone	Wear OS	Bluetooth to PC
Duclos 2016 [23]	Activity Monitoring	Samsung Gear Live	Acceleration vector variance, relative magnitude	Accelerometer ($\pm$2G)	Wear OS	Bluetooth to Mobile

	Application	Device	Features	Sensors	OS	Connectivity
Faye 2015 [24]	Activity Monitoring	Samsung Gear Live	Heart Rate, Wrist velocity (Average & Maximum)	Heart rate, pedometer, accelerometer (±2G)	Wear OS	Bluetooth to Mobile
Haescher 2015 [25]	Activity Monitoring	Simvalley Mobile AW-420 RX	Magnitude area, energy, mean crossing rate, dominant frequency, movement intensity	Accelerometer (±2G), gyroscope (±256°/s), microphone	Wear OS	Bluetooth to Mobile
Jeong 2015 [26]	Healthcare Education	Samsung Gear Live	Positive peak accelerations	accelerometer (±2G)	Wear OS	Bluetooth to PC
Kalantarian 2015 [27]	Self Care (During Chronic Health Conditions)	Samsung Gear Live	AF (Audio Frequency) distribution	Microphone	Wear OS	Bluetooth to Mobile
Lockman 2011 [28]	Self Care (During Chronic Health Conditions)	Smart Monitor Smart watch	Duration, frequency, intensity	Accelerometer (±4G)	Wear OS, iOS	Bluetooth to PC
Lopez 2014 [29]	Self Care (During Chronic Health Conditions)	Not Specified	Intensity, direction	Accelerometer (range not available)	Not Specified	Not Specified
Mortazavi 2015 [30]	Activity Monitoring	Samsung Gear Live	Difference, gyroscope intensity, mean, sum, dominant frequency	Accelerometer (±2G), gyroscope (±300°/s)	Wear OS	Bluetooth to PC
Neto 2015 [31]	Self Care (During Chronic Health Conditions)	Samsung Gear Live	Temporal Coherence & Facial Recognition	Camera, microphone	Wear OS	Bluetooth to Mobile
Panagopoulos 2015 [32]	Nursing or Home-Based Care	Not Specified	Not Specified	LCD screen, accelerometer (range not available)	Wear OS, iOS	Bluetooth to Mobile

(Continued)

Table 19.2 (Continued)

Article	Implementation Area	Smart watch Used	Classification Features	Sensors Available	Operating System	Connectivity
Sharma 2014 [33]	Self Care (During Chronic Health Conditions)	Pebble	Mean, energy, high frequency energy content, entropy	Accelerometer ($\pm$4G)	Wear OS	Bluetooth to Mobile
Thomaz 2015 [34]	Self Care (During Chronic Health Conditions)	Pebble	mean, variance, skewness, kurtosis, root mean square	Accelerometer ($\pm$4G)	iOS	Bluetooth to Mobile
Vilarinho 2015 [35]	Nursing or Home-Based Care	LG G Watch R^2	Vectorial acceleration (Absolute & Total), Fall index	Accelerometer, gyroscope (1G–3G)	Wear OS	Bluetooth to Mobile

Table 19.3 Available smart watches in 2016, their manufacturer, available sensors, OS, available Bluetooth version (where N/A means not available), rating of battery and price in 2016 on an average

Manufacturer	Edition	Price (USD)	Battery Rating (mAh)	Available Sensors	OS	Blue-tooth Version
Pebble Technology [36]	Pebble Classic Smart watch	100	140	Microphone, Compass, Accelerometer.	Android Wear	BLE 4.0
Samsung [37]	Samsung Gear Live	100	300	Accelerometer, gyroscope, heart rate, compass, camera.	Android Wear	BLE 4.0
Sony [38]	Sony Smart watch 3	130	420	Microphone, Compass, Accelerometer.	Android Wear	BLE 4.0
LG Electronics [39]	LG Gizmo Gadget	150	510	GPS	Android Wear, iOS	BLE 4.0
Pebble Technology [40]	Pebble Steel Smart watch	150	150	Microphone, Compass, Accelerometer.	Android Wear	BLE 4.0
Pebble Technology [41]	Pebble Time Round Watch	200	150	Pedometer, magnetometer, gyroscope, Microphone, Compass, Accelerometer.	Android Wear	BLE 4.0
Simvalley Mobile [42]	Simvalley Mobile AW-420	250	600	Camera, GPS, Compass, Gyroscope, Accelerometer.	Android Wear	BLE 4.0
LG Electronics [43]	LG G Watch R W110	300	410	Barometer, Heart rate, proximity, Gyroscope, Accelerometer.	Android Wear	BLE 4.0
Motorola [44]	Motorola Moto 360 Sport	300	300	GPS, Barometer, Heart rate, proximity, Gyroscope, Accelerometer.	Android Wear	BLE 4.0
BLOCKS [45]	BLOCKS Modular Smart watch	330	300	Microphone, Perspiration, temperature, ECG, SPO2, Altimeter; GPS, Heart Rate, Gyroscope, Accelerometer.	Android Wear (Custom)	BLE 4.0

(Continued)

Table 19.3 (Continued)

Manufacturer	Edition	Price (USD)	Battery Rating (mAh)	Available Sensors	OS	Blue-tooth Version
Huawei Technologies Co. Ltd. [46]	Huawei Watch GT	350	300	Heart rate, gyroscope, Altimeter, Accelerometer.	Android Wear	BLE 4.2
LG Electronics [47]	LG Watch Urbane W150	450	410	Heart rate, proximity, Gyroscope, Accelerometer.	Android Wear	BLE 4.1
LG Electronics [48]	LG Watch Urbane 2nd Edition LTE	500	570	LTE Communication, GPS, barometer, Heart rate, proximity, Gyroscope, Accelerometer barometer.	Android Wear	BLE 4.1
Apple Inc. [49]	Apple Watch First Generation	600	250	Heart rate, Microphone, Gyroscope, Accelerometer.	iOS	BLE 4.0
Smart Monitor [50]	Smart Monitor Smart watch	150 + $30 monthly fee	N/A	GPS, Accelerometer	Android Wear, iOS	BLE 4.0
Basis [51]	Basis B1 Band	Recalled	190	Heart rate, Accelerometer	N/A	BLE 4.0

provide an option to observe physical activities and behavior in situations where cell phones cannot be used, for example, while exercising or during hospital visits. Observation of one's physical activities can be quite helpful for the self-management of chronic diseases, especially in children [30] or bedridden patients who do not wear cell phones [28], [56]. In addition, under the circumstances LCD screens and speakers can be used to take user inputs and provide feedback. Smart watches not only allow observation of physical activities but can further be enhanced to classify more complex behaviors like face recognition during navigation [31] and type of food being eaten [21], [22], [27] using a variety of sensors like skin impedance [19], temperature [19], heart rate [57], cameras [31], and microphones [21], [22], [27], [31], [56], [58]–[60]. Clinical trials of multifaceted disease conditions that need individualized interventions can be greatly benefited from these complex classifications of diverse behaviours.

However, the use of smart watches in healthcare has been found to be only partial because of a few limitations. First, its function as a wearable computer has only been explored since 2014 and very little time has elapsed for them to become popular in the healthcare market. Also, since the environmental and physiological data collected by smart watches have very little to no validation to the clinical data its compliance with the Health Insurance Portability and Accountability Act (HIPAA) so far poses a serious challenge. The compliance requires data to be transmitted securely, maintaining its privacy. But, according to the current scenario, the majority of smart watches are still dependent on connected smartphones to transmit the data to secure servers. Along with the above design issues, there is a technological challenge of battery life, which further creates a challenge using the smart watch for interventions studies, as the user would like to use the same device without having to remember to charge it. Finally, advancements are required in creating human–machine interfaces to provide the feedback mechanisms that allow improved compliance to the intervention and hence the use of smart watches.

19.4 CONCLUDING REMARKS

It has been observed that only 25 articles out of the 1391 studies on smart watch utilization are directed towards its use in medical care. Moreover, these examinations had restricted applications, which included healthcare education, home-based care, nursing, self-management of chronic diseases and activity monitoring. All examinations were viewed as a possibility or convenience studies, and in this way had an extremely small number of study subjects tried out. Due to the lack of random clinical trial research, further examination on bigger populaces is recommended. This will evaluate the adequacy of utilizing smart watches in medical services intercessions and may at last prompt an inescapable selection of the innovation in this field.

REFERENCES

[1] T. C. Lu, C. M. Fu, M. H. M. Ma, C. C. Fang, and A. M. Turner, "Healthcare Applications of Smart Watches: A Systematic Review," *Appl. Clin. Inform.*, vol. 7, no. 3, p. 850, 2016, doi: 10.4338/ACI-2016-03-RA-0042

[2] E. M. Glowacki, Y. Zhu, E. Hunt, K. Magsamen-Conrad, and J. M. Bernhardt, "Facilitators and Barriers to Smartwatch Use Among Individuals with Chronic Diseases: A Qualitative Study." Presented at the annual University of Texas McCombs Healthcare Symposium. Austin, TX, 2016.

[3] B. Reeder and A. David, "Health at hand: A systematic review of smart watch uses for health and wellness," *J. Biomed. Inform.*, vol. 63, pp. 269–276, Oct. 2016, doi: 10.1016/J.JBI.2016.09.001

[4] D. C. S. James and C. Harville, "Barriers and Motivators to Participating in mHealth Research Among African American Men," *Am. J. Mens. Health*, vol. 11, no. 6, pp. 1605–1613, Nov. 2017, doi: 10.1177/1557988315620276

[5] T. de Jongh, I. Gurol-Urganci, V. Vodopivec-Jamsek, J. Car, and R. Atun, "Mobile phone messaging for facilitating self-management of long-term illnesses," *Cochrane database Syst. Rev.*, vol. 12, no. 12, Dec. 2012, doi: 10.1002/14651858.CD007459.PUB2

[6] R. Whittaker, H. Mcrobbie, C. Bullen, A. Rodgers, and Y. Gu, "Mobile phone-based interventions for smoking cessation," *Cochrane database Syst. Rev.*, vol. 4, no. 4, Apr. 2016, doi: 10.1002/14651858.CD006611. PUB4

[7] C. Smith, J. Gold, T. D. Ngo, C. Sumpter, and C. Free, "Mobile phone-based interventions for improving contraception use," *Cochrane database Syst. Rev.*, vol. 2015, no. 6, Jun. 2015, doi: 10.1002/14651858.CD011159. PUB2

[8] E. Fisher, E. Law, J. Dudeney, C. Eccleston, and T. M. Palermo, "Psychological therapies (remotely delivered) for the management of chronic and recurrent pain in children and adolescents," *Cochrane database Syst. Rev.*, vol. 4, no. 4, Apr. 2019, doi: 10.1002/14651858.CD011118. PUB3

[9] J. S. Marcano Belisario, J. Jamsek, K. Huckvale, J. O'Donoghue, C. P. Morrison, and J. Car, "Comparison of self-administered survey questionnaire responses collected using mobile apps versus other methods," *Cochrane database Syst. Rev.*, vol. 2015, no. 7, Jul. 2015, doi: 10.1002/ 14651858.MR000042.PUB2

[10] E. Ozdalga, A. Ozdalga, and N. Ahuja, "The smartphone in medicine: a review of current and potential use among physicians and students," *J. Med. Internet Res.*, vol. 14, no. 5, 2012, doi: 10.2196/JMIR.1994

[11] T. L. Webb, J. Joseph, L. Yardley, and S. Michie, "Using the internet to promote health behavior change: a systematic review and meta-analysis of the impact of theoretical basis, use of behavior change techniques, and mode of delivery on efficacy," *J. Med. Internet Res.*, vol. 12, no. 1, 2010, doi: 10.2196/JMIR.1376

[12] C. Free, G. Phillips, L. Felix, L. Galli, V. Patel, and P. Edwards, "The effectiveness of M-health technologies for improving health and health services: a

systematic review protocol," *BMC Res. Notes*, vol. 3, 2010, doi: 10.1186/ 1756-0500-3-250

[13] G. F. Dunton, Y. Liao, S. S. Intille, D. Spruijt-Metz, and M. Pentz, "Investigating children's physical activity and sedentary behavior using ecological momentary assessment with mobile phones," *Obesity (Silver Spring).*, vol. 19, no. 6, pp. 1205–1212, Jun. 2011, doi: 10.1038/ OBY.2010.302

[14] F. Ehrler and C. Lovis, "Supporting Elderly Homecare with Smartwatches: Advantages and Drawbacks," *Stud. Health Technol. Inform.*, vol. 205, pp. 667–671, 2014, doi: 10.3233/978-1-61499-432-9-667

[15] S. Mann, "Wearable Computing: A First Step Toward Personal Imaging," *Cybersquare Comput.*, vol. 30, no. 2, 1997, Accessed: Dec. 26, 2021. [Online]. Available: http://wearcam.org/ieeecomputer/r2025.htm

[16] H. Ali and H. Li,, "Designing a smart watch interface for a notification and communication system for nursing homes," in *Human Aspects of IT for the Aged Population. Design for Aging: Second International Conference, ITAP 2016, Held as Part of HCI International 2016, Toronto, ON, Canada, July 17–22, Proceedings, Part I 2*, pp. 401–411. Springer International Publishing.

[17] E. Årsand, M. Muzny, M. Bradway, J. Muzik, and G. Hartvigsen, "Performance of the first combined smartwatch and smartphone diabetes diary application study," *J. Diabetes Sci. Technol.*, vol. 9, no. 3, pp. 556–563, 2015.

[18] O. Banos *et al.*, "The Mining Minds digital health and wellness framework," *Biomed. Eng. Online*, vol. 15, no. 1, pp. 165–186, 2016.

[19] C. Boletsis, S. McCallum, and B. F. Landmark, "The use of smartwatches for health monitoring in home-based dementia care," in *Human Aspects of IT for the Aged Population. Design for Everyday Life: First International Conference, ITAP 2015, Held as Part of HCI International 2015, Los Angeles, CA, USA, August 2–7, 2015. Proceedings, Part II 1*, pp. 15–26. Springer International Publishing.

[20] P. Chippendale, V. Tomaselli, V. d'Alto, G. Urlini, C. M. Modena, S. Messelodi, ... and G. M. Farinella, "Personal shopping assistance and navigator system for visually impaired people," in *Computer Vision-ECCV 2014 Workshops: Zurich, Switzerland, September 6-7 and 12, 2014, Proceedings, Part III 13*, pp. 375–390. Springer International Publishing.

[21] H. Dubey, J. C. Goldberg, K. Mankodiya, and L. Mahler, "A multi-smartwatch system for assessing speech characteristics of people with dysarthria in group settings," in *2015 17th International Conference on E-health Networking, Application & Services (HealthCom)*, pp. 528–533. IEEE.

[22] H. Dubey, J. C. Goldberg, M. Abtahi, L. Mahler, and K. Mankodiya, "{EchoWear}: smartwatch technology for voice and speech treatments of patients with Parkinson's disease," 2015.

[23] M. Duclos, G. Fleury, P. Lacomme, R. Phan, L. Ren, and S. Rousset, "An acceleration vector variance based method for energy expenditure estimation in real-life environment with a smartphone/smartwatch integration," *Expert Syst. Appl.*, vol. 63, pp. 435–449, 2016.

[24] S. Faye, R. Frank., & T. Engel, "Adaptive activity and context recognition using multimodal sensors in smart devices," in *Mobile Computing, Applications, and Services: 7th International Conference, MobiCASE 2015, Berlin, Germany, November 12–13, 2015, Revised Selected Papers 7*, pp. 33–50. Springer International Publishing.

[25] M. Haescher, J. Trimpop, D. J. C. Matthies, G. Bieber, B. Urban, and T. Kirste, "aHead: considering the head position in a multi-sensory setup of wearables to recognize everyday activities with intelligent sensor fusions," in *Human-Computer Interaction: Interaction Technologies: 17th International Conference, HCI International 2015, Los Angeles, CA, USA, August 2–7, 2015, Proceedings, Part II 17*, pp. 741–752. Springer International Publishing.

[26] Y. Jeong, Y. Chee, Y. Song, and K. Koo, "Smartwatch app as the chest compression depth feedback device," in *World Congress on Medical Physics and Biomedical Engineering, June 7–12, 2015, Toronto, Canada*, pp. 1465–1468. Springer International Publishing.

[27] H. Kalantarian and M. Sarrafzadeh, "Audio-based detection and evaluation of eating behavior using the smartwatch platform," *Comput. Biol. Med.*, vol. 65, pp. 1–9, 2015.

[28] J. Lockman, R. S. Fisher, and D. M. Olson, "Detection of seizure-like movements using a wrist accelerometer," *Epilepsy Behav.*, vol. 20, no. 4, pp. 638–641, Apr. 2011, doi: 10.1016/J.YEBEH.2011.01.019

[29] W. O. C. Lopez, C. A. E. Higuera, E. T. Fonoff, C. de Oliveira Souza, U. Albicker, and J. A. E. Martinez, "Listenmee\textregistered{}and Listenmee\textregistered{}smartphone application: synchronizing walking to rhythmic auditory cues to improve gait in Parkinson's disease," *Hum. Mov. Sci.*, vol. 37, pp. 147–156, 2014.

[30] B. Mortazavi *et al.*, "Can smartwatches replace smartphones for posture tracking?," *Sensors*, vol. 15, no. 10, pp. 26783–26800, 2015.

[31] L. de S. B. Neto, V. R. M. L. Maike, F. L. Koch, M. C. C. Baranauskas, A. de Rezende Rocha, and S. K. Goldenstein, "A wearable face recognition system built into a smartwatch and the blind and low vision users," in *Enterprise Information Systems: 17th International Conference, ICEIS 2015, Barcelona, Spain, April 27–30, 2015, Revised Selected Papers 17*, pp. 515–528. Springer International Publishing.

[32] C. Panagopoulos, E. Kalatha, P. Tsanakas, and I. Maglogiannis, "Evaluation of a mobile home care platform," in *Ambient Intelligence: 12th European Conference, AmI 2015, Athens, Greece, November 11–13, 2015, Proceedings 12*, pp. 328–343. Springer International Publishing.

[33] V. Sharma *et al.*, "{SPARK}: personalized parkinson disease interventions through synergy between a smartphone and a smartwatch," in *Design, User Experience, and Usability. User Experience Design for Everyday Life Applications and Services: Third International Conference, DUXU 2014, Held as Part of HCI International 2014, Heraklion, Crete, Greece, June 22–27, 2014, Proceedings, Part III 3*, pp. 103–114. Springer International Publishing.

[34] E. Thomaz, I. Essa, and G. D. Abowd, "A practical approach for recognizing eating moments with wrist-mounted inertial sensing," in

Proceedings of the 2015 ACM International Joint Conference on Pervasive and Ubiquitous Computing, pp. 1029–1040.

[35] T. Vilarinho, B. Farshchian, D. G. Bajer, O. H. Dahl, I. Egge, S. S. Hegdal, … & S. M. Weggersen, "A combined smartphone and smartwatch fall detection system," in *2015 IEEE International Conference on Computer and Information Technology; Ubiquitous Computing and Communications; Dependable, Autonomic and Secure Computing; Pervasive Intelligence and Computing*, pp. 1443–1448. IEEE.

[36] "Pebble Classic | ▦ Full Specifications & Reviews." https://productz.com/en/pebble-classic/p/1Ymy (accessed Dec. 26, 2021).

[37] "Samsung Gear Live – Specifications." www.devicespecifications.com/en/model/5b742dad (accessed Dec. 26, 2021).

[38] "Sony SmartWatch 3 – Specifications." www.devicespecifications.com/en/model/07092f83 (accessed Dec. 26, 2021).

[39] "LG Blue GizmoGadget: Kid-Friendly Wearable GPS | LG USA." www.lg.com/us/cell-phones/lg-VC200-Blue-gizmo-gadget (accessed Dec. 26, 2021).

[40] "Pebble Steel Smartwatch (Brushed Stainless) 401SLR B&H Photo." www.bhphotovideo.com/c/product/1093229-REG/pebble_401slr_pebble_steel_with_leather.html/specs (accessed Dec. 26, 2021).

[41] "Pebble Time Round Detailed Specs." www.alphachooser.com/smart_watches--pebble_time_round--smartwatch-specs-and-profile (accessed Dec. 26, 2021).

[42] "simvalley MOBILE 1.5" Smartwatch AW-420. RX with Android 4.2, BT, WiFi, Black." www.simvalley-mobile.de/Android-Watch-IP67-PX-1795-919.shtml (accessed Dec. 26, 2021).

[43] "LG Watch R (W110): Design Comes Full Circle | LG USA." www.lg.com/us/smart-watches/lg-W110-lg-watch-r (accessed Dec. 26, 2021).

[44] "Motorola Moto 360 Sport (1st gen) Price in India, Specs & Features (26th December 2021)." www.giznext.com/smartwatches/motorola-moto-360-sport-1st-gen-gnt (accessed Dec. 26, 2021).

[45] "BLOCKS – The World's First Modular Smartwatch by BLOCKS Wearables – Kickstarter." www.kickstarter.com/projects/2106691934/blocks-the-worlds-first-modular-smartwatch (accessed Dec. 26, 2021).

[46] "HUAWEI WATCH GT Specifications – HUAWEI Global." https://consumer.huawei.com/en/wearables/watch-gt/specs/ (accessed Dec. 26, 2021).

[47] "LG Watch Urbane (W150) – Best Android Wear Watch | LG Electronics In." www.lg.com/in/smart-watches/lg-W150 (accessed Dec. 26, 2021).

[48] "LG Smart Watch Urbane 2nd Edition (W200V) | LG USA." www.lg.com/us/smart-watches/lg-W200V-lg-watch-urbane-2nd-edition-verizon (accessed Dec. 26, 2021).

[49] "Apple Watch Edition (1st generation) – Technical Specifications (IN)." https://support.apple.com/kb/SP737?viewlocale=en_IN&locale=en_IN (accessed Dec. 26, 2021).

[50] "SmartWatch by Smart Monitor | Epilepsy Foundation." www.epilepsy.com/deviceapedia/smartwatch-smart-monitor-0 (accessed Dec. 26, 2021).

[51] "Basis B1 review I 167 facts and highlights." https://versus.com/en/basis-b1 (accessed Dec. 26, 2021).

[52] Q. Jawhar, K. Thakur, and K. J. Singh, "Recent Advances in Handling Big Data for Wireless Sensor Networks," *IEEE Potentials*, vol. 39, no. 6, pp. 22–27, 2020.

[53] P. Sachdeva and K. J. Singh, "Automatic segmentation and area calculation of optic disc in ophthalmic images," *2015 2nd Int. Conf. Recent Adv. Eng. Comput. Sci. RAECS 2015*, Apr. 2016, doi: 10.1109/RAECS.2015.7453356

[54] K. J. Singh *et al.*, "Adaptive Flower Pollination Algorithm-Based Energy Efficient Routing Protocol for Multi-Robot Systems," *IEEE Access*, vol. 9, pp. 82417–82434, 2021, doi: 10.1109/ACCESS.2021.3086628

[55] K. Singh, K. J. Singh, and D. S. Kapoor, "Image Retrieval for Medical Imaging Using Combined Feature Fuzzy Approach," in *2014 International Conference on Devices, Circuits and Communications (ICDCCom)*, 2014, pp. 1–5.

[56] M. Velez, R. S. Fisher, V. Bartlett, and S. Le, "Tracking generalized tonic-clonic seizures with a wrist accelerometer linked to an online database," *Seizure*, vol. 39, pp. 13–18, Jul. 2016, doi: 10.1016/J.SEIZURE.2016.04.009

[57] A. Hosseini *et al.*, "HIPAA Compliant Wireless Sensing Smartwatch Application for the Self-Management of Pediatric Asthma," *Int. Conf. Wearable Implant. Body Sens. Networks. Int. Conf. Wearable Implant. Body Sens. Networks*, Jul. 2016, p. 49. doi: 10.1109/BSN.2016.7516231

[58] N. Micallef, L. Baillie, and S. Uzor, "Time to exercise! An aide-memoire stroke app for post-stroke arm rehabilitation," *Proc. 18th Int. Conf. Human-Computer Interact. with Mob. Devices Serv. MobileHCI 2016*, pp. 112–123, Sep. 2016, doi: 10.1145/2935334.2935338

[59] K. J. Singh, D. S. Kapoor, and B. S. Sohi, "The MAI: A Robot for/by Everyone," in *Companion of the 2018 ACM/IEEE International Conference on Human-Robot Interaction*, 2018, pp. 367–368.

[60] K. J. Singh, D. S. Kapoor, and B. S. Sohi, "Selecting Social Robot by Understanding Human–Robot Interaction," in *International Conference on Innovative Computing and Communications*, 2021, pp. 203–213.

Chapter 20

Nature inspired computing for optimization

Ashima Kalra and Gaurav Tewari

20.1 INTRODUCTION

In order to solve difficult or time-sensitive problems, *optimization* is the process of determining the best possible solution. These optimization algorithms can be stochastic or deterministic in nature. Different types of approaches are used for optimization, but nature is the best way to solve optimization problems as the mapping between the nature and the engineering problems is readily possible, because these nature-inspired algorithms mimic the behavior of nature in technology and therefore prove much better than the traditional or other approaches for solving complex tasks.

Nature is a marvelous teacher, and we human beings perhaps are the best learners on earth. The term "nature-inspired computing" (NIC) refers to a group of computing methodologies that have been inspired by natural systems and processes. These systems and processes can be seen in nature and can be modelled for computing applications. Table 20.1 lists few of the many computing techniques that have been inspired by nature. NIC mimics the natural processes and systems to develop algorithms that can be used by computing machines to solve highly complex and non-linear problems. A typical NIC system is a sophisticated, autonomous computer system run by a population of independent entities inside a context. The NIC system's autonomous entity is made up of two components: effectors, processing elements, and sensors. One or more sensors, processing elements (PEs), and effectors may be present. Sensors gather data about their wider surroundings and their immediate surroundings. The information obviously relies on the system being represented or the issue being handled. Effectors alter the internal present conditions, display certain behaviors, and alter the environment based on the input and output of PEs. In essence, the PEs and effectors enable information exchange between autonomous units. The NIC system provides a database of local behavior regulations. The behavior codes are essential to an autonomous unit.

Fuzzy logic, artificial neural networks, evolutionary computation, rough sets, granular computing, swarm intelligence, and physics and

DOI: 10.1201/9781003453406-20

Table 20.1 Nature effects

Cellular Automata	Self-Reproduction
Neural Computation	Structure of Brain
Fuzzy Logic	Philosophy/Psychology of Brain
Evolutionary Computation	Evolution
Swarm Intelligence	Artificial Life
Immuno-Computing	Membrane Computing
Big Bang Big Crunch	Evolution of the Universe
Optimization	

chemistry-based computing models like the simulated annealing (SA), big bang big crunch (BB-BC) algorithm are all included under the umbrella term "NIC intelligence". The development of NIC was made feasible by the outstanding advancements in computer science and the striking increase in processing power in terms of hardware and it has further added a powerful punch to computing power of the modern computing machines. Software agents may be used to simulate incredibly complex and dynamic systems, regardless of the models used – fuzzy, neural, swarming, colonies, or other natural metaphors. Control systems, computer science, model identification, routing, transportation, robotics, Very large-scale integration design, industrial applications, household appliances, and business are just a few areas where NIC has been extensively used. NIC approaches are used in physics, engineering, economics, management, and a host of other fields.

This chapter consists of five sections. Section 1 is the introduction. Section 2 presents the various constituents of NIC. Section 3 gives swarm intelligence based approaches. Section 4 presents physics or chemistry based search and optimization approaches. Section 5 concludes the chapter.

20.2 COMPONENTS OF NATURE-INSPIRED COMPUTING

This section presents the various constituents of NIC. We divide these constituents into three main groups: fuzzy logic based computing, artificial neural networks (ANNs), and search and optimization approaches. Fuzzy logic and ANNs are very well established and extensively used fields today. This section focusses on recent search and optimization approaches found in literature.

20.2.1 Fuzzy logic based computing

With his 1965 seminal paper [1] Prof. L.A. Zadeh, set the ball rolling, and fuzzy logic based systems have matured into an established and widely used field today [2]. Standard two-valued logic is generalized into fuzzy logic as a way to deal with uncertainty. In a larger sense, fuzzy logic encompasses

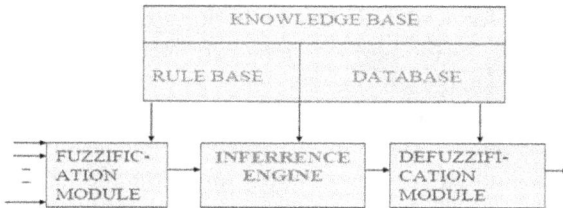

Figure 20.1 A fuzzy logic based computing model.

any theories and methods that make use of fuzzy sets .Thus we shall state that "Use of fuzzy sets in logical expressions is called fuzzy logic". Fuzzy logic based systems are modelled on the psychology and philosophy of the working of the human brain.

Figure 20.1's block diagram may be used to describe a computing paradigm based on fuzzy logic. The system consists of four main parts. The crisp input(s) are transformed into fuzzy values via the *fuzzification* module. Then, using the knowledge base (rule base and procedural knowledge) provided by the domain expert, the inference engine processes these values in the fuzzy domain (s). Finally, the defuzzification module converts the processed output from the fuzzy domain to the crisp domain.

Apart from fuzzy sets [1], rough sets [4], granular computing [5,6], Perception-Based Computing, Wisdom Technology, Anticipatory Computing [3] also find extensive applications in solution modelling for a complex problem.

20.2.2 Artificial neural networks

A massively distributed parallel processing system made up of straight-forward processing components, ANN has a built-in inclination for accumulating practical experience and subsequently making it accessible for usage. Because ANNs are streamlined representations of biological nervous systems, they were inspired by the type of computation that occurs in the human brain. In two ways, they resemble the brain. (1) The network learns from its surroundings through the learning process. (2) Acquired information is stored in synaptic weights, the strength of inter-neuron connections. Mapping skills, pattern association, and generalization capability for tolerance and improved dependability are only a few of the characteristics that define ANNs. Applications for manufacturing, marketing, healthcare, environmental applications, pattern identification, and control have all made extensive use of ANNs. [7–12].

The two major categories of ANN architectures are feed-forward networks and recurrent/feedback networks. Radial basis function networks, single-layer perceptron networks, and multi-layer perceptron networks are further categories for feed forward networks. Conversely, recurrent/feedback type networks include networks like competitive networks, Hopfield networks and Kohenen's self-organizing maps (SOM), as well as adaptive resonant theory (ART) models, and so forth [7–14].

Since ANNs have the ability to learn from examples, they may first be trained using a known set of examples for a specific issue before being evaluated for their ability to draw conclusions about unknowable instances of the same problem. As a result, they are able to recognize items for which they have not yet received training. The three types of ANN learning paradigms are supervised learning, unsupervised learning, and hybrid learning. Every input pattern used to train the network in supervised learning is linked to an output pattern that represents the goal or intended pattern. It is believed that a trainer or instructor will be present throughout the training or learning process. An error occurs when the network's calculated output is compared against the target. This miscalculation serves to change network parameters so as to reduce the error.

With unsupervised learning, the network is not shown the goal output. This is equivalent to a teacher not being there. By identifying and modifying the structural characteristics in the input pattern, the system learns on its own. Combining the two aforementioned forms of learning is termed hybrid learning. The algorithms that fall under the category of supervised learning include perceptron learning, back propagation, Adaline, Medaline Boltzman's learning algorithm, learning vector quantization, and so forth. The algorithms that fall under the category of unsupervised learning include principal component analysis (PCA), associative memory learning, vector quantization, Kohonen's self-organizing maps (SOMs), ART-1, ART-2, and so forth. The hybrid learning paradigm includes algorithms like radial basis function (RBF) learning methods. [7–14].

20.2.3 Search and optimization approaches

Search and optimization approaches are among the very important and integral constituents of computing system – specifically when the problems are non-deterministic polynomial (NP) hard or NP complete. When the problems are very complex and highly nonlinear, finding exact solutions may become too expensive. Under such circumstances researchers are looking to nature to provide some clues. Computing can be divided into two subfields: Hard Computing, where we use exact reasoning to find solutions, and soft computing, where we use approximate reasoning to find the desired solutions. Soft Computing offers very inexpensive solutions, where the best solution can be replaced with good enough solutions. Since, soft computing based search and optimization approaches are too important, and most of the approaches draw their inspiration from nature, we devote this section to a

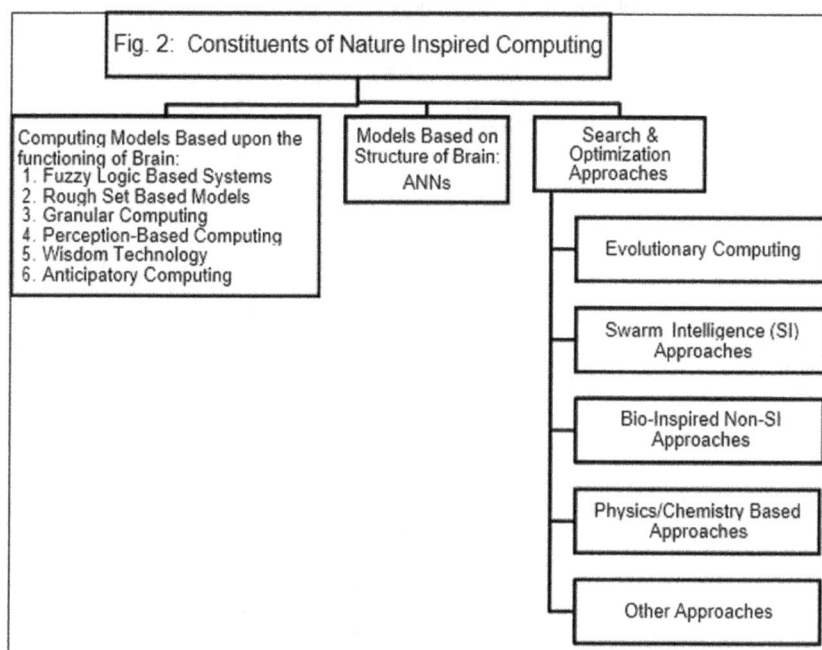

Figure 20.2 Constituents of nature-inspired computing.

third constituent of the nature-inspired computing constituent. Figure 20.2 groups these approaches into five categories: Evolutionary Computing [15–17], Swarm Intelligence (SI) Approaches [18–23], Bio-Inspired Non-SI Approaches, Physics/Chemistry Based Approaches and *another's* group. Tables 20.2, 20.3, and 20.4 list the most widely used nature-inspired search and optimization approaches.

Wolpert argues that all search algorithms perform equally well on average across all tasks [24]. In other words, when applied to all issues, the GA does not perform any better than a completely random search. Using the appropriate algorithm for the appropriate task is the notion. In order to do a search for an appropriate algorithm for an upcoming new problem, the research must go on to present new algorithms to meet this dynamic requirement.

20.2.3.1 Evolutionary computing

The term "Evolutionary Computing" (EC) [15–17] refers to a variety of computer-based problem-solving approaches that are founded on biological evolutionary concepts, including natural selection and genetic

Table 20.2 Swarm intelligence based algorithms

Accelerated PSO	Ant colony optimization	Artificial bee colony
Bacterial foraging	Bacterial-GA Foraging	Bat algorithm
Bee colony optimization	Bee system	Bee hive
Wolf search	Bees algorithms	Bees swarm optimization
Bumblebees	Cat warm	Consultant- guided search
Cuckoo search	Marriage in honey bees	Eagle strategy
Fast bacterial swarming algorithm	Firefly algorithm	Fish swarm/school
Good lattice swarm Optimization	Glowworm swarm optimization	Hierarchical swarm model
Krill herd	Monkey search	Particle swarm algorithm
Virtual ant algorithm	Virtual bees	
Weightless swarm algorithm		

Table 20.3 Physics and chemistry based algorithms

Big bang-big crunch algorithm	Parallel Big Bang-Big Crunch Algorithm	Black hole
Central force optimization	Charged system search	Electro-magnetism optimization
Galaxy-based search algorithm	Gravitational search GSA	Harmony search
Intelligent water drop	River formation dynamics	Self-propelled particles
Simulated annealing	Stochastic diffusion search	Spiral optimization
Water cycle algorithm	Pollination based optimization (PBO)	

Table 20.4 Bio-inspired (non-SI-based) algorithms

Atmosphere clouds model	*Biogeography-based optimization*	*Brain storm optimization*
Differential evolution	Dolphin echolocation	Egyptian vulture
Japanese tree frogs calling	Eco-inspired evolutionary algorithm	Fish-school search
Flower pollination algorithm	Gene expression	Great salmon run
Group search optimizer	Human-inspired algorithm	Invasive weed optimization
OptBees	Paddy field algorithm	Roach infestation algorithm
Queen-bee evolution	Shuffled frog leaping algorithm	Termite colony optimization

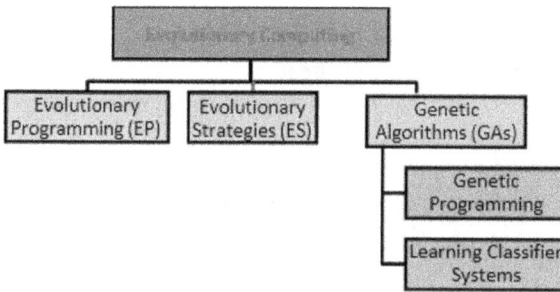

Figure 20.3 Evolutionary computing.

inheritance. These methods are being used more often to solve a wide range of issues, from cutting-edge scientific research to actual practical applications in business and industry. Evolutionary Computing is the study and application of the theory of evolution to an engineering and computing context.

In EC systems we usually define five things: A phenotype, a genotype, genetic operators (like combination/crossover and mutation), a fitness function, and selection operators. The phenotype is a solution to the problem we want to solve. The genotype is the representation of that solution, which will suffer variation and selection in the algorithm. Most often, but not always, the phenotype and the genotype are the same. A type of optimization approach called evolutionary computation (EC) [15–17] is motivated by the mechanics of biological evolution and the behaviors of living things. In general, EC algorithms include learning classifier systems, genetic algorithms, evolutionary strategies, and evolutionary programming (EP, GA, GP) (LCS). Differential evolution (DE) and the estimate of distribution method are also included in EC (EDA).

20.3 SWARM INTELLIGENCE

James Kennedy and Russell Eberhart initially put out the idea of swarm intelligence. This was inspired by various ant, wasp, bee, and other swarming behaviors. They lack intellect as a whole, but their capacity for coordinated action in the absence of a coordinator makes them appear clever. These agents communicate with one another to develop "intelligence", and they do so without any centralized control or supervision. Swarm intelligence-based algorithms are among the most widely used. A lot of them include bat algorithms, firefly algorithms, artificial bee colonies, cuckoo searches, and particle swarm optimization.

20.3.1 Particle swarm optimization (PSO)

Kennedy and Eberhart introduced this population-based optimization in 1995. Actually, this is a mimic of the flocking or schooling behavior of birds and fish. In order to find the location of the meal, birds follow the bird that has to go the smallest distance.

It is made up of a swarm of particles, each of which travels across the multidimensional search space with a certain velocity before eventually locating its best place. The best position of the particle (pbest) and the best position of the particle's neighbors (gbest) are continuously updated to determine a new place for the particle to travel on throughout each iteration. Each particle utilizes its own experience to solve the problem under consideration.

20.3.2 Ant Colony Optimization (ACO)

The foraging habits of ants served as the inspiration for ACO. Finding the quickest route between food and nest is one of their strongest skills. They continuously release a chemical called pheromone on their path as they go in quest of food, signaling some favorable way for other ant colony members. *Stigmergy* is the term for this form of indirect communication between ants that relies on pheromone production. With an increase in pheromone concentration along the path, the likelihood of choosing that way increases. Ants then make a probabilistic choice that is influenced by the pheromone concentration: the stronger the pheromone trail, the greater its attractiveness. As they are leaving pheromone deposits along their route, this behavior causes a self-reinforcing cycle that culminates in the creation of routes with a high pheromone concentration. In the early 1990s Dorigo introduced the first ant colony optimization, known as the Ant System (AS).

20.3.3 Artificial Bee Colony (ABC)

ABC was found in 2007 by Karaboga and Basturk, who were motivated by the honey bees' clever behavior. A popular algorithm that mimics the intelligent foraging activity of a honeybee swarm is called Artificial Bee Colony. Three types of bees make up the artificial bee colony in the ABC algorithm: workers, observers, and scouts. A bee travelling to the food source that it previously visited is known as an *employed bee*, whereas one waiting on the dance floor while deciding which food source to choose is known as an *observer*. The second kind of bee is called a scout bee, and it goes on erratic hunts for new resources. While a food source's location represents a potential resolution to the optimization problem, the nectar quantity of a food source relates to the quality (fitness) of the connected solution.

Apart from above-mentioned approaches, a number of SI-based and Non-SI based approaches are available. Non-SI based approaches are inspired by nature but are not due to any swarm movement or behavior. The various SI and non-SI based approaches are available in the literature and are shown in Tables 20.2 and 20.4.

20.4 PHYSICS OR CHEMISTRY-BASED SEARCH AND OPTIMIZATION APPROACHES

Certain physical or chemical principles, such as electrical charges, gravity, river systems, and so forth, have been imitated in some of the algorithms. These algorithms can be categorized as search and optimization methods based on physics or chemistry.

20.4.1 Intelligent Water Drops Algorithm (IWD)

Hamed Shah-hosseini initially presented it as a population-based optimization technique to address the issue of the travelling salesman (TSP). It is modelled after the activities and interactions that occur between water droplets in a river and the alterations to the environment through which the river flows in natural river systems. The quantity of dirt carried by and the speed of motion of this Intelligent Water Drop are two crucial characteristics. The environment that the water runs in depends on the issue at hand.

20.4.2 EM (Electromagnetism-like Mechanism) Algorithm

By treating each particle as an electrical charge and drawing inspiration from the electromagnetic theory of physics, Birbil and Fang originally suggested the EM method for unconstrained optimization problems. To move some points towards the ideal locations, it employs an attraction–repulsion mechanism based on Coulomb's Law. The method is known as an electromagnetism-like mechanism algorithm because it mimics the electromagnetic theory's attraction–repulsion mechanism. In the EM method, a solution may be thought of as a charged particle in the search space, and its charge is related to the value of the objective function.

20.4.3 Gravitational Search Algorithm (GSA)

This algorithm was put up by Rashiedi et al. and was motivated by Newton's theories of gravity and motion. The Newtonian theory of gravity is the basis of this algorithm, which states that "Every particle in the world is drawn to every other particle with an attraction force that is directly proportional to the product of their masses and inversely proportional to the square of their distance from one another". The searcher agents are a group of masses that communicate with one another using the principles of motion and gravity. The performance of the agents is determined by their mass, which is treated like an object. All items gravitate toward other objects with heavier weights due to the gravitational force. The algorithm's exploitation step is guaranteed, and excellent solutions correlate to the slower movement of heavier masses. Actually, the masses are obedient to gravity's laws.

Apart from the above-mentioned search and optimization approaches, a number of optimization approaches based on physics or chemistry laws are available in the literature and are shown in the Table 20.3.

20.5 CONCLUSION

This chapter presents an extensive survey of available nature-inspired search and optimization approaches in existing literature. These optimization approaches can be successfully further applied in the different fields of engineering, like wireless communication, control engineering, neural networks, and so forth. Depending upon the nature and the requirement of the problem, any of the optimization approaches can be chosen. It has been found that these nature-inspired optimization approaches are promoted by researchers due to their better results by comparison to the classical approaches.

REFERENCES

1. Zadeh, L.A. "Fuzzy sets", *Information and Control*, Vol. 8 (3), pp. 338–353, 1965.
2. Yen, J. and Langari, R. *Fuzzy Logic Intelligence, Control and Information.* Prentice Hall, Upper Saddle River, NJ, 1999, pp. 548.
3. Lavika Goel, Daya Gupta, V.K. Panchal and Ajith Abraham. "Taxonomy of Nature Inspired Computational Intelligence: A Remote Sensing Perspective", Fourth World Congress on Nature and Biologically Inspired Computing (NaBIC-2012), pp. 200–206.
4. Pawlak, Z. "Rough Sets", *International Journal of Computer and Information Sciences*, 11, pp 341–356, 1982.
5. Bargiela, A. and Pedrycz W. *Granular Computing: An Introduction*, Kluwer Academic Publishers, Boston, 2002.
6. Yao, Yiyu. "Perspectives of Granular Computing" Proceedings of IEEE International Conference on Granular Computing, Vol. I, pp. 85–90, 2005.
7. Bishop, Chris M. "Neural Networks and their applications", *Review of Scientific Instruments*, Vol. 65, No. 6, June 1994, pp. 1803–1832.
8. Soroush, A.R., Kamal-Abadi, Nakhai Bahreininejad A. "Review on applications of artificial neural networks in supply chain management", *World Applied Sciences Journal* 6 (supplement 1), pp. 12–18, 2009.

9. Jin, Yaochu, Jingping Jin, Jing Zhu. "Neural Network Based Fuzzy Identification and Its Applications to Control of Complex systems", IEEE *Transactions on Systems, Man and Cybernetics*, Vol, 25, No. 6, June 1995, pp. 990–997.

10. Simon Haykin. *Neural Networks: A Comprehensive Foundation*, Prentice Hall PTR, Upper Saddle River, NJ, 1994.

11. Jacek, M. Zurada. *Introduction to Artificial Neural Systems*, West Publishing Co., 1992.

12. Martin T. Hagan, Howard B. Demuth, Mark H. Beale, *Neural Network Design*, Martin Hagan, 2014.

13. Widrow, Bernard, Lehar, Michael A. "30 Years of Adaptive Neural Networks: Perceptron, Medaline and Back Propagation" Proceedings of the IEEE, Vol 78, No. 9, Sep 1990, pp. 1415–1442.

14. Jain, Anil K., Mao Jianchang, Mohiuddin, "Artificial Neural Networks: a Turorial", IEEE Computers, Vol. 29, No. 3, March 1996 pp. 31–44.

15. Back, T.. Evolutionary computation: comments on the history and current state. *IEEE Trans. Evol. Comput.* 1:3–17. 1997.

16. S. Kumar, S.S. Walia, A. Kalra. "ANN Training: A Review of Soft Computing Approaches", *International Journal of Electrical & Electronics Engineering*, Vol. 2, Spl. Issue 2, pp. 193–205. 2015.

17. A. Kalra, S. Kumar, S.S. Walia. "ANN Training: A Survey of classical and Soft Computing Approaches", *International Journal of Control Theory and Applications*, Vol. 9, issue 34 pp. 715–736. 2016.

18. Jaspreet, A. Kalra. "Artificial Neural Network Optimization by a Hybrid IWD-PSO Approach for Iris Classification", *International Journal of Electronics, Electrical and Computational System IJEECS*, Vol. 6, Issue 4, pp. 232–239. 2017.

19. *Comparative Survey of Swarm Intelligence Optimization Approaches for ANN Optimization*, Springer, ICICCD. AISC *Advances in Intelligent Systems and Computing* book series by Springer (scopus indexed). Vol. 624, pp. 305. 2017.

20. Mahajan, S., Abualigah, L. & Pandit, A.K. Hybrid arithmetic optimization algorithm with hunger games search for global optimization. *Multimed Tools Appl* 81, 28755–28778. 2022. https://doi.org/10.1007/s11042-022-12922-z

21. Mahajan, S., Abualigah, L., Pandit, A.K. *et al.* Hybrid Aquila optimizer with arithmetic optimization algorithm for global optimization tasks. *Soft Comput* 26, 4863–4881. 2022. https://doi.org/10.1007/s00500-022-06873-8

22. Run Ma, Shahab Wahhab Kareem, Ashima Kalra, Rumi Iqbal Doewes, Pankaj Kumar, Shahajan Miah. "Optimization of Electric Automation Control Model Based on Artificial Intelligence Algorithm", *Wireless Communications and Mobile Computing*, vol. 2022, 9 pages, 2022. https://doi.org/10.1155/2022/7762493

23. Soni, M., Nayak, N.R., Kalra, A., Degadwala, S., Singh, N.K. and Singh, S. "Energy efficient multi-tasking for edge computing using federated learning", *International Journal of Pervasive Computing and Communications*, Emerald. 2022.

24. Wolpert H. David and William G. Macready. "No free lunch theorems for optimization". *IEEE Transactions on Evolutionary Computation*. Vol. 1, No. 1, April 1997, pp. 67–82.

Chapter 21

Automated smart billing cart for fruits

*Tanuja S. Patankar, Abhishek Bharane, Vinay Bhosale,
Pranav Bongulwar, and Onkar Mulay*

21.1 INTRODUCTION

Retail outlets have taken on huge importance in ordinary life. People in metropolitan areas routinely go to malls to purchase their daily necessities. In such a situation, the environment ought to be uncontroversial. This system is planned for edibles like fresh food varieties and other consumable produce, normalized label stickers, and RFID names can't be used as they should be stuck on all of the things and the quality of everything should be freely assessed. This chapter proposes a system that contains a camera that itemizes the purchase using simulated intelligence methodology and a load cell that measures the item added to the shopping bag. This system also creates the shopper's final bill.

1. Machine learning

Machine learning uses data to create a model which is constantly adjusted by new data – simulating intelligence learning from experience. It is essential to allow the machines to adjust normally without any human intercession and change tasks as required. Man-made intelligence can be used for various applications, such as picture affirmation, conversation affirmation, thing idea, and so forth.

2. Internet of Things (IoT)

Real-world objects and things are connected through sensors, programming, and other technological advancements to communicate, exchange data with systems and other devices on the Internet.

3. Image processing

Image processing is a method for enacting a strategy on a picture. It also aids in the removal of some helpful information. It is a sort of sign, processing where the input is a picture and the outcome is a picture or its associated properties. Locating various types of vegetables and natural products is a demanding task in general stores, since the clerk should bring up the classes of a specific natural product to decide its cost. The utilization of standardized tags has generally resolved this issue for bundled items, but considering that most customers need to pick their items, they cannot be pre-bundled, and accordingly should be gauged. An answer is to have codes for each organic product, yet it is easy for the supermarket staff to make cost mistakes. Another arrangement is to give clerks a stock catalog with images and codes but, in any case, going through such a book is tedious. The programmed order of natural products by means of PC vision is as yet a convoluted assignment because of the different properties of many kinds of organic products. The natural product's quality-discovery strategy depends on the outer properties of organic products, like shape, size, and shading.

21.2 LITERATURE SURVEY

IoT is unquestionably driving humankind to a superior world, except that it is important to keep in mind factors like energy utilization, time required, cost factors, and so on. This features issues in eight categories:

(1) massive scaling plan and conditions
(2) creating data and colossal data
(3) robustness
(4) openness
(5) security
(6) privacy
(7) human all good

These topics are essential. From this we can arrive at a goal that IoT is the vision to what is to accompany extended intricacy in distinguishing, incitation, exchanges, control, and in making data from immense proportions of data, achieving an abstractly special and more direct lifestyle than is experienced today.

Taking care of pictures is an assessment and control of digitized pictures, especially dealing with the idea of picture processing. In picture processing, there are several stages: Picture Pre-processing, Image Segmentation, Image Edge Detection, Feature Extraction, and Picture Acknowledgment. The huge Picture Handling beneficiaries are Agriculture, Multimedia Security, Remote Sensing, Computer Vision, Medical Applications, and so forth. Using Otsu procedure picture thresholding is done, and a while later Pre-treatment of the picture is done and by using the K-Means algorithm, Fuzzy C Means [1] algorithm, TsNKM [4] computation further division is done and, subsequently, Altered Vigilant Edge Recognition estimation edges are recognized by lastly using Component Extraction Picture.

For anticipating the natural item we can use diverse AI computations. We inspected and pondered three estimations, SVM [5], KNN estimation, [5] and Arbitrary woodlands algorithm. The K-NN algorithm recognizes the similarity between new and open data and places the new data in an order that is essentially the same as the open categories. Random Woods is a classifier that uses several decision trees on distinct subsets of a dataset and figures the ordinary to extemporize the dataset's judicious precision. Do not depend on a single decision tree: rather, the subjective forest considers the larger part of each tree's assumption and predicts the final result. After comparing the accuracy of the three estimates, we discovered that the sporadic woods computation provides the most precision.

Concerning Cloud Administrations, there is no inescapable best; all that induces down to what exactly specifically best suits our requirement. The exact need for Cloud Administrations ought to be known to avoid extra chaos.

Shukla and Desai [5] discovered nine new types of natural things. Normal item photo datasets are obtained from the Internet in the same way that explicit pictures are obtained through the wireless camera. These photographs have been pre-processed to eliminate the unusual and focus on common elements. A blend of hiding, form, and surface components are employed to tend to the fruit's aesthetic qualities. These subset datasets are also fed into two distinct classifiers, SVM and KNN [5].

GLCM [1] transformed the image to grayscale (Dark Level Co-occasion Framework). In addition, the image has been transformed into a two-dimensional image. Morphological tasks are also employed and concentrate the largest bulk or article from the image, which is also called a natural item. Later, the greatest bulk is managed, and the same traits are surpassed with extraordinary power. In the vast majority of circumstances, the combination of concealing surface and form produces better or virtually indistinguishable results than using any two classes of components, according to the findings of the testing. Similarly, the second conclusion that can be drawn is that KNN produces favorable results in this scenario.

This investigation [2] has different steps of the ready cycle that is according to the accompanying: To retrieve the nature of natural items' image, first assemble a normal items picture, then combine extraction process utilizing FCH and MI approaches, then transform into vector feature structure that will be taken care of in the informational index. The vector of the natural objects picture in the informational collection is then clustered using the K-Means Grouping approach. The following is the methodology for the testing framework in this assessment: To sense normal stuff, open the record picture query. The next stage is to extract a portion of the facial image, which is then converted into a vector feature structure, similar to the getting ready cycle. Then, using the KNN approach, calculate the gap between new normal item image components and present item image features informational index using Euclidian distance, which is then planned using the gathering findings.

This chapter [3] relies on the usage of a speeded up solid component. The methodology removes the close by part of the separated picture and portrays the article affirmation. The fundamental advances are to make an informational collection of the picture to be portrayed. Then, picture pre-taking care of done through various picture taking care of strategies to chip away at the idea of the image and later a couple of channels are used to de-stretch the image. Finally, picture classifiers are utilized to determine how to proceed. Picture is changed over from RGB picture to constrain picture. Taking into account a speeded up lively technique area incorporate is removed and portrayed. To portray the outer layer of the data picture, verifiable assessment of inconsistency. Various components were eliminated, for instance, object affirmation, picture enrollment, seeing limit, and picture recuperation. Articles and cutoff lines of pictures are acquired by picture division. Then, incorporate extraction like shape, size, concealing, and the surface of normal not set in stone using computation. Then, for disease request configuration planning is applied. The system in like manner consolidates obvious surface flaws distinguishing proof estimations, not to solely to identify them, but to prevent their differences from hindering the creation of a standard.

On the shape and hiding ward on examination systems, two-layered normal item photos are required in this study. Using counterfeit neural association (ANN), a system was developed to increase the precision of the regular item quality area. The basic idea is to obtain the image of a common object. The image of the standard item tests is obtained using a standard contemporary camera with a white backdrop and a stand. The image of the second step is to prepare neural affiliation, natural items are layered into MATLAB® to fuse the component extraction of every single model in the dataset. The final phase eliminates aspects of the usual object testing. In the fourth step, neural affiliation is used to organize the data. The standard object test is chosen for testing in the fifth phase from an enlightening assortment. ANN preparation module button is used to execute a sixth testing under a concordance condition. Finally, ANN-based outcomes are obtained, with the client having the option of selecting the instance of the standard thing that must be obtained in total.

The typical item testing' features have been deleted. The data is prepared in the fourth stage using neural association. From the informational collection, a regular item test is chosen for testing in the fifth phase. ANN getting ready module button is used to do a sixth testing in a condition of harmony. Finally, ANN-based findings are obtained, with the consumer having the option of selecting a standard item case to test and ultimately purchase.

21.3 PROPOSED METHOD

21.3.1 System design

In this system (see Figure 21.1) will be the following main modules:

- Collection of Data Using IoT devices. (Raspberry PI)
- Measuring weight using load cell
- Capturing images through camera and processing them
- Transfer of data through Bluetooth module
- Bill generation

Our system will automatically identify the fruit or vegetable put into the cart. We are using the TensorFlow Lite object detection module for image processing. Collection of data is done with Raspberry PI. A load cell will measure the weight of the fruits added. A camera will capture images of fruits. Bluetooth sensor will send all the data to the mobile device. After adding all the fruits to the cart a bill will be generated on screen. After entering CVV and bank details, payment will be made.

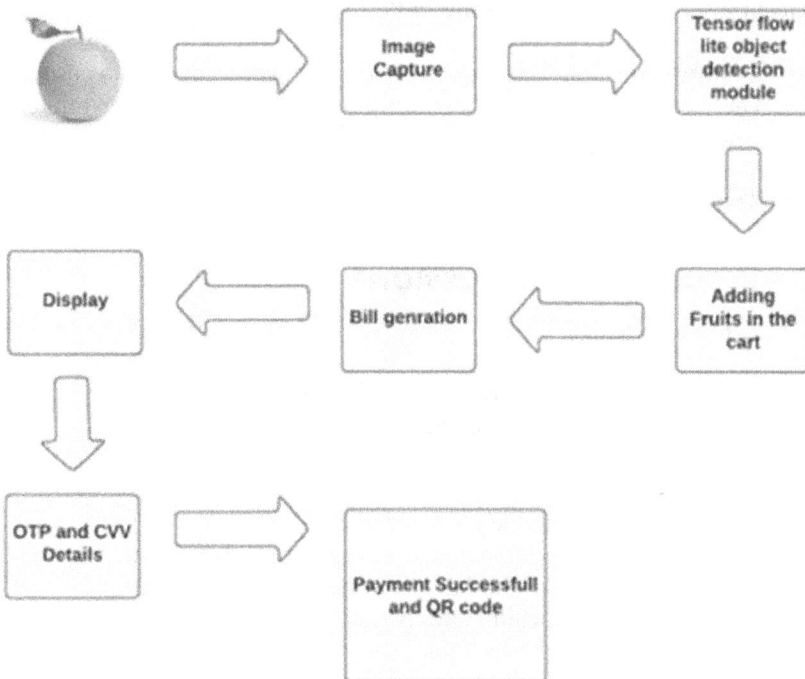

Figure 21.1 System Design of Automated Smart Billing Cart.

21.4 IMPLEMENTATION

- The user login in our application using the given design. The user enters personal information such as his or her name, phone number, and email address. The user can then login to the program after successfully registering. All the credentials are stored in the database.
- User has to put fruit in front of camera on the load cell. Load cell will weigh the fruits. Then the camera will recognize fruit in front of it using the TensorFlow Lite model. After the working of this model, the fruit is recognized. On the screen fruit is displayed with an image of the fruit.
- Once the fruit is recognized then the HC-05 sensor will transfer data from rpi to mobile app. Then all the fruits are added to the cart and the bill is generated on the screen. After clicking on proceed to payment button, a new window is displayed. A user must then enter their bank details with their card number and CVV.
- One-time password (OTP) is generated on the registered mobile number .After entering CVV details and otp QR code is generated on the app. A bill is also received on registered email ID.

Implementation procedure: Image processing with TensorFlow Lite Object Detection on Android and Raspberry Pi:

- A lightweight deep learning model deployment framework is used.
- These models use less processing power and have a faster inference time, allowing them to be employed in real-time applications.
- A custom object detection model can be run on Android phones or the Raspberry Pi.

21.5 RESULTS AND DISCUSSIONS

Implementation steps:

1. Customer registration on the application.
2. On successful registration, login with valid credentials.
3. Add your Credit/Debit card details.
4. Connect to respective trolley via the application.
5. Place a fruit / vegetable on the load cell.
6. Camera will capture the image and recognize the object.
7. Details of the object will be sent to the customer's mobile phone with the help of HC-05 (Bluetooth Module).
8. Data received can be seen once the view cart button is hit.
9. Further, proceed to checkout.
10. Enter CVV and receive an OTP.

11.   Enter the OTP and the process is completed and you receive a QR code (bill).

Image processing using TensorFlow Lite module gives fast results and accuracy. It also requires less processing power and offers more inference time. These models are used to obtain fast and accurate results in real time applications such as a smart cart system.

21.6 RESULTS

* User registers on the mobile application by providing personal details. After the successful registration, the user can login to the application. All the credentials are stored in the database (see Figures 6.1 and 6.2).

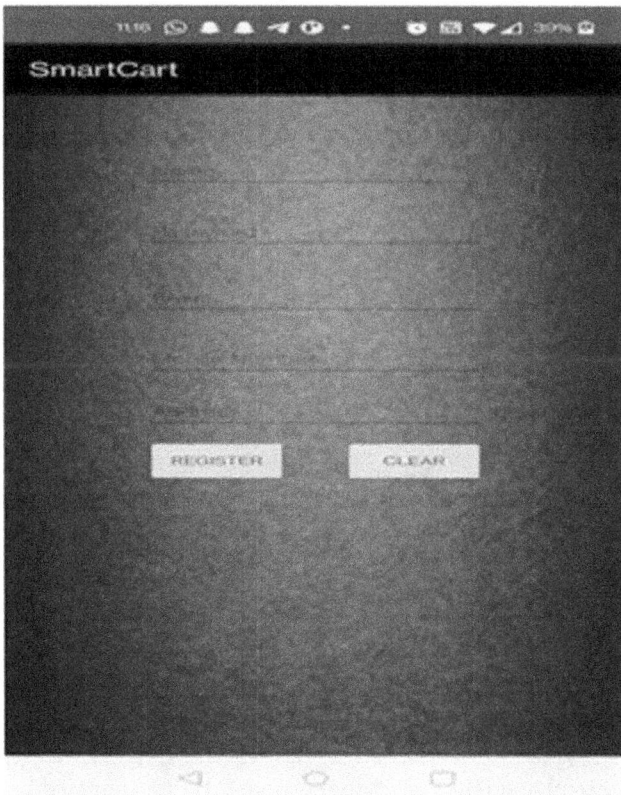

Figure 21.2 SmartCart Registration Form.

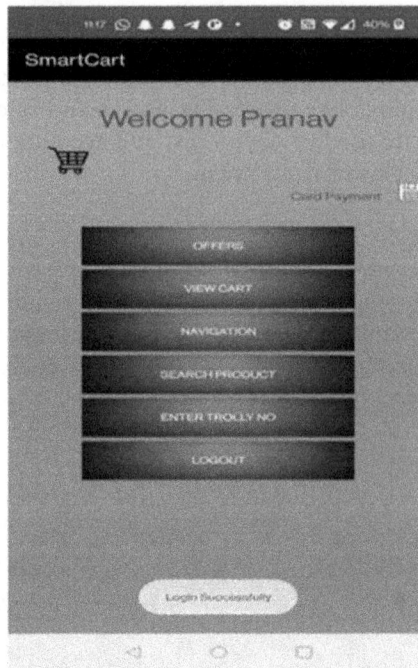

Figure 21.3 SmartCart User Home Screen after Registration.

- Users put fruit in front of camera on the load cell. Load cell will weigh the fruits. Then camera will recognize fruit in front of it using tensor flow lite model. After the working if this model, fruit is recognized. On the screen fruit is displayed with an accuracy of the fruit (see Figure 6.3).
- Once the fruit is recognized then HC-05 sensor will transfer data from rpi to mobile app. Then all the fruits are added to the cart and bill is generated on the screen. After clicking on proceed to payment button new window is displayed. A user must have to enter their bank details with their card number and CVV.
- OTP is generated on the registered mobile number. After entering CVV details and otp QR code is generated on the app (see Figure 6.4). A bill is also received on registered email ID.

21.7 CONCLUSION

This chapter explained the system that proposes the mechanized smart trolley that can be utilized by any shopping center and which will save time as well as decrease the number of customers close to the checkout. The proposed

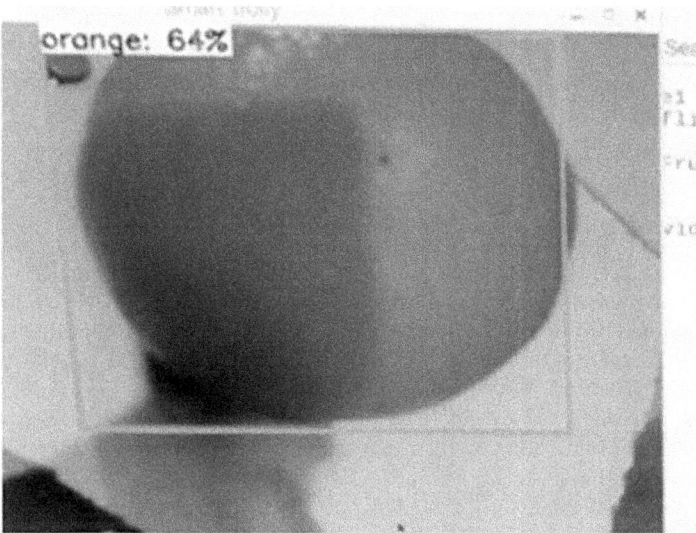

Figure 21.4 Orange Fruit Recognition on the Cart

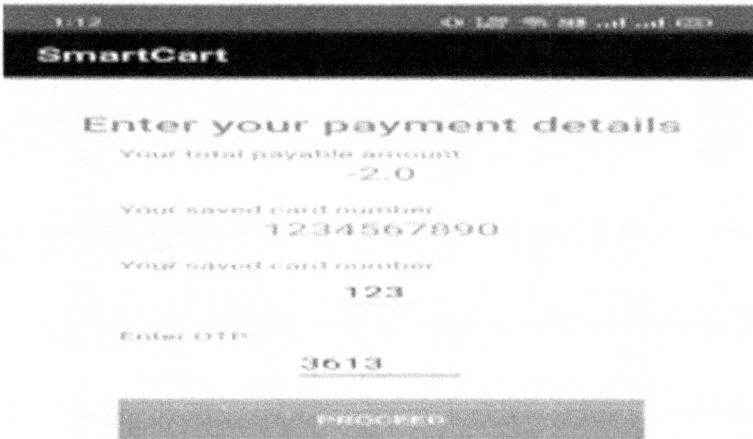

Figure 21.5 Payment Page Displayed on the Mobile App

trolley is not difficult to use and guarantees saving time and creating gains for shopping center proprietors. This framework is likewise truly reasonable for clients as it saves time and establishes a problem free climate. Tested recognition accuracy for oranges (65%), bananas (60%), apples (70%) and strawberries (68%) on this module. This automated smart shopping cart is user friendly and anyone can access it in supermarkets.

REFERENCES

[1] Yogesh, Iman Ali, Ashad Ahmed, "Segmentation of Different Fruits Using Image Processing Based on Fuzzy C-means Method," in 7th International Conference on Reliability, Infocom Technologies and Optimization. August 2018. DOI: 10.1109/ICRITO.2018.8748554

[2] Md Khurram Monir Rabby, Brinta Chowdhury and Jung H. Kim," A Modified Canny Edge Detection Algorithm for Fruit Detection and Classification," in International Conference on Electrical and Computer Engineering (ICECE). December 2018. DOI:10.5121/ijcsit.2017.9508

[3] Jose Rafael Cortes Leon, Ricardo Francisco Martínez-González, Anilu Miranda Medinay, "Raspberry Pi and Arduin Uno Working Together as A Basic Meteorological Station," In, *International Journal of Computer Science & Information Technology*. October 2017. DOI: 10.1109/ICECE.2018.8636811

[4] H. Hambali, S.L.S. Abdullah, N. Jamil, H. Harun, "Intelligent segmentation of fruit images using an integrated thresholding and adaptive K-means method (TsNKM)," *Journal Technology*, vol. 78, no. 6-5, pp. 13–20. 2016. DOI: 10.11113/jt.v78.8993

[5] D. Shukla and A. Desai, "Recognition of fruits using hybrid features and machine learning." In International Conference on Computing, Analytics and Security Trends (CAST), pp. 572–577. December 2016, DOI: 10.1109/CAST.2016.7915033

[6] M. Zawbaa, M. Hazman, M. Abbass and A.E. Hassanien, "Automatic fruit classification using random forest algorithm," 2014, in 14th International Conference on Hybrid Intelligent Systems, pp. 164–168. 2014. DOI: 10.1109/HIS.2014.7086191

[7] Y. Mingqiang, K. Kpalma, and J. Ronsin, "A Survey of shape feature extraction techniques." In *Book Pattern Recognition Techniques, Technology and Applications* pp. 43–98. 2008.

[8] D.G. Lowe, "Object recognition from local scale-invariant features." In *Computer Vision*, in the proceedings of the seventh IEEE international conference, Corfu, Greece. pp. 1150–1157. 1999. DOI: 10.1109/ICCV.1999.790410

Index

For Product Safety Concerns and Information please contact our EU
representative GPSR@taylorandfrancis.com
Taylor & Francis Verlag GmbH, Kaufingerstraße 24, 80331 München, Germany

www.ingramcontent.com/pod-product-compliance
Lightning Source LLC
Chambersburg PA
CBHW060811220326
41598CB00022B/2587